THE POLARIZATION METHOD OF SEISMIC EXPLORATION

SOLID EARTH SCIENCES LIBRARY

THE POLARIZATION METHOD OF SEISMIC EXPLORATION

E. I. GAL'PERIN

Academy of Sciences of the USSR, Institute of Earth Physics, Moscow

D. REIDEL PUBLISHING COMPANY

A MEMBER OF THE KLUWER ACADEMIC PUBLISHERS GROUP

DORDRECHT / BOSTON / LANCASTER

Library of Congress Cataloging in Publication Data

Gal'perin, E. I. (Evseĭ Iosifovich), 1920–
The polarization method of seismic exploration

(Solid earth sciences library)
Translation of: Poliarizatsionnyĭ metod seĭsmicheskikh issledovaniĭ
Includes bibliographical references and index.
1. Seismic waves. 2. Seismic reflection method. I. Title.
II. Series.
QE538.5.G3413 1983 551.2′2 83–4547
ISBN 90–277–1555–6

Translated from the 1977 edition of *Poliarisatsionnij metod seismicheskikh issledovanij*, Nedra, Moscow, by B. Kuznetsov and M. Samokhvalov

Published by D. Reidel Publishing Company
P.O. Box 17, 3300 AA Dordrecht, Holland

Sold and distributed in the U.S.A. and Canada
by Kluwer Academic Publishers,
190 Old Derby Street, Hingham, MA 02043, U.S.A.

In all other countries, sold and distributed by
Kluwer Academic Publishers Group,
P.O. Box 322, 3300 AH Dordrecht, Holland.

Printed in the Netherlands

To Grigorii Aleksandrovich Gamburtzev,
Member of the Academy of Sciences
of the USSR

TABLE OF CONTENTS

PREFACE TO THE ENGLISH EDITION

Five years have elapsed since this book was first published in Russian. Year after year geological practice poses increasingly difficult and variegated problems to be solved by seismic investigations not only in the field of oil and ore geology, but also in the field of seismology and prospecting. The major goal in seismic exploration at present is the transition from structural studies to direct studies of material composition. However, the potentialities of seismic methods fail to satisfy newly arising problems. This – to an appreciable extent – is the outcome of a scalar approach to the analysis of the wave field which is essentially a vector field, such an analysis being typical of modern seismic exploration. Because of such an approach, every type of wave employed in seismic studies requires its own 'personal' method (the method of longitudinal reflected waves, the method of transverse reflected waves, the method of converted reflected waves, the method of refracted waves, the method of converted transmitted waves, etc). In the course of field work, for each method an oscillation component peculiar to the method is recorded fixed in space (the vertical component for the longitudinal waves and the horizontal X and Y components for the transverse waves), and the analysis of wave fields is executed chiefly by distinguishing the waves according to two parameters: the frequency and the apparent velocity.

The immense volume of vertical seismic profiling carried out in the last decade has enabled the mode of propagation of seismic waves in media with greatly varying structures to be studied. The results of such studies have demonstrated in particular that for a substantial improvement of the effectiveness of seismic studies it is advisable to go over from the scalar analysis of the wave field to vector analysis. Studies along these lines have led to the creation of a general polarization method of seismic studies.

The generality and the effectiveness of the polarization method (PM) follow from its features:

– all the parameters of a complex wave field are utilized for its analysis, including the polarization;
– the vector analysis of the wave field enables waves of various types, irrespective of their velocity, propagation direction and polarization, to be discriminated and tracked;
– longitudinal, converted and transverse waves are utilized simultaneously to gain information about the Earth, this being the chief prerequisite for the study of its material composition and for predicting its cross-section;
– the method is applicable to all seismic studies both up- and downhole.

In the space of five years since the publication of the Russian edition of this book, great experience in the application of PM for the solution of all sorts of problems in variegated seismological conditions has been accumulated. The method has been developed further in two main directions:

– the perfection of the techniques and equipment of three-component field studies incorporating a continuous control of the identity of equipment both in deep boreholes and on the surface;
– the creation of efficient and straightforward software for the automatic processing of PM data.

Whereas the early stages of development of the method have been conspicuous chiefly by the use of polarization for the discrimination of the wave field, in recent years the emphasis shifts to the use of polarization parameters to obtain additional information about the Earth. This is especially important, as the polarization parameters are very sensitive to inhomogeneities in cross-section and enable those parameters of the Earth to be studies which cannot be obtained from kinematics. Here one should especially mention the polarization of transverse waves, which can act as one of the most reliable parameters for identifying and studying anisotropic properties of the Earth.

At present various modifications of PM are being developed applicable both to uphole – (PM RW = reflected waves method, PM CDP = common-depth-point method, PM refracted-wave method, PM of regional studies utilizing earthquakes) and to downhole observations (PM VSP = vertical seismic profiling). Not all the PM modifications are developing at the same rate. The method which has up to now been developed most extensively is PM VSP.

PM VSP has combined all kinds of downhole observations in the frequency band utilized in seismic exploration and has become the foundation of a new trend in seismic exploration; that of production seismics, i.e. of seismic studies at the stages of the exploration and exploitation of fields.

The efficiency and the effect of production seismics on the economy is determined by the opportunity it offers for substantial cuts in the volume of deep drilling required for the exploration of every field.

The most fundamental result of recent years, in the author's opinion, however, is that it has been possible to overcome the psychological barrier that arises every time the specialists are confronted with some new parameters, especially with such a parameter of the wave field as the polarization of oscillations.

The great exploratory potentialities and the high economic efficiency of PM have attracted the attention of a wide specialist audience. This was particularly noticeable during the course of a seminar conducted by the author (October, 1978, Bartlesville, Oklahoma, USA) and the author's report to the 49th Meeting of the International Society of Exploration Geophysicists (New Orleans, November, 1979).

Despite the fact that not all the principal potentialities of the method have been fully exploited, the results already obtained are proof of its good prospects. There is no doubt that, as the method and first of all the software for its processing, are progressively developed, its efficiency will improve continuously, and new opportunities may emerge that are hard to foresee at present. This justifies the generous financing of work aimed at testing and developing the method. We can already advise that the method be used extensively in all kinds of VSP observations. In uphole observations it is advisable to test the method in the studies of the material composition, as well as in the studies of complex media in various seismic and geologic situations where the effectiveness of the traditional methods is low.

The methods of analog processing described in this book should enable the reader to evolve a three-dimensional image required for a clear understanding of the results of digital data processing. Since we are dealing with a method which is totally new for seismic exploration, we have deemed it advisable to append a glossary of terms so as to guarantee unambiguous understanding of the many new terms that have had to be introduced during the course of the development of the method.

We are convinced that the subsequent development of seismic exploration will be associated with three-component recording and with a vector analysis of the wave field, i.e. with a transition to the polarization method. And we hope that the present book will encourage such a transition.

The author expresses his gratitude in advance for any remarks and advice the readers may care to provide when they become familiar with the book and employ the polarization method.

INTRODUCTION

At present, the most important problem in seismic prospecting is the need to increase the depth range and the efficiency of methods employed in the study of media with complex structures. In addition to this structural problem, there is another of great importance, which is the need for methods of studying the material composition of a section in such a manner that would enable forecasts to be made. Although in recent years, thanks to the introduction of systems-coverage techniques employing multichannel and computer data-processing, great progress has been achieved in increasing penetration, the efficiency of seismic prospecting in complex media is still low. This primarily applies to media with steep interfaces: salt domes, reefs, fracture zones in oil geology; intensely deformed media (intrusions, veins and dykes) in ore regions. Areas with such a structure with good prospects for oil and gas have not been adequately studied in traditional oil-producing provinces. Because of the complexity of geological structure, the efficiency of seismic methods of prospecting for ores still remains very low, and the growth of seismic prospecting studies in ore geology is extremely slow.

The efficiency of seismic prospecting is at present, as it has been during all the preceding stages of its development, determined mainly by the facilities available for the analysis of complex wave fields observed in the Earth itself. The development of such facilities led to gradual simplification of the recorded wave field through selection of oscillations based on their various characteristics. Frequency discrimination and phase correlation of the waves resulted in the development of the method of reflected waves which made seismic prospecting the leading method of geophysical prospecting for oil and gas. The discrimination of waves according to the direction of their propagation (or their apparent velocities) combined with gradual improvements in selectivity [(pattern techniques, controlled directional reception (CDR), common depth point method (CDP)], has been the principal method of increasing the efficiency of seismic studies for over 40 years. Consequently, the possibilities of modern seismic prospecting are based on the utilization of discrimination mainly based on two parameters (the frequency and the direction of propagation).

The introduction of an additional independent parameter would be expected greatly to increase the efficiency of analysis of complex wave fields and therefore of seismic prospecting of complex media. The polarization of seismic waves constitutes such a parameter. In contrast to optics and radar, where the polarization of oscillations has been extensively studied and has for a long time been in wide use, there has been little application of polarization of seismic waves in seismology, and the extent to which it has been studied is inadequate.

Of the two groups of parameters that define the wave field and are associated with the direction of motion of particles of the medium during the passage of waves through a point (directional characteristions of the first type) and with the direction of propagation of the wave in space (directional characteristics of the second type), only those of the latter group have been used, their application being limited to planar (as distinct from spatial) problems. Despite the fact that all problems in seismic prospecting are essentially spatial, three-dimensional analysis of wave fields has been practically non-existent.

It seems necessary to supplement the traditional 'line-plane' scheme in three-dimensional analysis of wave fields by two geometrical elements ('point' and 'space'). Research carried out in this direction at the Institute of Earth Physics of the Academy of Sciences of the USSR (IEP AS USSR) created the polarization method (PM) of seismic studies based on the combination of discrimination of waves at a point, related to the type of polarization of oscillations at this point, and discrimination, according to the direction of propagation of the wave, that is, the combination of systems of observations of the first and second types. Methodologically, this trend corresponds to the combination of three-component observations at a point with three-dimensional observations in space.

Initially, the main attention was focused on the three-dimensional analysis of waves at a point, that is, on the study of polarization of oscillations in seismic waves, which is one of the most sensitive elements of the wave field. The nature of polarization of oscillations is such that it requires three-dimensional analysis. In developing methods of analysis of the wave field at a point the same principles have been applied as form the basis of seismic prospecting (correlation methods and wave discrimination). This led to the formulation of two methods of analysis of oscillations at a point [polarization correlation and controlled directional reception based on the directions of particle displacement (CDR–I)]. The addition of discrimination according to the type of polarization of oscillations to the two former methods of wave discrimination based on their frequency and direction of propagation will probably mark the beginning of a new stage in the evolution of seismic prospecting in complex media.

For the analysis of complex wave fields, the polarization method employs polarization-position correlation (PPC) in which the components to be separated and tracked are not some fixed components (e.g., vertical components), but tracking (optimum) components characterized by maximum signal-to-noise ratio. PPC enables the regular waves to be separated and tracked, no matter what their polarization and velocities of propagation are. Therefore, in contrast to traditional methods which make use of waves of one definite type (e.g., the method of longitudinal reflected waves or of transverse reflected waves, etc.), PM simultaneously makes use of longitudinal and transverse waves excited by conventional sources, as well as secondary waves appearing in the medium (reflected and refracted waves, longitudinal, transverse, and converted waves). This, besides improving facilities for the solution of structural problems, opens up new prospects for the study of the material composition of a section. For this reason PM may turn out to be effective not only in regions with a complex geological structure, but also in such regions, where structural problems are being successfully tackled with the aid of the method of longitudinal waves.

The current literature contains scanty information on the polarization of seismic waves. There are only a few discrete papers dealing with individual problems involving the polarization of seismic waves. Since the polarization method is beginning to have wide application in practical prospecting, the author felt it necessary in this first monograph dealing with the method to present a systematic account of the available information on the study of polarization, and to dwell upon some general aspects connected with the polarization of oscillations, the methods of studying it, and of representing the results.

Although the waves observed in seismology and seismic prospecting are of a finite duration (i.e. they are a combination of monochromatic waves with different periods), the entire theoretical background for studying the nature of oscillations, including polar correlation, has been presented in the form applicable to monochromatic waves. This has been done for the sake of simplicity and clearer understanding, and also because it is the author's impression that the experience accumulated in the course of experimental studies of the polarization of seismic oscillations supports the conclusion that the results obtained in the theory of monochromatic oscillations are applicable, at least in the first approximation, to waves actually observed in seismic studies. It should also be kept in mind that the general information on polarization of seismic waves presented herein is meant to apply as a rule only to isotropic media.

This publication contains the results of studies carried out under the direction of the author mainly at IEP AS USSR, as well as at other institutions. The major part of the work has been carried out by G. I. Aksenovich, R. M. Gal'perina, A. V. Frolova (IEP AS USSR), M. S. Erenburg (Soyuzgeofizika), L. A. Pevsner, R. N. Khairutdinov, V. L. Pokidov, T. G. Chastnaya (Kaz VIRG), Yu. D. Mirzoyan, and V. S. Starodvorsky ('Krasnodarneftegeofizika'). Not all the aspects of the polarization method have been developed to an equally advanced stage, which was bound to affect the contents of the book.[1]

Although the processing of a substantial part of the data pertaining to the polarization method is currently performed by digital computers, the author deemed it advisable to consider methods of processing by analog computers, because this is important not only for rapid evaluation of the quality of the data and for supervising digital processing, but also for the understanding of physical sense and for interpretation of the results.

Data obtained in observations made on the oilfields of the Northeastern Cis-caucasian region (IEP AS USSR in collaboration with 'Krasnodarneftegeofizika') and on the ore deposits of Central Kazakhstan (IEP AS USSR in collaboration with Kaz VIRG) and the information from regional studies utilizing blasts and earthquakes, serve to illustrate the efficiency of the method.

[1] I consider it my pleasant duty to express gratitude to I. I. Gurvich and G. I. Petrashen for their suggestions and remarks made in the course of discussions of various stages of the research program and especially on reading the manuscript. I am also grateful to A. F. Frolova and R. M. Galperina for assistance in the preparation of the book.

CHAPTER 1

PHYSICAL AND GEOLOGICAL FOUNDATIONS OF THE POLARIZATION METHOD

The principal trends aimed at increasing the efficiency of seismic-prospecting methods employed in the study of media with complex structures are the development of three-dimensional analysis of wave fields and the simultaneous use of waves of different types. Both these trends are realized to an appreciable degree in the polarization method by the introduction of an additional independent parameter of the wave field-seismic-wave polarization.

The use of seismic-wave polarization in prospecting practice has been reserved on a very limited scale mainly for certain special problems involving the study of transverse and converted waves. Accordingly, we shall have to consider the polarization characteristics of the main types of seismic waves and to formulate the fundamental principles of utilization of polarization of oscillations. In addition, we shall have to discuss methods of representing the trajectories of motion of the particles of a medium in space, and describe methods of working with the stereographic projection conveniently used in the study of the seismic-wave polarization.

Physical Premises for Increasing the Efficiency of Seismic Methods Used in the Study of Media with Complex Structures

We shall begin our discussion of the principal trends in increasing the efficiency of seismic studies of complex media by analysing the wave field. One general rule stands out clearly in the evolution of seismic prospecting methods: at all stages of evolution the efficiency of seismic studies has always been determined mainly by the facilities available for the analysis of complex wave fields characteristic of the true Earth. The large amount of work completed in recent years on the vertical seismic-profiling method (VSP) has demonstrated that even now the facilities available for the analysis of wave fields remain the decisive factor in increasing the efficiency of seismic studies. Interfaces producing reflected waves of sufficient intensity have been located in practically all media at all observable depths. However, those waves could be tracked only up to the upper portion of the section, where their correlation became blurred by the superposition of a large number of regular and irregular unwanted waves of different types. The separation and tracking of such waves on the surface entails great difficulties and is not always possible.

The main trend in the evolution of seismic prospecting has been the improvement in the methods of analysing the wave field and the corresponding observation systems. The principal trend in the development of methods of analysing wave fields has been a gradual simplification of the wave fields, based on discrimination of the waves by several characteristics utilizing the parameters of the wave field with the aim of separating the useful waves against a background of varied noise made up both of regular and irregular unwanted waves. Both the systems of observation and the method of recording have in principle been adapted to the methods of analysing wave fields currently in use. The evolution of seismic prospecting can conventionally be subdivided into two principal stages. One involves the frequency discrimination of waves, and the other, the discrimination of waves according to their direction of propagation. The main results obtained in seismic prospecting may be said to be connected with these stages. Although the development of both types of discrimination proceeded practically along parallel lines, frequency discrimination was introduced earlier, and because of that it has, in particular, been instrumental in the development in the twenties of the reflection method. It is a well-known fact that the principal prospecting potentialities of reflected waves had been formulated long before such waves were identified on the seismograms. Only some 10–15 years later, frequency discrimination based on the phase correlation of waves, together with multi-channel recording, made it possible to analyse waves on the subsequent portions of the reords and, in particular, to identify reflected waves which greatly increased the prospecting potentialities of such studies and made seismic prospecting the principal method of geophysical prospecting for oil and gas.

As the geological problems became more complicated, the efficiency of the reflection method (RM) fell, and the need arose for a method of analysing more intricate wave fields. With this aim in view, frequency discrimination was supplemented by discrimination based on the direction of propagation, or in other words, by discrimination based on the value of apparent velocity. At first such discrimination was performed simply by seismometer grouping followed by source grouping (pattern shooting). As the problem became more complex, the selectivity of the receiving and analysing systems was increased mainly as a result of greater sophistication. Later in the 'forties, a new method (controlled directional reception) was invented and developed [113], and this greatly improved the system's selectivity and hence the efficiency of seismic prospecting. However, with increase in depth, the CDR-method became inadequate for many regions, and to improve the system's selectivity still further methods based on multiple coverage have been developed. Specifically, such methods include the method of a common depth point (CDP) based on the accumulation of signals received from the same element of the interface, this being the principal method at present in seismic prospecting. In the development of this method, an essential part has been played by digital-computer data processing which has enabled repeated summation of the signals (from 12 to 24 times) to be performed. The CDP brought about a rise in prospecting costs, but in many cases it enabled the penetration to be increased, and geological results to be obtained in regions formerly unsuitable for seismic prospecting. All the methods of interferential reception mentioned above (pattern techniques, CDR, and CDP) are based on the discrimination of waves based on the same parameter, that is, according to the wave's direction of propagation along the traverse, or in other words, according to the value of the apparent velocity along this line.

Hence, the potentialities of seismic prospecting are at present determined by the use of discrimination of waves based on two parameters only (the frequency and the direction of propagation). Under conditions of complex media both the direct and the converse problems of seismic prospecting are essentially three-dimensional in nature. This is true of oil prospecting, but much more true of ore prospecting, where one usually has to deal with steep interfaces of intricate shapes.

It should be pointed out that, despite some specific features peculiar to seismic prospecting for ores, the subdivision of seismic prospecting into prospecting for oil and for ores is at present largely a matter of convention. Now, when seismic prospecting for oil is required to handle problems involving complexly-constructed and intensely deformed media with steep interfaces (salt domes, reefs, diapirs, etc.), there is little difference from the methodological point of view between them and the problems facing seismic ore-prospecting. At this point, it would be appropriate simply to discuss seismic prospecting 'in media with complex structures in which the waves are propagated in different directions and in different planes, and in which three-dimensional analysis of the field should be possible. Whereas in the study of comparatively simple media one could, under appropriate conditions, ignore the fact of the three-dimensional nature of the wave fields and make use of two-dimensional interpretation, in media with complex structures this not only results in considerable mapping errors, but in many instances makes such studies worthless. For this reason one should in each case specially assess the validity of such simplifications and the errors introduced by them. However, the use of three-dimensional analysis has been totally inadequate.

The three-dimensional nature of the wave field is, as is well known, determined by parameters belonging to two groups (those determining the trajectory of motion of the particles in the medium at the time the wave passes through a point, and those determining the propagation of the wave in space). It has been deemed expedient to make use of both groups of parameters in order to improve the efficiency of studying media with complex structures. On the one hand it was decided to make use of an additional independent parameter for the purpose of wave separation (wave polarization), and on the other hand, to study the directions of wave propagation in space. As we shall presently see, both these trends determine the general approach based on a combination of three-dimensional systems of observations of the first and second types. The introduction of wave discrimination based on the third parameter of the wave field into the seismic-prospecting practice (wave polarization), and of polarization-positional correlation based on it, may mark the beginning of a new stage in the development of seismic prospecting, having extensive possibilities in analysing complex wave fields, just as in the case with the discrimination of waves based on frequency and direction of propagation.

To improve the efficiency of seismic prospecting in complex media, in addition to perfecting methods of analysing wave fields, one should simultaneously employ not only waves of the longitudinal type, but also other types (transverse and converted) as well. This trend, like the first, is not a new one. Research involving the use of waves of different types has been carried out at various institutions for a number of years. I. S. Berzon has directed work on the method of converted reflected waves at IEP.

A great volume of research aimed at developing the method of transverse reflected waves has been completed in the last 15 years under the direction of I. N. Pusyrev [17, 104–108, 130, 131]. This method is based on the excitation and recording of the transverse *SH* waves. A large amount of work has been devoted to studying the conditions of excitation of the *SH* waves and developing effective blast sources, in particular inverse directional sources possessing an azimuthal directivity that may, by changing the direction of action, be employed to improve relative efficiency in recording of *SH* waves. This has made it possible to start using the method of transverse reflected waves in practice, and has proved quite effective in solving a number of problems. However, difficulties have been experienced in the development of the method, apart from those connected with the excitation of transverse waves. They involve unwanted Love surface-type waves which are excited together with the SH waves and have the same horizontal polarization perpendicular to the direction of propagation as the SH waves. Such waves may in individual regions appreciably limit the depth of investigation.

In discussing transverse reflected waves it should be pointed out that normal blasts used for seismic-prospecting may, in addition to longitudinal waves, also excite transverse waves [33, 36, 102]. A great deal of work on the vertical seismic-profiling method completed in recent years and drill-hole recording of waves, have made a more detailed study of conditions of excitation and recording of transverse waves possible. It has specifically been demonstrated that transverse waves are practically universally observed, and that their intensity depends greatly on the conditions of excitation, but in a large number of cases this intensity is comparable with that of longitudinal waves. In some regions intense transverse waves recorded in surface observations made with the aid of vertical seismometers are sometimes classified as belonging to surface-type waves. The use of such transverse waves is connected with difficulties in analysing complex wave fields, owing to a large number of wave types involved and to possible instabilities in the conditions of their excitation and polarization. For this reason, such waves are usually regarded in seismic-prospecting practice as unwanted waves, and special observation systems and processing methods are employed to suppress them.

The polarization–positional correlation of waves based on the combination of waves according to the direction of displacement of the particles at a point, with the discrimination of the waves according to their direction of propagation in space, makes it possible to separate regular waves with different polarizations propagating in different directions and with different velocities. This makes it feasible to replace traditional methods based on the study of waves of a single type (e.g. the method of longitudinal reflected waves, or the method of transverse reflected waves, or the method of converted reflected waves or transmitted waves) by methods making use simultaneously of waves of various types connected with the same inhomogeneities of the section and excited by the same blast. Such a method provides maximum information about the medium and extends the potentialities of seismic exploration, especially studying the material composition of a section. The ratio of the velocities of the longitudinal and transverse waves and the absorption coefficients for these waves may be of special interest. That is why the use of the polarization method may prove valuable even in regions with comparatively simple conditions, where structural problems can be successfully solved exclusively by the use of longitudinal waves and traditional methods. In conditions of media with complex structures, where efficiency of seismic-prospecting is at present low, the polarization method holds out the promise of becoming the major method of seismic prospecting.

It should be pointed out that the use of transverse waves excited by conventional blasts in no way precludes the use of transverse waves excited by controlled sources. On the contrary, there can be no doubt that the efficiency of transverse waves, having the natural radiation pattern obtained during blasts, cannot be equally adequate in all regions. In some regions it will probably not be sufficient to separate the waves according to their polarization alone at the point of reception, and it will be necessary to introduce additional discrimination by employing directional sources. However, such discrimination should be introduced only after the potentialities of discriminating waves at the point of reception have been fully exploited. The combination of directional and non-directional sources will facilitate better exploitation of the great prospecting potentialities of transverse waves.

PARTICLE MOTION AT A POINT (DIRECTIVITY OF THE FIRST KIND)

The first group of parameters that determine the three-dimensional properties of the wave field is connected with the polarization of seismic waves. The polarization of oscillations is a space-time characteristic of the wave field. The term polarization of seismic waves applies to certain characteristics of the trajectory described by a particle of the medium (or by the end of the displacement vector) at each point in the medium.

Quantitatively, the polarization is characterized by parameters that determine the trajectories of motion of the particles of the medium. Such trajectories differ for different waves. For instance, for simple three-dimensional (longitudinal and transverse) waves in homogenous media, they take the form of straight segments, whereas surface (Rayleigh) waves at the surface are elliptically polarized in the vertical plane. Since the wave field in the true Earth is a very intricate one and may be due to the superposition of a large number of waves of various types and origins, the trajectories of particle motion are in most cases represented by complex three-dimensional curves.

At the same time only the vertical oscillation component is usually recorded in seismic prospecting practice, especially in the method of reflected waves. This was legitimate as long as the conditions of observation were comparatively simple, since the longitudinal reflected waves of interest arrive at the seismometers from directions close to the vertical. However, the directions of arrival in the case of surface observations in conditions of complex media, or in case of drill-hole observations in conditions of horizontally-layered media, may differ appreciably from the vertical. Moreover, for the same wave they may change greatly from point to point [36].

The situation is aggravated by increased deviation of the complete vector of interferential oscillation from the vertical as a result of superposition of unwanted waves at each point of observation. Under such conditions, the vertical oscillation component loses its special importance and turns into a random component like any other. Hence, the separation and tracking of a definite wave may better be performed by recording (instead of the vertical component), some oblique component that permits optimum identification of the specific wave, that is, with a maximum signal-to-noise ratio. The recorded component may be located in space on some basis or may generally even be unlocated. Such a component will in future be termed tracking or optimum.

For longitudinal waves recorded outside the interference area, the tracking component will coincide with the vector of oscillations. The direction of the vector of oscillations for each simple wave is determined by the mutual disposition of the excitation and reception points, as well as by the structure of the medium. Usually such directions change regularly and smoothly from point to point. In complex wave fields formed as a result of superposition of different waves the orientation of the tracking components in space will be appreciably affected by the interference of the waves. In the case of a composite multi-polarized oscillation the tracking component of each wave may, depending on the polarization diagram, deviate from the oscillation vector, the change in the orientation of tracking components of a chosen wave from point to point along the line of the profile being possibly more abrupt than the change in the orientation of the vector of the same wave. The tracking components may be determined accurately with the aid of a digital computer or approximately in an analog form by visual discrimination from fixed components.

The tracking components of each wave are found by combining the discrimination of waves based on the direction of propagation with that according to their type of polarization. The discrimination of waves according to the apparent velocity in the vicinity of the point being studied helps to separate regular waves, whereas the discrimination based on polarization parameters enables tracking components that facilitate optimum separation of the particular wave to be found. Discrimination according to the parameter V_a has been extensively studied, is widely used, and will not be considered here. Discrimination based on the polarization parameters is also being used in seismologic practice. There are various polarization filters in existence that discriminate on the basis of various polarization parameters. In order to study the trajectories of particle motion, we have developed polar correlation and the discrimination of waves based on the direction of particle motion in a medium in space. Such discrimination is performed with the aid of the method of controlled directional reception ('directional' refers to the directions of displacements) [27, 36] termed CDR–I by analogy with CDR–II which employs discrimination according to V_a. CDR–I is one of the possible modifications of polarization filtering that can easily be realized in an analog form. At present two main trends in using polarization of oscillations are being developed: one for the analysis of complex wave fields and the other for the extraction of direct information about the Earth.

The analysis of wave fields is performed with the aid of the methods of polar and polarization-positional correlation. Polar correlation is used mainly for the analysis of wave fields recorded at individual discrete observation points. The effective means for analysing wave fields recorded along a profile or across an observation area is the polarization-positional correlation method based on the separation and tracing of the tracking components of oscillations along a line or across the plane of observations. Polarization filtering based on the discrimination of waves according to the nature of their polarization may be regarded as a further development of CDR–I, aimed at improving the selectivity of the system. As will be demonstrated below, apart from the analysis of wave fields, polarization may be employed to study the variation in the velocity of propagation of seismic waves with depth. It provides more facilities for the solution of certain problems than the exclusive use of the kinematic characteristics of the waves. This is true in particular of the identification of wave refraction. Polarization is also helpful in mapping structures and in determining stratigraphic detail both of refracting and reflecting interfaces.

In short, the use of polarization at every point involves the discrimination of the waves in accordance with the polarization parameters, and hence the existing seismic-prospecting 'line-plane' scheme should be supplemented by another geometrical element – the 'point', that is, the wave field should be analysed at every point.

To ensure the discrimination of waves at every point, efforts have also been made to design seismometers with improved selectivity characteristics as compared with those of conventional seismometers. In this connection, we may mention combination gauges consisting of a pressure gauge and a displacement velocity gauge. Such gauges are known [46] to posses a cardioid-shaped characteristic, which, under certain conditions, makes it possible in principle to record waves arriving at the seismometer only from below and to disregard multiple waves arriving from above. However, such methods have not yet been developed beyond the experimental stage.

PROPAGATION OF WAVES IN SPACE (DIRECTIVITY OF THE SECOND KIND)

The second group of parameters connected with the three-dimensional properties of a wave field include the parameters of the waves that control their direction and propagation velocity. At present all plotting is usually performed in the vertical plane, this, strictly speaking, being correct only in case of media with a symmetry axis. In case of oblique interfaces this is permissible only, if the line of the profile coincides with the line of slant of the interface.

The study of complex media requires the use of three-dimensional observation systems and three-dimensional interpretation. True, individual elements of three-dimensional systems have sometimes been used previously for the solution of certain problems [11, 12]. For instance, sections of surface travel-time curves from one or more shot points have been plotted. Three-dimensional interpretation in resolving numerous problems of seismic prospecting usually begins at the stage of drawing of structural maps of sections. However, in contrast to this the study of complex media involves a three-dimensional study of the wave field and the development of complete three-dimensional observation systems featuring multiple coverage and signal accumulation. In such systems it is frequently advisable to supplement areal surface observations with VSP observations and thus obtain three-dimensional stereo systems. Initially they may be made up of separate stereo elements to be combined later into a comprehensive stereo system which will make possible direct transformation of observation data into structural maps of interfaces and will appreciably improve the efficiency of seismic prospecting in the study of complex media.

Thus, the traditional 'line-plane' scheme should be supplemented by the geometrical 'space' element which envisages the use of three-dimensional systems for observations and for processing and interpretation of experimental material.

Hence, the polarization method of seismic prospecting is based on a scheme which may be represented in the form 'point-line-plane-space' (Figure 1). Such a scheme is the most complete, and any other scheme may be regarded as its particular case.

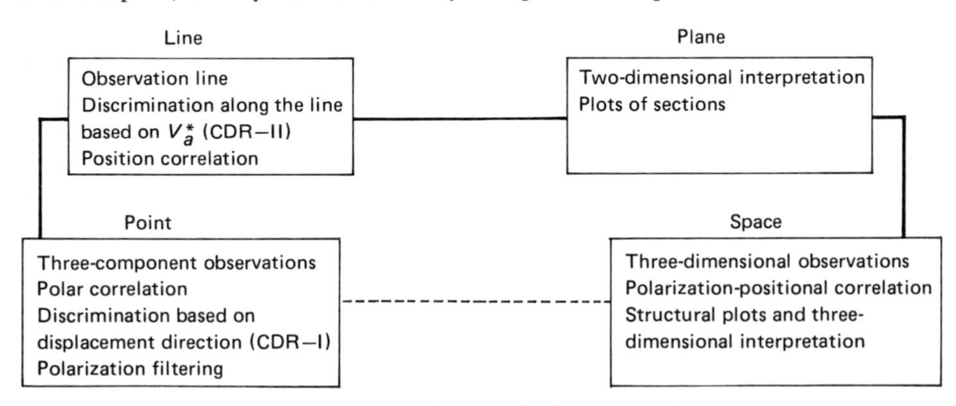

Fig. 1. Schematic diagrams of seismic observations.

Whereas the observations conforming to the 'line-plane' scheme, which employs the discrimination of waves based on two parameters alone (frequency and apparent velocity), required only one component of the oscillation to be measured, the operation of the general scheme 'point-line-plane-space' based on the discrimination of waves according to three parameters (polarization, frequency, and apparent velocity) requires the recording of three oscillation components and the study of the vector of oscillations.

Naturally, in the study of sundry problems, depending on their complexity, only some elements of the general scheme may be used instead of the scheme as a whole. For example, if the efficiency of observations conforming to the traditional 'line-plane' scheme (Figure 1) is not adequate for the solution of problems aimed at increasing penetration in the study of slanting platform structures, it is advisable to introduce, in addition, the three-dimensional analysis of oscillations at a point, that is, to make use of the 'point-line-plane' scheme. The best scheme for studying steep or bent interfaces (e.g. domes, reefs, etc.) is often the 'point-line-space' scheme. With increase in facilities for perfecting stereo observation systems, when multi-channel recording with the aid of a large number of seismometers uniformly distributed over a large area becomes a technical possibility, the study of complex media will, probably, conform to the 'point-space' scheme, because in this case the elements 'line' and 'plane' are intermediaries. However, owing to the imperfection of the observation systems such elements facilitate the transition from 'point' to 'space'. Present-day data processing in most cases conforms to the 'line-plane' scheme. The 'point-line-plane-space' scheme makes it possible to unify the observation system and to set up a comprehensive universal technological data processing scheme that makes use of the polarization-positional correlation of waves.

To sum up, the folowing principal features determine the efficiency of the polarization method in studying complex media.

1. It employs the discrimination of a complex wave field according to all its principal parameters (frequency, polarization of oscillations, and apparent velocity) and three-dimensional analysis of the wave field at a point and in space.

2. It involves the separation and tracking of regular waves irrespective of their polarization and propagation velocity. Therefore, in contrast to the traditional methods based on the use of waves of one type (e.g. the method of longitudinal reflected waves or the method of transverse reflected waves, etc.), the polarization method can make simultaneous use of the principal wave types (longitudinal and transverse) excited by conventional sources employed in seismic prospecting and of secondary waves connected with them. The discrimination of waves at the source (the source's radiation pattern) is utilized, as required, in cases when the discrimination of waves at the receiver does not prove effective enough.

3. It makes use of observation systems based on the combination of three-component observations at a point with three-dimensional observations in space.

It should be pointed out that the separation in the polarization method of a large number of regular waves of different origin and type for joint interpretation greatly enhances the importance of VSP which should make it possible to sort out waves of different types related to the same inhomogeneities of the Earth from a great number of other waves and to correlate them stratigraphically.

General Information on the Polarization of Seismic Waves

The particles of the medium involved in the wave process leave their equilibrium positions and perform oscillatory motion in space describing definite trajectories about their equilibrium points. After the elastic wave has passed the particles again return to their original equilibrium positions. The trajectories of particle motion are not the same for different types of waves, that is, different waves may have different polarization. At this point we should define more precisely the concept 'nature of polarization' to mean the qualitative type of the trajectory (linear, elliptical, three-dimensional, and 'polarization parameters' that quantitatively describe the trajectory of motion itself (the direction in space for linearly polarized oscillations, the shape of the ellipse, its position in the polarization plane, and the orientation of the latter in space for elliptically polarized oscillations). The simplest trajectories are linear. However, the trajectories may also be of a more complex shape (two- or three-dimensional curves). The polarization parameters may generally depend not only on the type of the wave, but also on the inhomogeneities of the Earth and the source. The seismic motion of points both at the surface and within the Earth is generally determined by vectors variable in magnitude and direction. In many cases scalar quantities may conveniently be used to study trajectories instead of vectors.

Depending on the choice of the coordinate frame, the position of a particle on the trajectory at every instant of time may be described by different parameters. In a rectangular coordinate frame such parameters are the projections A_x, A_y, and A_z of the oscillation vector A on the x, y and z axes. If the positive direction of the x axis is set at North, y at East, and z towards the zenith, then

$$A_x(t) = |A(t)| \sin \varphi(t) \cos \omega(t), \qquad |A_y(t)| = \sin \varphi(t) \sin \omega(t),$$
$$A_z(t) = |A(t)| \cos \varphi(t), \tag{1}$$

where $|A(t)|$ is the magnitude of the vector A or the displacement at the instant t; $\omega(t)$ is the azimuth of the direction of oscillation; and $\varphi(t)$ is the angle between the vector A and the vertical at each instant of time.

Without dwelling here on the methods of describing the trajectories that are to be discussed later, let us consider the polarization of the principal wave types.

LONGITUDINAL WAVES

A simple longitudinal wave recorded both on the present surface and at points within the Earh, in the absence of interference with the other waves, is a linearly-polarized wave. This means that the direction of the displacement vector remains unchanged, its magnitude changing so that the particles of the medium oscillate about their equilibrium positions, describing rectilinear trajectories. In this respect a linearly-polarized oscillation may be regarded as scalar. Characteristic features of a linearly-polarized oscillation are the constant form of its trace and the equal phase of its components no matter how they are oriented in space. On account of this the only difference between, say, the components $A_z(t)$ and $A_x(t)$ may be in their amplitudes as described by the relation

$$A_z(t) = A_x(t) \cot \varphi,$$

where φ is the angle between the direction of motion and the vertical.

At internal points in a homogenous and isotropic medium the direction of particle displacements coincides with the direction of propagation of the longitudinal waves. Such coincidence is not observed in anisotropic media.

The nature and the parameters of polarization of longitudinal waves are independent of the source and of the forces acting in it. On account of this, the nature of polarization of a *P* wave excited by a blast or by an earthquake will be the same at every individual point ofobservation and cannot provide any information about the source (its radiation pattern). At the same time it should be pointed out that the sources of elastic waves may be different with respect to the mechanisms of the forces acting in them. This is primarily true of blasts and earthquakes. A blast in a homogenous medium may, to a first approximation, be regarded as a dilatation centre. In this case only longitudinal waves of intensity equal in all directions are excited. The forces acting at the focus of an earthquake result in fractures and displacements causing both longitudinal and transverse waves. Their relative intensity greatly depends on the direction of propagation and is determined by the orientation of forces acting at the focus. Thus, the difference in the mechanism of the source will show up, specifically when the polarization of a *P* wave in a plane (at the present surface) is being studied, in the first that, in case of a blast, the first to be recorded in the entire plane will be a compressional wave, and therefore the first particle movements will coincide in direction with that of the wave propagation. In the case of an earthquake, a compressional wave will be the first to be recorded in one part of the plane, and a dilatational wave in the other part. For this reason, at some observation points the first motion of the particles will coincide in direction with that of wave propagation, and at other points it will be directed opposite to it. The fields of opposite sign are separated by the so-called nodal line [114].

Stable linear polarization of the *P* wave is often used to improve the useful recording selectivity by separating the *P* wave against the background of microseismic oscillations polarized at random, with the aid of polarization filtering. The nature of the polarization usually becomes more complex as the first event is followed by subsequent phases. The interval in which the recording of pure linearly-polarized oscillations is possible depends on the conditions of superposition of subsequent waves on the direct wave. In the practice of seismic prospecting this interval is usually very short. In the majority of cases, the first wave on the records is immediately followed by waves excited by inhomogeneities at the sites of the source and the seismometer, and the first longitudinal wave on the seismogram ceases to be linearly polarized.

When observations are made close to interfaces, the seismometer records total oscillations formed as a result of superposition of the initial incident waves and the secondary waves excited by them. This is true both of VSP and surface observations. In surface observations, the particle motion, because of the superposition of secondary reflected waves on the longitudinal incident wave, will move in a direction that does not coinicde with the direction of arrival of the longitudinal wave.

The waves recorded in seismology have much longer periods than those recorded in seismic prospecting and are much less distorted at inhomogeneities in the supper part of a section. On account of this, a much more rigorous linear polarization of oscillations in the first wave is observed on earthquake records than on shot records in seismic prospecting.

TRANSVERSE WAVES

Most general laws governing the propagation of seismic waves are true both for longitudinal and transverse waves. However, there is a substantial distinction between the polarization of transverse and longitudinal waves. The action of a transverse wave on the seismometer depends on the orientation of the seismometer, and in which of the planes passing through the direction of propagation of the wave the transverse oscillation takes place. In a homogeneous isotropic medium the transverse waves are polarized in a plane tangent to the *P* wave-front, that is, the vector of particle motion lies in the above-mentioned plane and is perpendicular to the ray. In contrast to longitudinal waves for which the polarization of oscillations is determined solely by the direction of wave propagation, the polarization of oscillations in case of a transverse wave depends also on the type of the source.

The nature of particle motion is determined by the orientation of forces acting at the source. The waves excited in a medium with a vertical axis of symmetry by sources in which the force acts in a horizontal plane will be of the transverse *SH* type, and they will be propagated in a direction perpendicular to that of the force and have a horizontal polarization, that is, parallel to the force. Such waves are of special interest, because their propagation in axially symmetrical media does not result in the formation of converted waves at the interfaces, and the wave field retains a comparatively simple pattern. This is the reason why the method of transverse reflected waves is based on the use of *SH* waves. A characteristic property of such waves is that the direction of particle motion varies with change in the direction of the acting force. This property is widely used to improve the signal-to-noise ratio. This is done by subtracting signals from oppositely-directed perturbations. However, even blasts in shallow holes normally used in seismic prospecting also always produce, besides longitudinal waves, transverse *SV* waves. The predominant plane of polarization of the *SV* waves is the vertical plane, and in contrast to the *SH* waves they may be excited by sources of various types.

Even in homogereous layered media containing non-parallel interfaces the refraction of a transverse wave may result in the formation of two waves propagating along a single ray, if the polarization of the original wave does not lie in the plane of incidence and is not normal to it. If refraction takes place at an angle exceeding the maximum, a phase difference appears between the waves, and the resultant wave will be transverse elliptically-polarized. Similar phenomena are observed in inhomogeneous as well as in anisotropic media. In the latter the nature of elliptical polarization changes as the result of variation in the phase difference along the line of wave propagation.

Transverse waves excited by earthquakes depend greatly on the forces acting at the source and on the nature of their polarization. It is worth noting that whereas, during interference of two simple longitudinal waves propagated in the same direction, the directions of their oscillations also coincide, the interference of two transverse waves propagated in the same direction may take place under conditions when the directions of their oscillations do not coincide, since in a transverse wave the oscillations may generally take place in any direction perpendicular to the direction of propagation.

SURFACE (RAYLEIGH) WAVES

Rayleigh waves on the surface are elliptically polarized. The polarization plane is vertical, there being a phase shift of a quarter of a period between the components $A_z(t)$ and $A_x(t)$. The ratio of the vertical axis of the ellipse to its horizontal axis in surface observations, with a velocity ratio $v_p/v_s = 1.73$ (Poisson's coefficient $x = 0.25$), is equal to 1.47. The ellipse is elongated in the z direction, and the particles of the medium move in their trajectories in the direction of the source of oscillations. The amplitude of the Rayleigh wave decreases with depth. The depth of penetration of excitation caused by a Rayleigh wave greatly depends on the wavelength, the longer waves penetrating deeper. The variation in the amplitudes of the horizontal and the vertical components with depth is not the same. As a result, not only the trajectories of particle motion change with depth, but the direction of their motion, as well. At a certain depth, the amplitude of the horizontal component vanishes, and the Rayleigh wave has a vertical component only. Such patterns hold only in homogeneous media. In the true Earth the nature of the variation in the trajectories with the depth may be much more complicated.

SURFACE (LOVE) WAVES

In contrast to the Rayleigh waves, Love waves are linearly polarized in the horizontal plane at right angles to the direction of propagation. Hence, the polarization of Love waves is the same as that of transverse *SH* waves, and this can make the tracking of reflected transverse *SH* waves much more difficult, and limit the depth range of the method.

INTERFERENCE OF WAVES

The characteristic features of polarization of various waves discussed above can be observed only if the waves are recorded in a pure form, that is, without other waves superimposed on them. This situation is quite rare in the true Earth which, as a rule, has a very complex structure. In the majority of cases the wave field is the result of interference of waves of different types. This is to a large extent responsible for the difficulties involved in the interpretation of wave fields. The polarization, being a very sensitive element of the wave field, is greatly affected by the parameters of the interfering waves and may display great variety and complexity. This makes polarization a more rigorous criterion for testing the 'purity' of waves than kinematics, and enables it to be used to detect even negligible deficiencies in such purity.

MICROSEISMIC OSCILLATIONS

Microseismic oscillations are characterized by unstable polarization. This means that the phase difference between various recorded components is a random quantity. The records of different components in individual intervals of the seismogram may be in phase, and the records themselves be of a similar form, but this will be limited to a comparatively short interval. During a large time interval the correlation coefficient will tend to zero. This does not apply to microseismic oscillations from concentrated sources, in which case the oscillations are represented by regular waves with quite stable polarization.

Brief Review of Papers Dealing with the Study of Polarization of Seismic Waves

The first results obtained in the study of polarization of seismic waves are connected with the name of the prominent seismologist B. B. Golitsyn who in 1909 suggested a method for determining the direction towards the epicentre of an earthquake from the direction of particle motion in the first longitudinal wave [59]. To this end he first made use of a three-component cluster of seismometers. The first experiment in the use of polarization of seismic waves had a very marked effect on the evolution of seismology, since it enabled the location of an earthquake to be determined from observations performed at a single station. At present, too, the overwhelming majority of the world's seismic stations are equipped with three-component seismometer clusters of various designs with a variety of parameters. Another reason why three-component observations are needed in seismology is that not only longitudinal, but also transverse waves are used in the study of earthquakes. During the first stages of development, seismic prospecting employed observation methods borrowed from seismology, that is, it also made use of three-component observations. However, in the 'thirties frequency discrimination and multi-channel recording resulted in the development of phase correlation, and this made it possible to introduce correlation of wave phases instead of tracking times of first arrivals and thus to use the subsequent part of the record for the purpose of seismic prospecting. This served as the corner stone for the method of longitudinal reflected waves. Three-component observations at disconnected, comparatively remote points have been replaced by continuous tracking of the phases of longitudinal reflected waves along the line of the profile. The wave travel-time curves still remain the principal element in the analysis of wave fields. The subsequent development of seismic prospecting methods on the whole followed the line of perfecting the methods of separation and tracking of vertical components of longitudinal reflected waves along the traverse.

In the 'fifties some comparatively small-scale work on two- and three-component recording of waves was carried out in conjunction with research aimed at developing the method of converted transmitted waves [21]. However, because of difficulties involved in the identification of converted waves, research in that direction made little progress at that time.

AZIMUTHAL OBSERVATIONS

The idea of an extensive use of wave polarization to improve the efficiency of seismic prospecting and seismological exploration was first suggested in the Fifties by Gamburtsev [51–57]. He invented and developed, together with the present author, a new type of wave correlation (azimuthal phase correlation) based on the tracking of phases of seismic waves as functions of orientation in space of the oscillation components for a constant position of points of observation. Azimuthal phase correlation was performed with the aid of polarization seismograms obtained with special multi-component seismometer clusters. The development of such research resulted in the creation of the azimuthal method of seismic observations [24]. Fundamental equations of azimuthal phase correlation of seismic waves have been deduced and criteria to be used in the selection of waves of the principal types according to their polarization, have been formulated. Methods of processing azimuthal seismograms and determining the polarization parameters of seismic waves have been devised. The azimuthal method made possible the study of polarization of seismic waves on a substantially larger scale. Azimuthal-phase correlation, which makes it possible to assess the state of polarization of waves visually, using primitive instrumentation, has helped to obtain in a comparatively short time voluminous information about the polarization of seismic waves in the true Earth.

It has been demonstrated in numerous azimuthal observations performed in the course of seismic prospecting and seismological studies carried out under various seismological conditions, that the polarization of seismic waves is one of the most senstive characteristics of the wave field. The overwhelming majority of seismic waves are not linearly polarized, owing to superposition of waves of different types. Usually the least distorted is the polarization of the first arrivals on the seismograms and waves with dominant amplitudes on the subsequent part of the record. The polarization of seismic waves is unstable along the line of observation and is very sensitive to all inhomogeneities. It has also been demonstrated that the wave fields observed in the true Earth are represented by an extensive class of waves of different origin and type, and that, apart from longitudinal waves, a large number of transverse waves of different origin are usually recorded. Moreover, and this is very important, this work has proved that it is possible on the basis of polarization studies, simultaneously to separate, track, and use all the principal wave types for the exploration of the Earth.

The azimuthal method of seismic studies has been further developed by Bondarev [14–16]. Specifically, he devised analytical methods for determing the directions of particle displacement and the parameters of elliptically-polarized oscillations and estimated the accuracy of determination of the direction of the vector of particle motion and of the spread in the amplitude characteristics of the channels.

Azimuthal observations have demonstrated that the low-velocity layer may have a marked effect on the directions of particle motion. Hence, for the study of wave polarization in the true Earth it was first of all necessary to make drill-hole observations.

OBSERVATIONS IN DRILL-HOLES

In 1959, IEP carried out drill-hole studies of wave polarization. Three-component observations in drill-holes have been carried out in regions with different structures: a foothill downward (Western Ukraine), the Russian Platform (the Kuibyshëv District and the northern part of the Krasnodar Krai (province), and an intermontance depression (the Fergana Valley). The trajectory of motion has been studied together with the process of propagation of waves in space [28–32, 42, 43]. The method of tracking waves in drill-holes has resulted in the creation of the method of vertical seismic profiling (VSP) which has become the main method of experimental studies of seismic waves in the true Earth in the frequency range employed in seismic prospecting, and which has helped to improve the efficiency of all the modifications not only of drill-hole, but also of surface seismic observations.

Studies of particle-motion trajectories have received a great impetus from the investigations of converted transmitted waves along vertical profiles carried out in the Volga Basin near Volgograd [36, 50]. Polar and polarization-positional correlations were first employed here [1]. The combination of wave discrimination, based on the direction of particle motion (CDR–I) with discrimination based on wave-propagation direction

(CDR–II), performed with the aid of a digital computer, has made it possible to separate converted transmitted waves in a complex wave field and to study them in media made up of a succession of high-velocity carbonate and low-velocity terrigenous deposits, characteristic of the southern part of the Russian Platform. From the methodological point of view these studies may be regarded as the beginning of the polarization method.

However, studies of polarization in VSP observations, as has previously been the case with surface observations, have subsequently been dropped, and the study of wave propagation in space has become the leading trend. Accordingly, three-component observations have everywhere been dropped in favour of recording the vertical components of oscillations only. One explanation for this is that even the recording of the Z components alone makes it possible to obtain information needed to improve the efficiency of seismic prospecting comparatively easily and rapidly. Another explanation is that specific difficulties were involved in three-component observations in deep drill-holes, and the technology of such observations was far from perfect.

POLARIZATION OF TRANSVERSE (REFLECTED AND CONVERTED) WAVES

Recent years have witnessed substantial progress in the study of polarization, especially of transverse waves. This has largely been due to long-term large-scale research aimed at developing the method of transverse reflected waves carried out at the All-Union Scientific Research Institute for Geophysics (VNII Geofizika) and at the Siberian Branch of the Academy of Sciences of the USSR (SO AN SSSR) under the direction of Puzyrev [6, 8, 9, 82, 92–94, 102, 104–108, 130–132]. The polarization of converted transmitted waves has been studied as part of the development program of the method based on the utilization of waves of this type from remote earthquakes. Extensive research in this direction has been carried out at VNII Geofizika by Pomerantseva, Egorkina *et al.* [168–71, 100, 101, 112]. It should be pointed out that much greater attention has been paid to the study of polarization of waves in seismology than in seismic prospecting, which is based mainly on the study of longitudinal reflected waves. This situation still persists, mainly on account of special conditions of seismological observations with the stations, as a rule, far apart, so that it becomes necessary to obtain data directly from observations made at each individual station. To this end transverse waves are also used in data processing.

The development of the method of transverse waves and its introduction into prospecting practice for the purpose of regional exploration in the last 10 years has greatly extended the scale of field operations employing three-component recording of remote earthquakes. The 'Zemlya' ('Earth') apparatus, designed by Moszhenko, featuring magnetic recording, has greatly increased the potentialities for studying the transmitted PS waves.

In developing tests for the identification of P and PS waves, information has been obtained on the polarization of converted waves and other dynamic characteristics. Since both methods (of reflected transverse and of transmitted converted waves) deal with the polarization of transverse waves, let us discuss the results of such studies together.

A result that may be regarded as important in this respect is the method of discrimination of transverse reflected SH waves, based on their polarization [58, 82]. This method is based on the summation of excitations of opposite signs to suppress waves without phase inversion, primarily longitudinal waves (in some field operations an increase in the ratio A_S/A_P as high as 10 has been obtained). Such discrimination is now used in practically all operations employing the method of transverse reflected waves.

In the course of studies of propagation velocities of transverse waves needed to interpret the reflected waves it has turned out in many cases that the propagation velocities of the *SH* and *SV* waves greatly depend on the direction in which they are propagated [8, 74, 85, 132]. Three-component drill-hole observations [8, 9] have demonstrated that the polarization of transverse waves is frequently not linear, that the polarization of the *S* waves is very senstive to inhomogeneities in the Earth, and that the main factors determining the polarization of transverse waves in space are the orientation of the interfaces and the anisotropy of elastic properties of the Earth.

In the period from 1972 to 1974 VNII Geofizika carried out a large amount of work in regions of salt-dome tectonics (Gur'evo District) with the aim of studying the anisotropy of the true Earth and the polarization of waves in complex media. In this work surface and drill-hole three-component observations have been combined with the use of sources with controlled artificial directivity and with the polarization discrimination of waves at the source. Similar work has later been carried out in the same region by the Siberian Geophysical Expedition [121]. In those studies the indicatrices of ray velocities of the longitudinal *P* and the transverse *SV* and *SH* waves were obtained. It was also demonstrated that the polarization of the converted *PS* waves (both reflected and transmitted) measured in the holes does not remain constant, but varies along the vertical profile both continuously (in homogeneous strata) and abruptly across sharp interfaces with an increase in the ratio A_y/A_x usually accompanying the progress of the wave. The large values of A_y of converted waves observed, exceeding the theoretical values by an order of magnitude, are due not only to the location of the conversion interfaces (their dips) and to the properties of the Earth through which the waves are propagated, but also to the specific conditions of wave formation at the actual interfaces. The studies have resulted in the discovery that waves with specific polarization may be inherent to specific interfaces, no matter what their inclination, and that conversion is accompanied by the appearance of an intensive component in the immediate vicinity of the interface.

The polarization of waves is greatly dependent on their frequency: the higher the frequency, the more intense the tangential component. A difference in the spectra of the components is already noticeable in the vicinity of the conversion interface. Variation in the polarization of waves of different frequency with depth follows a different pattern. Anomalous polarization is, as a rule, characteristic of waves connected with sharply-defined interfaces which usually are erosional surfaces, angular nonconformities, etc.

Similar results were obtained during studies of converted transmitted waves on records of remote earthquakes [18, 19, 68, 69]. In the majority of cases the observed intensity of the waves on the *Y* components appreciably exceeded the theoretical value. Different explanations for this have been advanced. The anisotropy of the elastic properties of the Earth's crust was considered to be the main reason for the anomalous polarization. It has been theoretically demonstrated [78] that the combined effect of the anisotropy of velocities and the dips of the interfaces is responsible for intensification of oscillations on the *Y* components, and this agrees with the experimental date. Tsibul'chik [2, 127] classified such waves as belonging to lateral waves of the surface type connected with the partial transformation of energy of the three-dimensional wave during the process of its interaction with the inhomogeneities of the upper part of the section. Papers by

Puzyrev, Obolentseva, Brodova, and Bakharevskaya [6, 65, 92, 94, 105, 130] contain theoretical and experimental data on the polarization of longitudinal and transverse waves reflected from an oblique interface. On the basis of those studies, it has been demonstrated that anomalous polarization of transverse waves may be explained as a result of the combined effect of the anisotropy of velocities of propagation of the *SH* and *SV* waves and the presence of oblique interfaces.

The fact that, without taking account of anisotropy, it has been difficult to interpret dynamic parameters conditioned the study of the kinematics and dynamics of waves in anisotropic media. Obolentseva has made theoretical studies of the polarization of various types of longitudinal, transverse, and converted waves in space (reflected, head, and transmitted) for typical models of geological media, in particular, for media containing oblique interfaces, both isotropic and anisotropic [65, 78, 92–94]. With the aid of the ray method, a methodology has been devised in zero approximation for finding the displacement fields in three-dimensional media, and the effect of various factors (dip of the interface, its curvature, inhomogeneities in the overlying sequence, and anisotropy of elastic properties of the medium) has been studied. A comparison of the experimental data on polarization of a converted wave in space reflected from an oblique interface with theoretical data for isotropic and anisotropic models favours the latter. It has been discovered that the polarization of transverse waves may be quite unstable, since it is greatly affected by the presence of thin strata not greater than a wavelength thick. Such studies have demonstrated that the use of polarization offers good prospects for evaluating the correspondence of the true Earth to the isotropic thick strata model.

In recent years a tendency has become apparent to employ wave polarization to gain direct information about the structure and the properties of the Earth. Egorkina [68–70] has studied the effect of a planar oblique interface on the polarization of waves as a function of the parameters of the interface and the incident wave, and has devised a method of determining the parameters of the interface (the orientation of the interface in space and the angle of incidence of the wave on the interface from below) based on the polarization of the *P* and the *PS* waves. Such information may not only substantially improve the accuracy of data on the depths of conversion interfaces obtained from the time differences between the *P* and the *PS* waves, but also helps to evaluate the errors arising out of the situation that the dips of the interfaces have not been taken into account in the process of interpretation. The polarization of seismic waves has been used for mapping structures in resolving problems in drill-hole seismic prospecting [31, 103].

Apart from experimental studies, there have also been theoretical studies of wave polarization. Rudnitsky [112] has carried out a theoretical analysis of polarization of converted seismic oscillations in the form of pulses as a function of their parameters, thereby making an important contribution to the separation of converted transmitted waves on the seismograms. Despite much field work involving the method of converted transmitted waves excited by distant earthquakes, carried out by industrial and scientific research organizations, the origin of the waves that constitute the initial part of the seismograms is still not quite clear. The prospecting potentialities of this method, too, are still open to discussion, owing largely to the absence of sufficiently physically-based tests for the identification of the *P* and *PS* waves associated with a common interface.

A paper [95], containing the results of processing of observational data on converted transmitted waves recorded in the Altai-Sayany region, demonstrates that the azimuthal direction of particle motion excited by the passage of converted waves may differ from that towards the source. Generally there is no connection between these directions. The existing criteria for separating converted waves (dominant amplitudes on the horizontal components, orientation of oscillations in the plane of the ray of the longitudinal wave, coincidence of the signs of arrival of the *PS* and *P* waves recorded on the same horizontal component) did not enable the converted *PS* waves from the Mohorovičić interface to be separated. On the contrary, proof is presented that the horizontal components in the initial part of the seismogram are, on the whole, not *PS* waves, since they are characterized by independence of the amplitudes on the distance to the epicentre; absence of correlation between the direction of displacement and that towards the source; and substantially larger values for the observed amplitudes of the horizontal components (3 times and more) as compared with the maximum theoretical values. The authors of [95] insist that the horizontal components obstruct the separation of the converted *PS* waves, but do not discuss their origin.

MICROSEISMIC OSCILLATIONS

Wave polarization is widely used in the study of microseisms. Extensive research has been carried out both in the USSR and abroad. Walzer [157], who has studied the polarization of microseismic oscillations in the range of periods from 2 to 10 s, has demonstrated that most microseisms caused by storms and recorded by continental stations are plane-polarized oscillations, the polarization plane coinciding with the direction of propagation of the wave in microseismic oscillations. The proportion of linearly-polarized oscillations is less than 1%. This has made it possible to devise a method for improving the useful sensitivity of the instruments (the signal-to-noise ratio), based on the difference between the polarization planes of the microseisms and three-dimensional waves from earthquakes. A great deal of research on the polarization of microseisms has been carried out by Strobach [153, 154].

Various pieces of equipment have been devised to facilitate the processing of data from three-component observations. Specifically, Strobach has developed a system of stereographic fan-recording [156, 157]. Various types of polarization sets have also been used, intended to record the horizontal component of the oscillation vector or to obtain azimuthgrams [83]. A set-up for the mathematical processing of three-component seismograms has also been proposed [91, 128].

EVOLUTION OF THE POLARIZATION METHOD

The above analysis demonstrates that, up to the present time, the polarization of seismic waves has been studied mainly with the aim of devising methods based on the recording of transverse (uniform or converted) waves. Polarization has practically never been studied or used with the aim of improving the efficiency of seismic prospecting based on the study of longitudinal waves.

This was the reason why, after VSP has begun to be extensively employed in the practice of seismic observations, IEP continued three-component observations aimed at improving the efficiency of seismic studies of complex media. This required first of all the perfection of instrumentation and technology of the three-component observations themselves. Such research was carried out in the period from 1969 to 1972 by IEP in collaboration with KazVIRG in the ore deposits of Central Kazakhstan (Nura-Taldy, Zhairem, and Sayak). The structure of ore-rich regions is more intricate than that of oil-rich regions. They are characterized by high-velocity media differentiated comparatively little with respect to the propagation velocities of seismic waves, as well as by steeply-dipping interfaces and those with a complex relief, by the prevalence of high-velocity rock in the section, by the frequent absence of a low-velocity layer, by anisotropy of velocities not only in the vertical, but also in the horizontal direction, and by the instability (and often the absence) of datum horizons. Under such conditions the waves are propagated in different planes and in different directions. An additional complication is that ore bodies which are of geological interest are usually located at comparatively shallow depths. For this reason the useful waves are recorded in the initial part of the seismogram, and the entire interval of useful recording may be limited to 1 s. As a rule, a complex wave field is observed within this interval.

During the process of research a new technology of three-component observations, and a new method of studying polarization of seismic waves have been developed [36–41, 47, 49, 78]. Methods of obtaining continuous seismic-instrumentation control [40], oriented records [116, 120, 125] and multi-component polar seismograms based on the pattern technique of the first kind [20, 47, 124, 126] have been proposed. The results obtained have demonstrated that wave polarization could be also used to good effect in seismic oil-prospecting. With this aim in view IEP, together with the 'Krasnodargeofizika' trust in 1973–1976, carried out observations in areas with good prospects for oil and gas. Such areas had a complex structure and were located along the southern margin of the Western Kuban' downward, characterized by high dips, and on the Taman Peninsula, characterized by widespread diapir tectonics where up to now seismic prospecting has been almost ineffective.

During the first stages of the work wave polarization was studied mainly in the context of VSP problems [1, 75]. The fact is that although the VSP method began to be widely used and became one of the main methods for studying seismic waves in the true Earth, which enabled to resolution of problems largely concerning the efficiency of seismic studies (choice of conditions of excitation, determination of the origin of the recorded waves, their stratigraphic correlation, evaluation of potentialities of seismic studies and choice of optimum methods in specific situations), the principal potentialities of VSP have not been fully realized. The study of polarization of seismic waves in drill-hole observations when the direction of particle motion changes at every point along the profile has created new possibilities for improving the efficiency of the VSP method [1, 75].

Subsequently extensive research on wave polarization has been carried out in surface observations as well. The accumulated experience has proved that wave polarization can be used in a variety of seismic studies in the study of excitation, in improving wave correlation, in the separation of converted waves, in the study of anisotropy of velocities and of the properties of the Earth, in mapping structures, etc. Polarization presents considerable interest in the study of complex media. It is our opinion that the combination of the discrimination of waves based on polarization and that based on direction of propagation constitutes one of the main tendencies towards improvement of efficiency of studies of complex media. We are also of the opinion that the simultaneous use of all types of waves envisaged by the polarization method and capable of drastically increasing the amount of information not only about the structure, but also about the material composition of the Earth, offers good prospects from the point of view of improving the efficiency of seismic studies.

It should be pointed out that the progress of polarization studies in seismic prospecting was hindered largely by the high labour costs of processing three-component observations. To overcome these difficulties and to computerize data processing, IEP in collaboration with the Computer Center of SB AS USSR in 1967 launched a polar correlation program [1] and in 1973 a program for calculating trajectories of motion [80]. In 1974 IEP in collabotation with the 'Krasnodarneftegeofizika' trust developed an automatic data-processing system (ADPS) called 'Polarizatsiya'. The results of pilot observations and the availability of ADPS 'Polarizatsiya' has enabled a regular program of studies of complex media to be launched.

In recent years polarization has been used to improve the useful sensitivity of instruments. This is especially important when both the signal and the noise lie in the same frequency range and in the same range of apparent velocities. To this end methods of polarization filtering have been devised based on the discrimination of waves according to polarization [61, 73, 134, 139, 140, 149, 150, 152, 157, 159]. This trend is at present making rapid progress thanks to the possibilities offered by digital data-processing.

To sum up, an adequate amount amount of information has accumulated to prove the expedience of studying polarization of waves for the purpose of improving the efficiency of various seismic studies.

Selected Results from the Theory of Harmonic Oscillations

Let us recall some results obtained in the theory of harmonic oscillations that will be required in the study of polarization of seismic waves. When analysing vector oscillations, it is convenient to operate with their projections on the coordinate axes. This makes it possible to replace vector oscillations by the sums of scalar oscillations. Each vector oscillation in a plane is described by two and, in space, by three scalar oscillations. A linearly-polarized oscillation characterized by a rectilinear trajectory, as a rule, is a simple oscillation and is not the result of interference of several oscillations. However, a linearly-polarized oscillation may also be the result of the superposition of two or more oscillations with different frequencies, but with identical displacement directions, or with different displacement directions, but with identical frequencies and phases. The sum of two oscillations with identical frequencies and displacement directions, but with different phases

$$A_1 = A_{01} \sin\left(\frac{2\pi}{T} + \beta_1\right), \qquad A_2 = A_{02} \sin\left(\frac{2\pi}{T} + \beta_2\right),$$

is a harmonic oscillation of the same frequency and displacement direction whose amplitude A and phase β are expressed in terms of amplitudes A_{01} and A_{02} and of the phases β_1 and β_2 of the component oscillations by the equations:

$$A^2 = A_{01}^2 + A_{02}^2 + 2A_{01}A_{02} \cos(\beta_1 - \beta_2), \tag{2}$$

$$\tan \beta = \frac{A_{01} \sin \beta_1 + A_{02} \sin \beta_2}{A_{01} \cos \beta_1 + A_{02} \cos \beta_2} \ . \tag{3}$$

The the case of superposition of an arbitrary number of oscillations with an identical frequency

$$A_1 = A_{01} \sin\left(\frac{2\pi}{T} + \beta_1\right); \qquad A_2 = A_{02} \sin\left(\frac{2\pi}{T} + \beta_2\right), \ldots,$$
$$A_n = A_{0n} \sin\left(\frac{2\pi}{T} + \beta_n\right),$$

the resultant oscillation is a harmonic oscillation with the same frequency

$$A = A_0 \sin\left(\frac{2\pi}{T} + \beta\right).$$

Here

$$A_0 = A_{01}^2 + A_{02}^2 + \ldots + A_{0n}^2 + 2A_{01}A_{02} \cos(\beta_1 - \beta_2) + \ldots + 2A_{0n-1}A_{0n} \cos(\beta_{n-1} - \beta_n) \tag{4}$$

$$\tan \beta = \frac{A_{01} \sin \beta_1 + A_{02} \sin \beta_2 + \ldots + A_{0n} \sin \beta_n}{A_{01} \cos \beta_1 + A_{02} \cos \beta_2 + \ldots + A_{0n} \cos \beta_n} \ . \tag{5}$$

The summation of two or more oscillations with different displacement directions may also produce a vector oscillation with a variable displacement direction. Hence, during the passage of seismic waves the particles of the Earth move not only in rectilinear trajectories, but also much more often describe more intricate trajectories. Very often such trajectories are represented by a second-order curve (ellipse). A wave with an elliptical trajectory is termed an elliptically-polarized wave.

An elliptically-polarized oscillation is generally the result of superposition of two oscillations with the same frequency polarized in different directions. The interference of two mutually perpendicular oscillations, depending on their phase difference, produces the following specific cases:

1. The oscillations are in phase, $\beta_1 - \beta_2 = \pm 2\pi n$. In this case the trajectory of motion is a straight line.
2. The oscillations are in opposite phases, $\beta_1 - \beta_2 = \pm(2n + 1)\pi$. The resulting oscillation is also linearly-polarized, but perpendicular to the trajectory considered in the first case.
3. The phase difference is $\beta_1 - \beta_2 = (\pi/2) \pm 2\pi n$. The oscillation is elliptically-polarized (the coordinate axes coincide with the principal axes of the ellipse) with the particle moving clockwise.
4. The phase difference is $\beta_1 - \beta_2 = (\pi/2) \pm (2n + 1)\pi$. The trajectory coincides with that of the preceding case, but the particle moves counter-clockwise.

We may show that in the case of interference of two linearly-polarized oscillations with an identical period, but with arbitrary phases, amplitudes and directions of arrival we may always find two mutually-perpendicular directions in which oscillations take place with a phase difference of 90°.

Consider two oscillations $A = A_0 \sin [(2\pi/T)t]$ and $B = B_0 \sin [(2\pi/T)t]$ making an angle γ (Figure 2a). We shall find directions of the rectangular coordinate axes ξ and η in

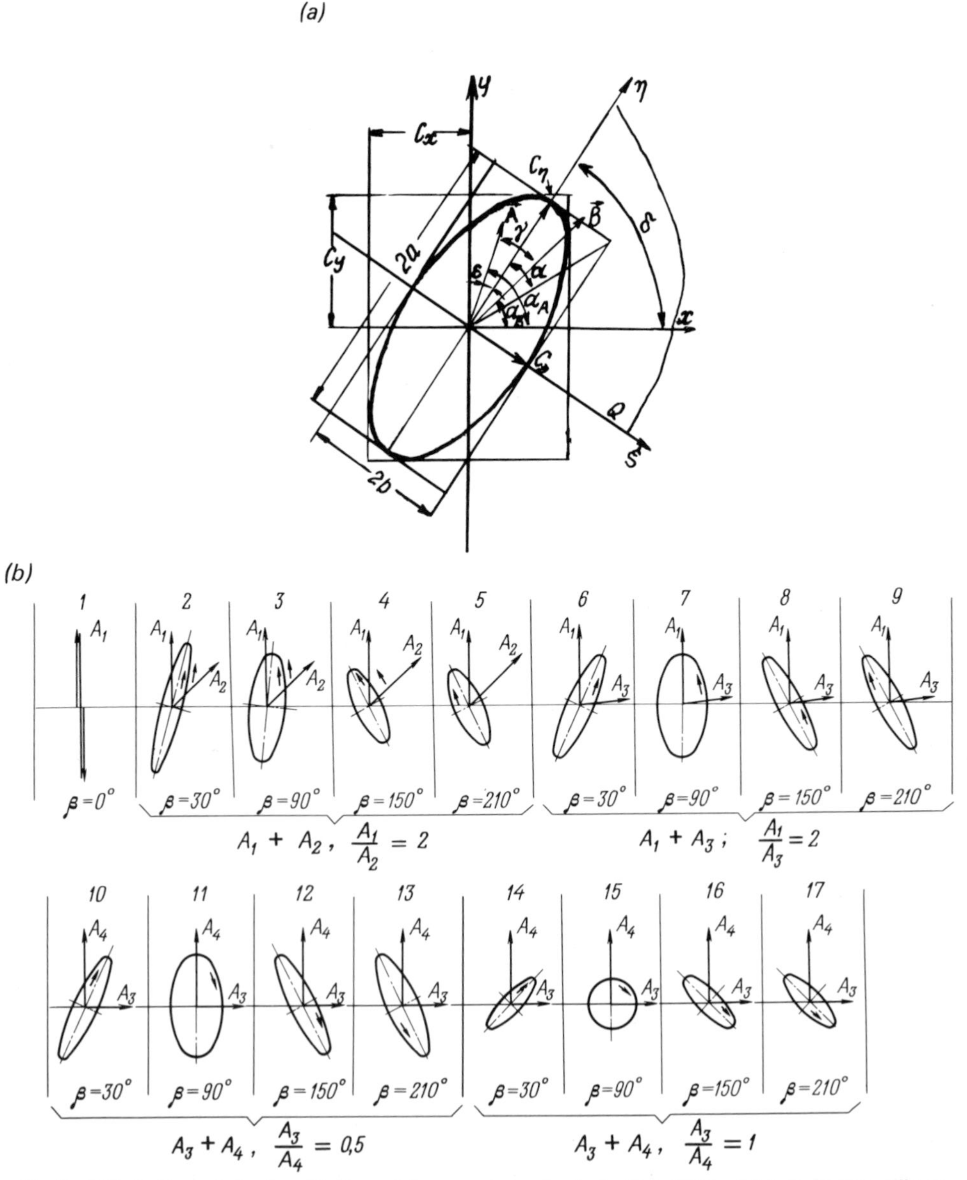

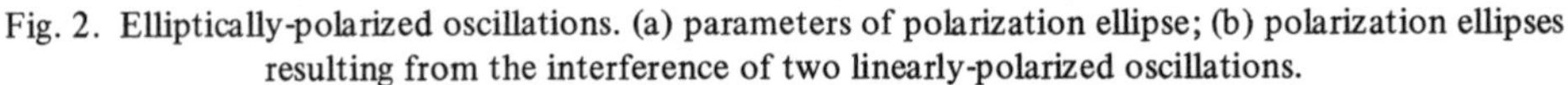

Fig. 2. Elliptically-polarized oscillations. (a) parameters of polarization ellipse; (b) polarization ellipses resulting from the interference of two linearly-polarized oscillations.

the plane Q of the oscillation vectors such that the vectors **A** and **B** projected on them would have a phase difference of 90°. The sums of projections of the vectors **A** and **B** in the directions ξ and η will be equal to

$$C_\eta \sin\left(\frac{2\pi}{T} + \beta_\eta\right) = A_0 \sin\frac{2\pi}{T} t \sin\epsilon + B_0 \sin\left(\frac{2\pi}{T} t + \beta\right) \cos(\epsilon + \gamma),$$

$$C_\xi \sin\left(\frac{2\pi}{T} + \beta_\xi\right) = A_0 \sin\frac{2\pi}{T} t \sin\xi + B_0 \sin\left(\frac{2\pi}{T} t + \beta\right) \cos(\epsilon + \gamma),$$

where ϵ is the angle between the vector **A** and the η axis in the plane Q.

The plase shifts β_η and β_ξ may be determined from the formulae

$$\tan\beta_\eta = \frac{B_0 \cos(\epsilon + \gamma) \sin\beta}{A_0 \cos\epsilon + B_0 \cos(\epsilon + \gamma) \cos\beta},$$

$$\tan\beta_\xi = \frac{B_0 \sin(\epsilon + \gamma) \sin\beta}{A_0 \sin\epsilon + B_0 \sin(\epsilon + \gamma) \cos\beta}.$$

Since, as presumed, the phase difference $\beta_\eta - \beta_\xi = 90°$, it follows that $\tan\beta_\eta = -\cot\beta_\xi$, or after the expressions are simplified

$$\tan 2\epsilon = \frac{-(B_0^2/A_0^2) \sin 2\gamma - 2(B_0/A_0) \cos\beta \sin\gamma}{B_0^2/A_0^2 \cos 2\gamma + 2(B_0/A_0) \cos\beta \cos\gamma + 1}. \tag{6}$$

This expression determines one of the directions of the rectangular coordinate axes η sought. In the specific case of $A_0/B_0 = 1$

$$\tan 2\epsilon = -\frac{\sin\gamma + \cos\beta}{\cos\gamma + \cos\beta}.$$

Note that the axes ξ and η are the axes of an interferential elliptically-polarized oscillation.

It can be demonstrated that the sum of an arbitrary number of linearly-polarized oscillations with an identical period can always be reduced to the sum of *two* linearly-polarized oscillations, and that therefore the interference of any arbitrary number of oscillations linearly polarized in one plane with arbitrary directions of arrival, arbitrary amplitudes and phases, but with equal periods will always produce interferential elliptically-polarized oscillations.

ELLIPTICALLY-POLARIZED OSCILLATIONS

The quantitative characteristics of an elliptically-polarized oscillation are the geometrical parameters of the polarization ellipse: its shape and its orientation in space, as well as the direction of particle motion alone the elliptical trajectory.

The shape of the ellipse is characterized by its form-factor (ellipticity) the magnitude of which is determined by the ratio of the semi-axes $m = b/a$, where b and a are the major and minor axes of the ellipse (Figure 2a). The form-factor is related to its excentricity l through the expression $l = \sqrt{m^2 - 1}$. The sign of the form-factor m is determined by the direction of particle's motion along the ellipse. If, when looked at in the direction of wave propagation, the particle is seen to move clockwise, m is positive, and the oscillation is right-polarized; a counter-clockwise motion means negative m and a left-polarized oscillation. This definition is not complete and applies only to waves whose direction of propagation does not lie in the polarization plane. In some cases it will be convenient to to use the angle of ellipticity $\alpha = \operatorname{arc\,tan} m$.

The angle α is equal to one half of the acute angle between the diagonals of the rectangle in which the ellipse may be inscribed, and whose sides are parallel to the principal axes of the ellipse. The angle of ellipticity uniquely determines the shape of the ellipse, and the sign indicates the direction of the radius of rotation of the vector. The factor m and the angle α are independent of the choice of the coordinate frame. The position of the ellipse in the polarization plane is determined by the angle δ between the major axis of the ellipse and the abcissa of the specified rectangular coordinate frame $x\,O\,y$ (Figure 2a).

The shape of the polarization ellipse and its orientation in the plane depend substantially on the parameters of the interfering oscillations. Consider the orientation of the polarization ellipse with respect to the directions of two interfering linearly-polarized oscillations with an identical period, written in the form

$$A = A_0 \sin\left(\frac{2\pi}{T} t + \beta_1\right), \qquad B = B_0 \sin\left(\frac{2\pi}{T} t + \beta_2\right).$$

and making angles α_A and α_B with the abcissa of the rectangular coordinate frame $x\,O\,y$.

The principal axes of the ellipse coincide in direction with the coordinate axes ξ and η, which make an angle δ with the axes $x\,O\,y$, as determined by the equation

$$\tan 2\delta = \frac{\sin 2\alpha_B + M^2 \sin(2\alpha_B + 2\gamma) + 2M \cos(\beta_2 - \beta_1) \sin(2\alpha_B + \gamma)}{\cos 2\alpha_B + M^2 \cos(2\alpha_B + 2\gamma) + 2M \cos(\beta_2 - \beta_1) \cos(2\alpha_B - \gamma)}, \tag{7}$$

where $\gamma = \alpha_A - \alpha_B$ is the angle between the interfering oscillations, and $M = B_0/A_0$ is the ratio of their amplitudes.

Hence, in the case of interference of two harmonic oscillations the position of the polarization ellipse in the polarization plane is uniquely determined by the direction, the amplitudes and the phase differences of the interfering oscillations.

The factor m can be determined with the aid of the formula

$$m^2 = \frac{\cos^2(\alpha_B - \delta) + M^2 \cos(\alpha_A - \delta) + 2M \cos(\alpha_B - \delta) \cos(\alpha_A - \delta) \cos(\beta_2 - \beta_1)}{\sin^2(\alpha_B - \delta) + M^2 \sin(\alpha_A - \delta) + 2M \sin(\alpha_B - \delta) \sin(\alpha_A - \delta) \cos(\beta_2 - \beta_1)}.$$

It follows from this expression that for arbitrary values of the angle γ between the directions of the interfering oscillations and those of the phase difference $(\beta_2 - \beta_1)$, an increase in the amplitude of one of the interfering oscillations brings one of the principal axes of the ellipse closer to the direction of the dominant oscillation:

$$\delta \to \alpha_A \quad \text{for} \quad M \to 0, \qquad \delta \to \alpha_B \quad \text{for} \quad M \to \infty.$$

When the amplitudes are almost equal ($M = 1$), one of the principal axes of the ellipse takes up a position approximately midway between the directions of the interfering oscillations:

$$\gamma = \frac{\alpha_A - \alpha_B}{2}$$

The greater the difference between the ratio of the amplitudes of the interfering oscillations and unity, the more elongated the shape of the ellipse will be. The shape of the ellipse depends to a great extent on the angle between the interfering oscillations.

Figure 2b depicts polarization ellipses formed in different cases (1–17) of superposition of two linearly polarized oscillations. The three-component seismograms corresponding to these cases are depicted in Figure 8. If the angle between the interfering oscillations remains unchanged, the shape of the ellipse and the direction of particle motion will be determined by the phase shifts β between the oscillations. The interference of two mutually-perpendicular oscillations of equal amplitude, but with a phase shift of 90°, results in a circularly-polarized oscillation (Figure 2b, case 15). For phase shifts between the interfering oscillations $\beta > 180°$, the particles of the medium rotate along the ellipse in one direction, and for $\beta < 180°$ they rotate in the opposite direction [see $\beta = 150°$ (8), and $\beta = 120°$ (9)]. If β changes sign, the ellipse retains its shape.

The particles in elliptically-polarized waves move in ellipses with a periodically variable velocity. A complete revolution takes a complete oscillation period. The interference of two or more elliptically-polarized oscillations also produces an interferential elliptically-polarized oscillation. To determine the polarization ellipse resulting from the superposition of two waves with arbitrary elliptical polarizations and with different amplitudes and phases, one may make use of the procedure for the synthesis of polarization ellipses widely employed in radar [76].

OSCILLATIONS POLARIZED IN SPACE

The interference of three or more oscillations with different periods whose trajectories do not lie in a single plane produces oscillations polarized in space (see Figure 33). The parameters of the spatially-polarized waves depend on the parameters of the interfering waves, and there may be a great variety of them. The trajectories of spatially-polarized waves are distinguished by their complexity. Accordingly, their directions of displacement may conveniently and graphically be represented in the form of stereographic projections. The individual sections of the surfaces described by the radius-vector can be approximated by planes. However, in general such a surface may be more complex.

WAVES OF PULSED SHAPE

In the presentation of general results from the theory of oscillations we have made use of steady-state oscillations and their superposition. In considering the theory of polar correlation, stationary harmonic vector oscillations are employed. It is characteristic of the superposition of such oscillations that the shape and the position of the ellipse (its orientation) in space do not change with time. Actually, the waves observed in seismic prospecting belong rather to the pulsed type and contain different frequencies. On account of this, strictly speaking, not only can the trajectories of particle motion have shapes slightly different from elliptical, but their shape and dimensions as well as the orientation of the ellipses in space may continuously change with time. Such distinctions are determined by the relationships existing between the oscillation periods and the envelopes of seismic pulses. If the period of the envelopes of seismic pulses greatly exceeds the period of the oscillations themselves, then the amplitudes of adjacent extrema are almost equal, and in this case the trajectory of particle motion consists of separate ellipses of almost regular shape, distinguished from each other by their dimensions and orientation with respect to the coordinate axes. The variations in shape and orientation of the ellipses may be regular or random. An example of superposition of two identical Berlage pulses with a phase shift of $\beta = \pi/2$ between them is shown in Figure 3.

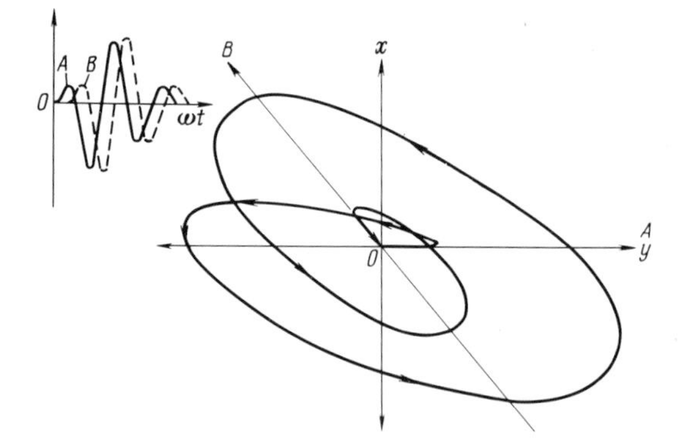

Fig. 3. Superposition of two Berlage pulses. Original pulses: $A = at^b e^{-c\omega t} \sin(\omega t + \beta_1)$; $B = at^b e^{-c\omega t} \sin(\omega t + \beta_2)$; $a = 1$; $b = 3$; $c = 0.5$; $\beta_1 - \beta_2 = 30°$. The plot shows the trajectory of particle motion.

Since polar correlation is based on the analysis of oscillations at a point, it is equally applicable to waves of pulsed shape, provided separate sections of such oscillations (up to one period) can be approximated by a section of a sinusoid. The validity of such an assumption is fully substantiated by experimental data.

Stereographic Projections

Before dealing with problems relating to polarization, we shall have to dwell on stereographic projections. This is necessary for the study of wave polarization, because we shall come across space trajectories the representation of which on a plane involves great difficulties.

For more than two thousand years, stereographic projection (the projection of a sphere on a plane) has been used in astronomy to depict the dome on a plane. Later, stereographic projection has been employed for the same purpose in cartography. Towards the end of last century, the stereographic projection began to be successfully used in studying the angles between lines and planes in space. It is this application of stereographic projection that is of interest to us, since it facilitates the solution of problems connected with the study of particle trajectories at a point and the propagation of seismic waves, which are in essence three-dimensional. We shall not dwell on the theory of stereographic projections presented in several publications [26, 86, 109], but shall cite only their principal properties that enable directions in space to be studied on a plane with the aid of their intermediate transformation on to a sphere.

1. There is a direction in space described by two angles: the angle ω in the horizontal plane and the angle φ in the vertical plane corresponding to each point in a plane.
2. The entire upper or lower hemisphere may be represented inside a circle.
3. The angles between the arcs of great circles on a sphere are equal to the angles between the arcs of their projections, that is, between the lines in a plane.
4. The arcs of both small and great circles are projected as arcs of circles or in specific cases as straight lines (for general purposes, the latter may be regarded as arcs of circles of infinite radius).
5. The stereographic projection is an equiangular projection, that is, the angle between the projections of spherical lines is equal to the angle between the lines on the sphere themselves. This property of stereographic projections which is shared by some other projections is a necessary and sufficient condition for the projection to be a congruent one, that is, spherical figures having infinitesimal dimensions in all directions are projected as congruent infinitesimal figures.
6. The projection of a circle is a circle.

THE STEREO-NET

The term 'stereo-net' implies the projection of a degree graticule on a sphere on to a plane. The stereo-net, depending on the position of the projection plane, may either be polar (the projection plane coincides with the equator, and the point of observation with the nadir (the lower pole)) or meridional, when the point of observation lies on the equator, and the projection plane coincides with the meridian 90° from the observation point. The meridional nets are the most interesting for the purpose of seismic investigations. Let us consider them in more detail. Figure 4a shows the construction of a meridional stereo-net. The plane of the figure coincides with the meridian ZEZ_1 and the projection plane Q with the meridian ZMZ_1. The observation point lies on the equator EMS at point S. The equator EMS and the meridian ZEZ_1 are represented by two mutually perpendicular diameters MN and ZZ_1 of the projection circle, with the projection of the meridian coinciding with the polar axis of the sphere ZZ_1. The other meridians are represented by circles passing through the poles Z and Z_1 and intersecting the equator.

Consider the meridian ZM_1Z_1 φ degrees away from the point E along an arc of the equator. The distance from the projection of the point m at the intersection of the equator with the meridian M_1 from point O will be equal to

$$O_m = R \tan\frac{1}{2}\varphi,$$

where R is the radius of the sphere.

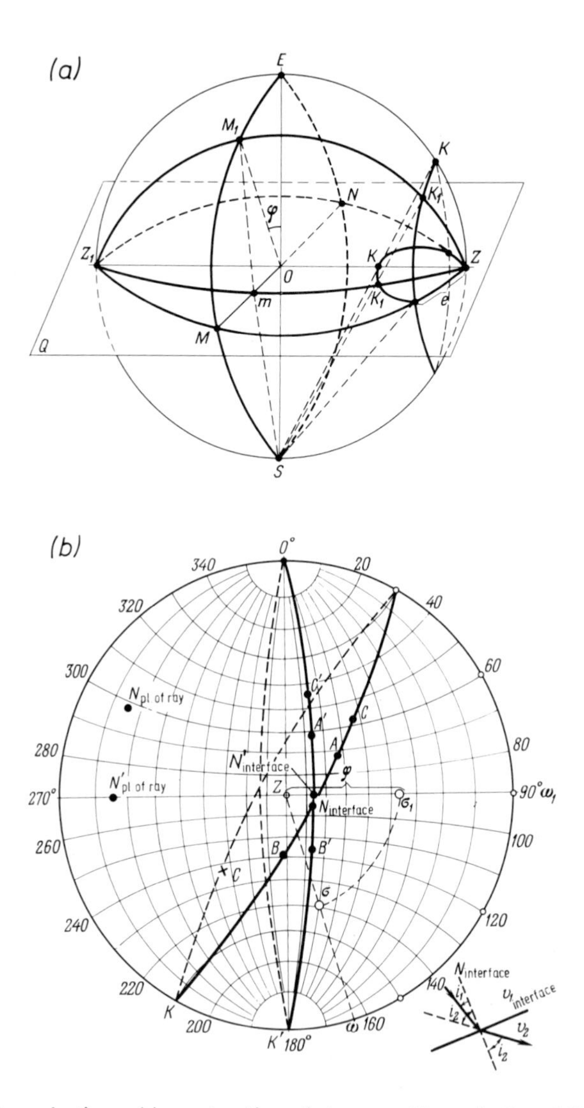

Fig. 4. Stereographic projections. (a) construction of stereographic projections [109]; (b) examples of plots on Wulff net.

The distance of the parallel ZK with the coordinate $e = 15°$ from the pole is equal to 15° along any meridian. Therefore, if one marks points on the projection circle 15° on both sides of the pole Z and a segment $ZK = 15°$ from the point Z along the straight line ZO on the stereographic scale, one will obtain three points lying on the parallel. One will then be able to draw the entire parallel.

To project directions in the upper hemisphere, that is, for $\varphi < 90°$, the point of observation is placed at the south pole, and *vice versa*. The projections of directions in the upper hemisphere are usually designated by dots, and in the lower by crosses. The vertical direction is projected in the centre of the projection plane. The stereographic projection of a plane passing through the centre of the sphere will be an arc of a great circle, that is, an arc terminating at a diameter. The stereographic projection of a horizontal plane will coincide with the circumference of the main projection circle. An oblique plane is projected as two symmetrical arcs of a great circle (in the upper and in the lower hemispheres). The curvature of the arcs (of the projection) decreases with increase in the inclination of the plane, and the stereographic projection of a vertical plane will be a straight line. The projection of a plane that does not pass through the projection centre will take the form of an arc of a small circle. Azimuths on the stereogram are measured along the circumference of the main projection circle from the northerly direction (in a clockwise direction), and the angle with the vertical is measured from the projection centre along a horizontal or vertical diameter.

In constructing projections one may conveniently use a meridional stereographic projection of the graticule on to a meridional plane. Such a graticule for a sphere of radius $R = 10$ cm with a 2° grid was first used in crystallography by Wulff and is named after him.

PRINCIPAL USES OF THE WULFF NET

The Wulff net is a lined guide-sheet, with the aid of which all drawings are made on transparent paper (waxed or tracing paper). Compass and/or ruler are not required. The tracing paper is centered, and dashes (indices) are drawn on it to mark the end of the meridian, which serves as the origin for measuring azimuths. In this way the initial position of the tracing paper is fixed. This index is subsequently used to return the tracing paper to its initial position. In seismic exploration, the main importance attaches to the study of directions in space and their variations.

Any direction in space is uniquely determined by two angles: the angle ω (azimuth), measured clockwise from north from 0° to 360°, and the angle φ the direction makes with the vertical, its range being from 0° to 180°. Consider first how to plot coordinate rays on the net and how to read the direction coordinates from the net.

Plot a direction with specified coordinates on the net. Let the coordinates of the direction be $\omega = 162^\circ$ and $\varphi = 54^\circ$. On the tracing paper measure on the net's outer circle in a clockwise direction an angle equal to the azimuth of our direction and mark point ω (Figure 4b). Rotating the tracing paper, make the point obtained coincide with one of the diameters of the projection circle. For example, let it be the horizontal diameter marked 90° (point ω_1). Measure the angle $\varphi = 54^\circ$ on the scale along the diameter from the projection centre to the circumference (point σ_1). With the tracing paper returned to its original position, the point σ_1 will assume the position σ. Determine the coordinates of the direction specified by point σ on the net (see Figure 4b). To this end draw a straight line from the centre line through point σ until it intersects the projection circle and measure the azimuth ω on the circumference. Next transfer the point to the equator (σ_1) and measure the angle φ from the center.

Now let us consider problems that occur in geometrical seismics.

Problem 1. Knowing the directions of the ray and the normal to the interface, find the position of the plane of the rays in space. The incident, the reflected, and the refracted rays together with the normal to the interface are known to lie in the same plane, that is, in the plane of the rays. Make use of this property to find the plane of the rays. For instance, the direction of the ray coming from a source is specified by the coordinates $\omega = 234°$ and $\varphi = 31°$, and the direction of the normal to the interface specified by the coordinates $\omega = 112°$ and $\varphi = 14°$. Imagine the projection centre to be placed on the interface at the point of incidence of the ray and plot the direction of the ray and the normal on the net. It is worth noting that the sign of the ray's azimuth will thereby be reversed, since the projection centre is placed at the point of incidence of the ray. The point A in Figure 4b with the coordinates $\omega = 54°$ and $\varphi = 31°$ corresponds to the direction of the incident ray, and the point N_{in} with the coordinates $\omega = 112°$ and $\varphi = 14°$ corresponds to the normal to the interface at the point of incidence.

To find the plane of rays, rotate the tracing plane, bringing both points into coincidence with the same meridian (A', N'_{in}) and draw this meridian on the tracing paper (in Figure 4b the lines plotted on the tracing paper are shown on the net as solid lines). Mark the normal to the plane of the rays on the tracing paper (point $N'_{p.r.}$). Now return the tracing paper to its initial position. This will enable us to determine the position of the plane of rays in space; in our case the coordinates of the normal $N'_{p.r.}$ to the plane of rays will be $\omega = 300°$ and $\varphi = 86°$.

If one of the directions lies above the plane of the plot, and the other below it, the points should be placed on symmetrical (equidistant from 0) meridians. This is correct, since the polar projection of the lower half of the meridian will be in the form of a symmetrical arc.

Problem 2. Find the angle between two directions in space. plot both directions with the aid of the net on the tracing paper (points A and B – see Figure 4b), and rotate the paper. Determine the plane in which both points lie. The points A and B will assume the positions A' and B', respectively. With the aid of the net, measure the angle between points A' and B' in this plane; in our case it will be 54°. This angle will be the angle between the directions specified in space.

Problem 3. Knowing the directions of the incident ray and of the normal to the interface, find the directions of the reflected, refracted, and grazing rays. To find the directions of these rays in space, we will have first to find the plane of rays, that is, to solve Problem 1, after which we shall be able to plot the directions of the rays of interest in the plane of rays. Let the point N_{in} (see Figure 4b) ($\omega = 112°$ and $\varphi = 14°$) correspond to the direction of the normal to the interface, and the point A ($\omega = 54°$ and $\varphi = 31°$) correspond to the direction of the inclined ray (with an azimuth of reversed sign). Then the plane of rays will be determined by the points A' and N_{in}. Find the directions of the reflected, refracted, and grazing rays in the plane.

The direction of the reflected ray is determined as follows. Since the reflection angle is equal to the angle of incidence, find with the aid of the net the angle between the incident ray and the normal (between points A' and N'_{in}), which will be equal to 27°, and measure this angle in the plane of rays from the normal in the direction opposite to that of the incident ray (point B'). Next return the tracing paper to its initial position; the point B' will accordingly move to point B, this point corresponding to the direction in which the ray is reflected. In our case the direction of the reflected ray is determined by the coordinates $\omega = 185°$, $\varphi = 30°$.

The direction of the refracted ray is determined in a similar fashion, the only difference being that in this similar fashion, the only difference being that in this case the angle measured from the normal will not be the reflection angle equal to the angle of incidence, but the refraction angle. The refraction angle is calculated from the angle of incidence and from the ratio ν_2/ν_1 of the velocities ν_1 in the first and ν_2 in the second layers in accordance with Snell's formula

$$i_2 = \arcsin i_1 \frac{\nu_2}{\nu_1},$$

where i_2 is the refraction angle; i_1 is the angle of incidence. For example, let $\nu_2/\nu_1 = 1.5$, then for $i_1 = 27°$ the refraction angle will be $i_2 = 43°$. Since the refracted ray will be below the plane of the drawing, plot a direction opposite to it on the net. As can be seen from the diagram of rays (in the lower right corner of Figure 4b), in order to do this, it suffices to plot an angle on the plane of rays equal to the refraction angle from the normal in the direction of the incident ray. The direction of the refracted ray is designated in Figure 4b by the point C'. Returning the tracing paper to the original position, we obtain point C with the coordinates of the refracted ray ($\omega = 43°$ and $\varphi = 47°$). However, it should not be forgotten that this direction is opposite to the direction of the refracted ray, and therefore when reading the coordinates off the net we must take the inverse azimuth $180° + \omega$ and the angle $90° + \varphi$. Then the direction of the refracted ray will be $\omega = 223°$, and $\varphi = 137°$, and it will be denoted on the net not by a dot but by a cross.

The refraction angle for a grazing ray should be 90°, therefore to determine the direction of the grazing angle it suffices to plot an angle of 90° with the normal N_{in}, and returning the tracing paper to its original position to read off the net coordinates of the grazing ray (point K).

In the example considered here, the grazing angle is practically horizontal.

The information presented here should be adequate for the understanding and application of stereographic projections. We advise the reader to become familiar with these methods. This will not only make the understanding of the material contained in the book appreciably easier, but will also facilitate the future use of wave polarization in the practice of seismic studies, even if computers are to be used in data processing.

Controlled Directional Reception Based on Displacement Directions (CDR–I)

The method of controlled directional reception based on displacement directions was developed for the study of polarization of seismic waves and for their discrimination according to the direction of particle motion. Since this method makes use of the directional diagram of the first kind, in order to distinguish it from the traditional directional reception which makes use of a radiation pattern of the second kind, we have termed it CDR–I [27, 36, 38, 40].

SEISMOMETER DIRECTIONAL DIAGRAM

The term seismometer directional diagram with respect to the displacements of the Earth, applies to the relationship between the sensitivity of a seismometer whose maximum

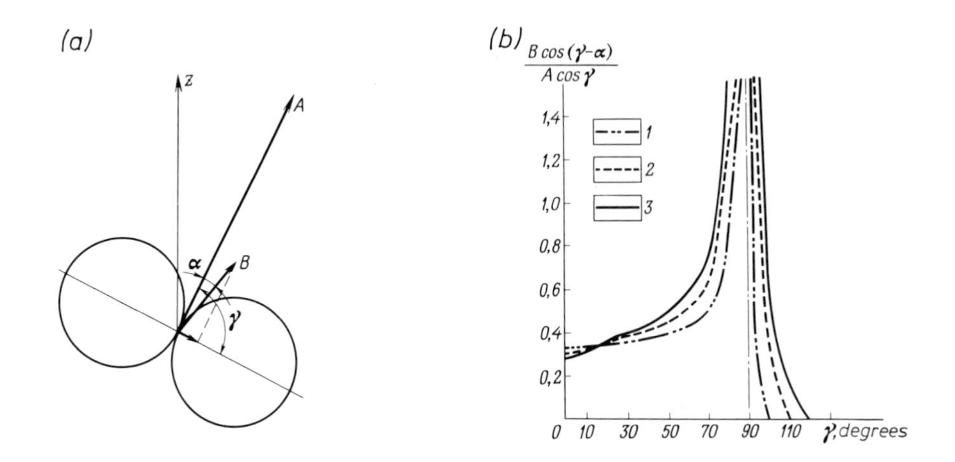

Fig. 5. Diagram explaining the efficiency of CDR–I. (a) directional diagram; (b) dependence of signal-to-noise ratio on orientation of directional diagram, for the case of interference of two linearly-polarized oscillations making an angle α equal to 10° (1), 20° (2), and 30° (3).

sensitivity axis is fixed in space and the orientation in space of displacements of equal amplitude. Since the seismometer is a system with a single degree of freedom, this diagram in polar coordinates takes the form of two tangential spheres (Figure 5a) described by the equation

$$r = K \sin \psi_i \sin \psi_j + K \cos \psi_i \cos \psi_j \cos (\omega_i - \omega_j) ,$$

where r is the radius-vector of the diagram in the direction determined by the coordinates ψ_i, ω_i (ψ_i is the angle with the horizon and ω_i is the azimuth); ψ_j, ω_j are the coordinates that determine the direction of the seismometer's maximum sensitivity axis; K is the coefficient that determines the seismometer's sensitivity in the direction of this axis.

The direction of the seismometer's maximum sensitivity axis coincides with the direction of displacement of its inertial mass. The equation of the directional diagram for a vertical seismometer takes the form

$$r = K \sin \psi_i ,$$

The selectivity of the directional diagram in different directions is not the same. For directions close to the axis of maximum sensitivity of the seismometer, the diagram is very poor, but for directions making an angle of 90° with the axis, the diagram's selectivity increases sharply, because the sensitivity of the seismometer becomes zero in the plane orthogonal to the axis of the directional diagram. For example, the recorded amplitudes of displacements in directions making angles up to 20° with the diagram's axis will be almost equal to (not less than 0.94) the amplitude of displacement, coinciding in direction with the axis of maximum sensivity. At the same time the decrease in the amplitude within the range of 70° to 80°, that is, following a change in the angle of only 10°, is already double (from 0.34 to 0.17). Therefore, to obtain maximum sensitivity of directional reception, the orientation of the directional diagram should be such that the displacements of the wave assumed to be unwanted would be in the plane of zero displacements of the diagram. In that case there will be optimum recording of the useful wave.

We may cite an example to illustrate the potentialities of CDR–I. Suppose two linearly-polarized oscillations A and B, making an angle α equal to 10°, 20°, and 30° interfere at a point. Oscillation B with an intensity three times less than that of oscillation A will be assumed to be the useful signal. Then, owing to a rapid decrease in the amplitude of oscillation A, the signal-to-noise ratio will increase rapidly for angles γ close to 90° (Plots 1–3 in Figure 5b), the gain in the signal-to-noise ratio depending greatly on the orientation of the directional diagram. Naturally, the smaller the angle between the oscillations, the greater should be the accuracy of orientation of the directional diagram.

CDR–I, like CDR–II, is achieved by the aid of summation of signals from a group of seismometers. However, in contrast to CDR–II, in which the seismometers making up a group are arranged along the traverse, those making up a group in CDR–I are located at a point, but with different orientations in space (in a cluster). Moreover, in CDR–II the summation of signals from individual seismometers making up the group takes place with a shift in time, whereas CDR–I employs in-phase summation, the variable parameter being the relative sensitivity of the seismometers making up the cluster.

The directional diagram for a cluster of seismometers is determined by its shape and its amplitude and by the orientation in space of its maximum sensitivity axis [27, 36].

DIRECTIONAL DIAGRAM FOR A SEISMOMETER CLUSTER

Let us consider the two-dimensional and the three-dimensional (spatial) problems.

Two-dimensional Problem. The equation for the directional diagram of a cluster made up of m seismometers arranged in a plane at different angles ψ_n to the horizontal (the so-called planar arrangement) in polar coordinates takes the form

$$R = \sin \psi_i \sum_{n=1}^{n=m} K_n \sin \psi_n + \cos \psi_i \sum_{n=1}^{n=m} K_n \cos \psi_n \;, \tag{8}$$

where $\Sigma_{n=1}^{n=m} K_n \sin \psi_n$ and $\Sigma_{n=1}^{n=m} K_n \cos \psi_n$ are constants for the specific cluster.

It follows from Equation (8) that the directional diagram for a cluster coincides in shape and direction with that of a single seismometer whose maximum sensitivity axis makes an angle ψ with the horizontal determined by the equation

$$\tan \psi = \frac{\Sigma_{n=1}^{n=m} K_n \sin \psi_n}{\Sigma_{n=1}^{n=m} K_n \cos \psi_n}\;, \tag{9}$$

In this instance, the axis of the cluster's directional diagram remains in the plane of all the seismometers of the cluster. In the case of a planar arrangement, any desired diagram (in amplitude and direction), arbitrarily arranged in the plane, may be obtained simply by changing the sensitivity of a single seismometer in the arrangement.

Denoting

$$\sum_{n=1}^{n=m-1} K_n \cos \psi_n = A_1 ; \qquad \sum_{n=1}^{n=m-1} K_n \sin \psi_n = B_1 ,$$

we may conveniently represent equation (9) in the form

$$\tan \psi = \frac{B_1 + K_m \sin \psi_m}{A_1 + K_m \cos \psi_m} . \tag{10}$$

For $K_m = 0$, $\tan \psi = B_1/A_1 = \text{const.}$ for the given cluster. For $K_m = \infty$, $\tan \psi = \tan \psi_m$ or $\psi = \psi_m$. If the sensitivity of one seismometer greatly exceeds that of the others, the direction of the combined directional diagram will approach that of this seismometer's axis. In the extreme case, when the cluster consists of one horizontal and one vertical seismometer, $\tan \psi = K_2/K_1$, where K_1 and K_2 are the sensitivities of the horizontal and the vertical seismometers, respectively. For $K_1 = K_2$ the axis of the directional diagram makes an angle of 45° with the horizon[1].

Three-dimensional Problem. The directional diagrams of a cluster of m seismometers, the directions of whose axes in space are specified by the angles $\psi_1, \omega_1 ; \psi_2, \omega_2 ; \ldots ; \psi_m, \omega_m$ is determined by the expression

$$R = \cos \psi_i \sum_{n=1}^{n=m} K_n \cos \psi_n \cos (\omega_i - \omega_n) + \sin \psi_i \sum_{n=1}^{n=m} K_n \sin \psi_n , \tag{11}$$

where R is the radius-vector in the direction ψ_i, ω_i.

It follows from Equation (11) that the directional diagram of a cluster is equivalent in shape to that of a single seismometer whose orientation and sensitivity are determined by the arrangement of the seismometers in the cluster and by their sensitivity. In future, we shall call such a seismometer a 'conventional' seismometer. The direction of the cluster's directional diagram is determined by the expressions

$$\tan \omega = \frac{\sum_{n=1}^{n=m} K_n \cos \psi_n \sin \omega_n}{\sum_{n=1}^{n=m} K_n \cos \psi_n \cos \omega_n} , \tag{12}$$

$$\tan \psi = \frac{\sum_{n=1}^{n=m} K_n \sin \psi_n}{[(\sum_{n=1}^{n=m} K_n \cos \psi_n \sin \omega_n)^2 + (\sum_{n=1}^{n=m} K_n \cos \psi_n \cos \omega_n)^2]^{1/2}} . \tag{13}$$

By varying the sensitivity and the arrangement of the seismometers in the cluster, one can obtain any desired orientation of the cluster's directional diagram. The orientation of this diagram may be conveniently varied by controlling the sensitivity of the seismometers. Consider the dependence of the direction of the axis of the directional diagram on the sensitivity of just one (the mth) seismometer. In this case the direction of the

[1] A cluster of two seismometers (z and x) has been proposed by Bereza [10].

axis of the directivity diagram is determined by the expression

$$\tan \omega = \frac{A_2 + K_m \cos \psi_m \sin \omega_m}{B_2 + K_m \cos \psi_m \cos \omega_m} = \frac{A_2 + aKm}{B_2 + bKm}, \tag{14}$$

$$\tan \psi = \frac{C_1 + K_m \sin \psi_m}{[(A_2 + K_m \cos \psi_m \sin \omega_m)^2 + (B_2 + K_m \cos \psi_m \cos \omega_m)^2]^{1/2}}$$

$$= \frac{C_1 + cKm}{[(A_2 + aKm)^2 + (B_2 + bKm)^2]^{1/2}}, \tag{15}$$

where

$$\left.\begin{aligned} A_2 &= \sum_{n=1}^{n=m-1} K_n \cos \psi_n \sin \omega_n, \\ B_2 &= \sum_{n=1}^{n=m-1} K_n \cos \psi_n \cos \omega_n, \\ C_1 &= \sum_{n=1}^{n=m-1} K_n \sin \psi_n, \end{aligned}\right\} \quad \text{for constant-sensitivity seismometers,}$$

$$\left.\begin{aligned} a &= \cos \psi_m \sin \omega_m, \\ b &= \cos \psi_m \cos \omega_m, \\ c &= \sin \psi_m. \end{aligned}\right\} \quad \text{for a single variable-sensitivity seismometer.}$$

The equation of the plane passing through the direction of the axis of the cluster's directional diagram and the maximum sensitivity axis of the mth (variable-sensitivity) seismometer in polar coordinates has the form

$$A' \sin \omega + B' \cos \omega + C' \tan \psi = 0. \tag{16}$$

If we analyse equation (16), we can demonstrate that by varying the sensitivity of just the mth seismometer the direction of the cluster's directional diagram will change within the limits of this plane.

An arbitrary directional diagram in space may be obtained by changing the sensitivity of two seismometers in the cluster. A definite orientation of the plane of the diagram's axes in space corresponds to each fixed value of the sensitivity of one of the seismometers. The position of the axis in this plane is determined by the sensitivity of the other seismometer. Let the sensitivities of all the seismometers in the cluster, except two (the mth and $(m-1)$th), have fixed values; and the orientation in space of all the axes of the seismometers in the cluster be fixed, as well. Then the direction of the axis of the cluster's directional diagram will be determined by the expressions

$$\tan \omega = \frac{A_3 + K_{m-1} \cos \psi_{m-1} \sin \omega_{m-1} + K_m \cos \psi_m \sin \omega_m}{B_3 + K_{m-1} \cos \psi_{m-1} \cos \omega_{m-1} + K_m \cos \psi_m \cos \omega_m},$$

$$\tan \psi = \frac{C_2 + K_{m-1} \sin \psi_{m-1} + K_m \sin \psi_m}{\left[\begin{aligned} &(A_3 + K_{m-1} \cos \psi_{m-1} \sin \omega_{m-1} + K_m \cos \psi_m \sin \omega_m)^2 + \\ &+ (B_3 + K_{m-1} \cos \psi_{m-1} \cos \omega_{m-1} + K_m \cos \psi_m \cos \omega_m)^2 \end{aligned}\right]^{1/2}},$$

where $A_3 = \Sigma_{n=1}^{n=m-2} K_n \cos \psi_n \sin \omega_n$; $B_3 = \Sigma_{n=1}^{n=m-2} K_n \cos \psi_n \cos \omega_n$; $C_2 = \Sigma_{n=1}^{n=m-2} K_n \sin \psi_n$ are constants for the specific cluster.

A change in the direction of the axis of the cluster's directional diagram also changes the sensitivity of the cluster.

The Amplitude of the Directional Diagram (Sensitivity) for a Seismometer Cluster. The maximum magnitude of the radius-vector of the cluster's directional diagram $R = R_{max}$ corresponds to the direction of the axis of the directional diagram (ψ, ω), and is determined from Equations (11) and (13) by the expression

$$R_{max} = \frac{\Sigma_{n=1}^{n=m} K_n \sin \psi_n}{\sin \psi_n} . \tag{17}$$

The formula obtained makes it possible to compute the cluster's sensitivity as a function of the sensitivities of the seismometers in the cluster. The theoretical seismograms for a cluster may be computed with the aid of formulae for the line-up and the amplitudes presented in [24]. These equations will also be used to evaluate the required accuracy of sensitivity adjustment of the set's seismometers.

It should be pointed out that the orientation and amplitude of the cluster's directional diagram may, with a degree of accuracy adequate for practical purposes, be determined with the aid of a method of graphic analysis which makes use of stereo-nets [86].

Analysis of Three-component Sets. Let us analyse the equations obtained above for the case of two types of seismometer clusters of greatest practical interest, which are used as three-component sets.

The direction of the axis of the directional diagram of a three-component XYZ set is determined by the expressions

$$\tan \omega = \frac{K_2}{K_1} , \qquad \tan \psi = \frac{1}{(K_1^2 + K_2^2)^{1/2}} .$$

The direction of the axis of the diagram may be changed by altering the sensitivity of any two of the set's seismometers: two horizontal seismometers or one horizontal and one vertical seismometer. In the first case the sensitivity of seismometer Z is equated to unity ($K_Z = 1$). The relation between the direction of the axis of the cluster's diagram and the sensitivities of the seismometers in the cluster for this case is depicted in Figure 6a. It will be seen from Figure 6 that to obtain, for example, a cluster diagram with its axis pointing in the direction $\omega = 225°$, $\psi = 29°$ (point A), the relative sensitivities of the seismometers should be equal to $K_X = -0.44$ and $K_Y = 1.64$. It follows from the dependence of the diagram's amplitude on the direction of its axis that the amplitude of the diagram (sensitivity) of a cluster decreases with decrease in the angle between the vertical and the axis of the diagram. All the cluster's diagrams whose axes are located on the surfaces of cones having a vertical axis, have equal amplitudes. Similar relationships can also be obtained for the second case, when $K_X = 1$, and the cluster's directional diagram is varied by controlling the sensitivities of the Z and Y seismometers.

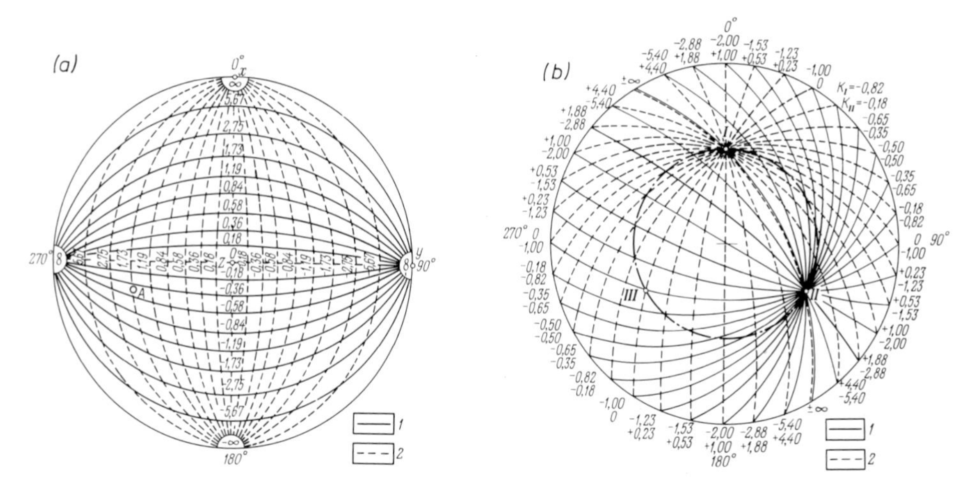

Fig. 6. Directions of axes of directional diagrams of three-component cluster groups. (a) XYZ set: $1 - K_X$, $2 - K_Y$; (b) symmetrical set I, II, III: $1 - K_I$, $1 - K_{II}$.

Consider in addition the directional diagram for a three-component set, the axes of all of whose seismometers I, II, and III are inclined to the horizontal at an angle of 35°20′, the difference in azimuth of adjacent seismometers being 120°. Such sets are termed symmetrical. They have substantial advantages over the XYZ sets (see also Chapter 2).

For a symmetrical three-component set, the dependence of the direction of the axis of the cluster's directional diagram on the sensitivity of two of its seismometers may be obtained from the general equations and is depicted graphically in Figure 6b. The amplitude of the cluster's directional diagram depends greatly on its direction [27, 36], and this is taken into account in the realization of CDR–I.

CDR–I is realized with the aid of pattern technique I on both analog and digital computers. The first program was drawn up by Alekseev at the SB AS USSR [1]. In many cases CDR–I makes it possible to improve the effective sensitivity of instrumentation used to separate useful waves. The correct orientation of the directional diagram produces a sharp rise in the signal-to-noise ratio.

Polarization Filtering

The CDR–I method described above achieves the discrimination of waves based on the direction of particle displacement and is derived from the variation in orientation of the directional diagram without any change in its amplitudes. The selectivity of the diagram is determined by the cosine law of variation of amplitudes as a function of direction of displacement. Such selectivity may in many cases prove inadequate. This applies primarily to the analysis of multi-polarized waves and waves with a time-dependent polarization direction. In this case, the system's selectivity may be impaired at the moment in time when the displacement vector of the unwanted wave is outside the zero-sensitivity plane of the directional diagram. For this reason, it may be necessary to employ more selective systems to analyse the wave field at a point. Here we may cite on example of wave discrimination based on other traditional parameters (frequency and ν_a), where the principal tradition has been a continuous improvement in selectivity (e.g. in the case of discrimination according to ν_k, pattern techniques, CDR, and CDP).

Recent years have witnessed the development in seismology of polarization filtering based on the use of the properties of polarization of seismic waves. First experience in the use of such method was connected with noise suppression. Owing to the circumstance that, in most cases, the microseismic background is formed by Rayleigh surface waves elliptically polarized at the surface with the three-dimensional *P* and *S* waves having a predominantly linear polarization, polarization filters proved quite effective, and this brought about a rapid progress in this trend. Up to the present, several types of polarization filters have been devised which employ the difference in the parameters and the nature of polarization of oscillations in various types of waves.

Polarization filtering may be performed either with constant numerical polarization parameters or with parameters determined directly from the observed field. The latter case involves self-adjusting polarization filtering. CDR–I is a particular case of such filtering with the particle-displacement direction serving as the filtering parameter. Polarization filters are employed for resolving the following problems: detection and separation of linearly-polarized oscillations from among oscillations with polarization of a more complex type; determination of displacement directions in linearly-polarized oscillations; separation of oscillations linearly-polarized in a specified direction against a background of noise or regular waves with other polarization parameters; separation of waves with different types of polarization; determination of oscillation parameters of elliptically-polarized waves.

Polarization filters may be subdivided into two groups: linear and non-linear. A brief review of the principal papers dealing with polarization filters follows. The polarization filter proposed by Flinn [138, 139, 140] was first used in seismology to improve the effective sensitivity of recording weak signals. The filter is a time-variable operator. Filtration involves the computation of a quadratic form of the covariation matrix whose surface is an ellipsoid from a three-component seismogram in a time interval. The ratio of all semi-axes determines the degree of linearity of motion and orientation, which coincides with the direction of polarization. The filter enables waves with different types (linear, elliptical, circular) and directions of polarization to be separated.

In a paper by White [159], a method is proposed for transforming signals to suppress irregular unwanted waves and to determine the direction of arrival of an arbitrary regular wave caused by a blast or by an earthquake. The method is based on phase relationships between different oscillation components observed at a point. The same paper deals with phase relationships for various types of waves with the object of assessing the potentialities of polarization filters. Each horizontal component of motion is multiplied by the vertical component, and the two resulting products are represented in the form of a vector indicating the direction of the source of seismic waves. For in-phase waves (*P*, *SV*), multiplication of the components is performed without phase reversal, and for out-of-phase waves, with a phase shift of $\pi/2$. In the first case, the output signal from the system is accumulated so long as a wave with in-phase *H* and *V* components is being received. In the second case, the signal is accumulated only for waves, whose components of movement have been displaced in phase by $\pi/2$ relatively to one another (e.g. a Rayleigh wave). Similar methods are discussed in [155].

Paper [152] deals with separation of a signal against background noise from the records of a single three-component set. The method uses the difference between the degree of linearity of the signal's polarization and the elliptical polarization of the noise. The vector product of the vertical and radial components averaged over time is multiplied by the initial signal. Among various computed functions, the simple product of the trace multiplied by the vector product of the vertical and radial components is best suited for separating and identifying the phases. This paper deals also with the determination of the excentricity of the ellipse, its principal axis, and the angle it makes with the vertical for a frequency corresponding to oscillations of maximum intensity. These parameters are subsequently used to draw up tests for identifying P and SV type waves.

In 1968, the Remodo phase filter was proposed. A similar filter was used in 1969 for the study of multiple waves from nuclear blasts. The polarization filter facilitates the identification of numerous longitudinal waves that follow in the wake of a P wave.

One design of a non-linear polarization phase filter is described in [151]. The filter makes use of the linearity of polarization and the directions of particle motion. Covariance matrices for three orthogonal components of motion are calculated for this purpose. The parameters of the filter change slowly with time, and the form of the recorded signal changes less than in the case of the Remodo filter.

Improvement in the filter's selectivity for separating linearly-polarized oscillations in specified directions increases the useful sensitivity of recording of displacements in specified directions, in the same way as narrow-band frequency-filtering of waves does for waves with specified frequencies. One possible design of filters for such a purpose is described in [73]. The paper deals with two types of two-component systems: the optimum type, when the correlation properties of noise and signal are known, and the self-adjusting type, when the properties of the noise are assessed from the seismograms as they are processed. The paper contains definitions of two-component systems of types I and II. The parameters in the system of type I are chosen so as to obtain a maximum amplitude of the useful signal for constant power of the combined signal + noise oscillation. In the system of type II the parameters are determined from the condition of minimum power of combined oscillations at the system's output for a useful signal of fixed amplitude.

The polarization filters enumerated above perform time-variable non-linear operations which may distort the form of the record. Whereas such distortions are unimportant in the study of arrival times of linearly-polarized oscillations, they are undesirable when the dynamic characteristics of the record are being studied. Moreover, time-variable operations require much greater computer time than filters with stationary parameters. In this respect the filters based on linear phase-transformations offer considerable advantages. Such filters include, in particular, the linear phase filter proposed by Mercado [149]. A set of linear filters was employed for separating longitudinal reflected and Rayleigh waves on the basis of phase relationships between their horizontal and vertical components. The separation method involves sequential filtering of the orthogonal oscillation components. As a result, practically sinusoidal oscillations with a frequency equal to the filter's resonance frequency appear at the output. The records of various components of linearly-polarized oscillations at the filter's output are in-phase, the only difference being in their amplitudes. The records of the same components of elliptically-polarized waves (e.g. Rayleigh waves) characterized by a $\pi/2$ phase shift at the output of the same filter, display a phase shift which depends on the filter's resonance frequency. This enables a corresponding two-channel velocity-filter to be effectively used for each frequency component. Moreover, the summation of outputs in a definite frequency band makes it possible to separate linearly-polarized waves from elliptically-polarized Rayleigh waves. Hence, the linear phase filter described above is in effect a combination of a two-channel velocity and a narrow-band frequency filter.

Computerization facilitates wide use of various types of polarization filtering, which is bound to become the principal method of separating tracking components in polarization-positional correlation of waves.

Methods of Representing Trajectories of Particle Motion

Polarization is quantitatively characterized by parameters that determine the trajectory of particle motion. In general, moving particles describe intricate trajectories in space, and for this reason methods of representing such trajectories are of great importance in their analysis and understanding.

The trajectories may be represented in various coordinate frames: rectangular and spherical. The choice of the method of representation depends in each specific case on the objective of the study and the ability of the representation to present a clear picture.

THE RECTANGULAR COORDINATE FRAME

The traditional and most widely used method is to represent the trajectory of motion in the form of three mutually-perpendicular oscillation components. This is largely due to the fact that in three-component observations, the trajectory of particle motion is automatically recorded in a rectangular coordinate frame, since each seismometer records the projection of the oscillation vector on the direction in space of its axis. The axes of a coordinate frame may be arbitrarily oriented in space. Two systems are in use in the practice of seismic observations: the xyz and the symmetrical systems. The first corresponds to the XYZ arrangement and is widely used in surface observations in which the directions of arrival of the waves are close to the vertical, with the longitudinal waves being recorded mainly on the vertical (Z) component, and the transverse waves, on the horizontal components: in the plane of rays (X) and perpendicular to it (Y). The symmetrical system whose axes are mutually perpendicular and inclined at an angle of $35°20'$ to the horizontal, with the difference between azimuths of adjacent seismometers being $120°$ [24, 30, 42, 55], is more effective as an instrument for analysing complex wave fields and in the study of complex media. The directions of the axes of both systems at all points of observation are usually fixed in space.

When studying polarization, it is often convenient to use a special reference frame for each observation point instead of a coordinate frame fixed in space. A convenient and frequently-used frame is that whose axes are oriented not in the direction of the points of the compass, but in the direction of the first arrival at the specific point. We shall in future use the term 'local frame' to describe such a rectangular coordinate frame. In this frame the positive direction of the p axis coincides with the direction of particle motion in the first longitudinal wave, and the r and q axes lie in a plane tangent to the wave front P. The q axis is horizontal and perpendicular to the plane of rays. The

r axis may be both inclined and vertical. To determine the axes of a local coordinate frame, one has to find the direction of the oscillation vector of the first longitudinal wave and to draw a plane tangent to the wave front. Here it is convenient to make use of the stereographic projection. Since the orientation in space of the plane of rays is not determined by the direction of the ray of a longitudinal wave, the orientation of the r axis may be determined only with the aid of some additional assumptions. Specifically, if the plane of rays is presumed to be vertical, the direction of the r axis will coincide with the line of intersection between the vertical plane of rays and the plane tangent to the wave front P.

The positive directions of the axes in a local coordinate frame are determined with the aid of the left-hand rule. If the left forefinger points in the direction of the p axis, the thumb will point in the positive direction of the r axis, and the middle finger in the direction of the q axis. The stereogram (Figure 7) depicts the directions of the axes of a local coordinate frame and of some oblique components connected with the local frame. The orientation of such components in space changes from point to point, but remains constant with respect to the axes of the local coordinate frames.

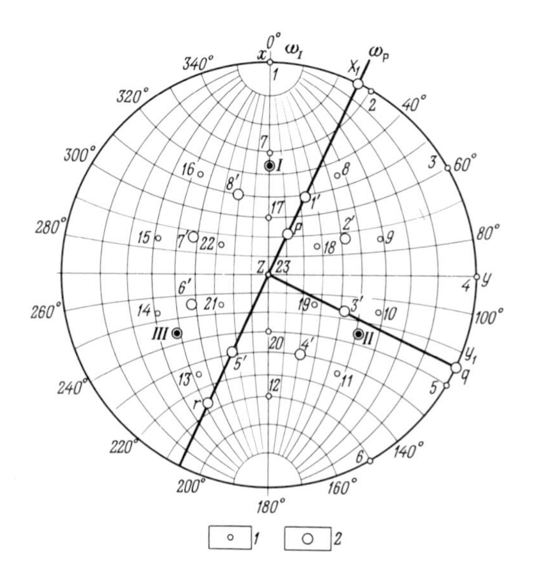

Fig. 7. Directions of rectangular coordinate frame axes of symmetrical set I–III, an XYZ set, and a local set prq. (1) directions of polar-seismogram components in the coordinate frame connected with axes of set I, II, III; (2) the same for a local coordinatė frame.

The R and Q oscillation components in a local coordinate frame are expressed in terms of rectangular frame components A_x and A_y, as follows

$$
\begin{aligned}
&R(t) = A_x(t)\cos\alpha + A_y(t)\sin\alpha,\\
&Q(t) = A_x(t)\sin\alpha + A_y(t)\cos\alpha,\\
&P(t) = A(t),
\end{aligned}
\tag{18}
$$

where α is the angle between the x axis and the projection of P on the horizontal plane.

If the x axis points North, the angle α will be the azimuth of the vector P (ω_P). Generally, $\omega = \omega_x + \alpha$, where ω_x is the azimuth of the x axis. The local coordinate frame is of interest, because all simple waves in a homogeneous or an axially-symmetrical medium are polarized either in the plane of rays or in the plane of the wave front. Moreover, the P component is always an optimum component of the first wave at every point.

THE SPHERICALLY SYMMETRICAL COORDINATE FRAME

The position of a particle in a spherical coordinate frame is determined by the radius-vector characterized by its modulus $|A|$ and its direction in space. The latter is specified with the aid of two angles: in the horizontal plane by the azimuth ω measured clockwise from North, and in the vertical plane by the polar angle φ measured from the vertical. The trajectory of particle motion is fully determined by the collection of curves determining the variation of those parameters in time (the records of the modulus $|A(t)|$, the azimuth $\omega(t)$, and the angle $\varphi(t)$), just as it was determined by the three-component seismogram. By way of illustration, let us consider the records of the azimuth, the angle, and the modulus for plane-polarized oscillations with different displacement directions, phase shifts, and intensities. The displacement directions in interfering oscillations are shown on the stereogram (Figure 8a). The trajectories of the combined oscillation in the polarization plane have been considered earlier (see Figure 2b). Note that in order to construct a trajectory of particle motion in the case of harmonic oscillations it suffices to determine φ_1, ω_1, and $|A_1|$ at Δt intervals in the first half-period of an oscillation. For the second half-period, we obtain the following values: $\varphi_2 = \varphi_1 + 90°$; $\omega_2 = \omega_1 + 180°$; $|A_2| = |A_1|$, and hence the scales on the $\varphi(t)$ and $\omega(t)$ plots coincide. On the φ scale 0° coincides with 90°, and 90° coincides with 180°. On the ω scale 0° coincides with 180°, and 180° with 360°.

It is evident from seismograms in spherical coordinates (Figure 8b) that:

1. For a linearly-polarized oscillation (Figure 8b, 1), the records of the polar angle $\varphi(t)$ and the azimuth $\omega(t)$ are straight lines parallel to the t axis, because the direction of motion remains constant. This may serve as a test for the separation of a linearly-polarized oscillation.

2. In the case of a plane-polarized oscillation, the records of the polar angle and the modulus depend greatly on the orientation in space of the polarization plane.

For the plane Q_1, which makes a comparatively small angle with the vertical (Figure 8b, 2–5), the values of $\varphi(t)$ vary from 0° to 180°. At the same time the azimuth mainly assumes two values (ω and ω + 180°), deviating little from each of these values. The values for the polarization plane Q_3, which is close to the horizontal (Figure 8b, 10–17) are approximately equal to 90° and vary little with time. On the contrary, the azimuths $\omega(t)$ vary in a wide range from 0° and 360°.

3. In the case of circular polarization, when the intensities of the interfering oscillations are equal, and the phase shift is $\beta = 90°$ (case 15), the record of the modulus is represented by a straight line.

The reversal of the line-ups observed on the three-component seismograms of symmetrical sets (Figure 8b, 8, 9) is due to the change in direction of particle motion.

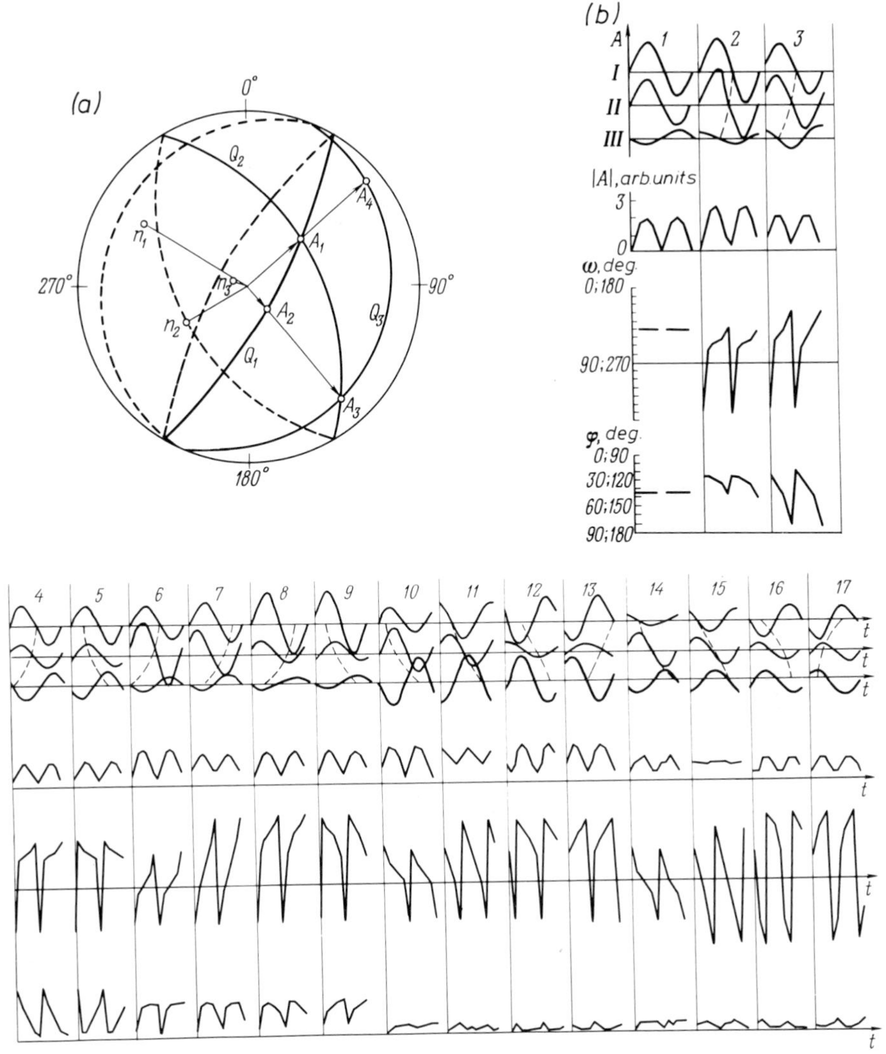

Fig. 8. Representation of trajectory of particle motion in rectangular and in spherical coordinate frames. (a) Direction of interfering oscillations ($A_1 - A_4$), orientation of planar polarization ($Q_1 - Q_3$), stereographic projection; (b) seismograms in rectangular coordinates (symmetrical set I–III) and in spherical coordinates.

The conditions of wave interference on the seismograms recorded in the true Earth (number of waves, their frequencies, intensity, phase shifts, and directions of arrival) may vary continuously, and as a result the records of the azimuth $\omega(t)$ and the polar angle $\varphi(t)$ may be quite intricate.

Figure 9 shows three-component seismograms of a blast in rectangular-symmetrical and spherical coordinate frames. As will be seen from the figure, rectilinear segments of plots corresponding to linearly-polarized oscillations are observed for the direct longitudinal P wave during short time intervals (points 1–5, 10–13). The transverse S wave (points 20–31) is complexly-polarized. Oscillations close to the linearly-polarized type

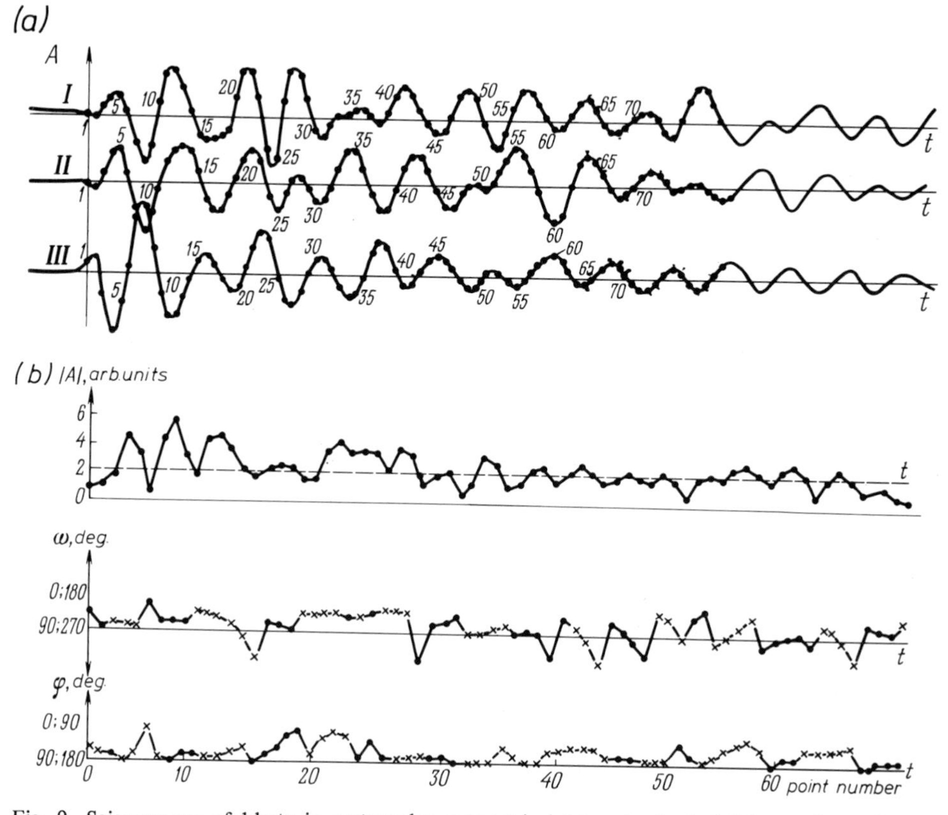

Fig. 9. Seismograms of blasts in rectangular symmetrical (a) and spherical (b) coordinate frames. (Depth of observation point H = 80 m, distance from firing point to drill-hole site l = 125 m, depth of blast h = 9 m).

are also observed in the interval of points 30–39. If some experience is acquired in working with seismograms represented in spherical coordinates, it will be found that they present a clearer picture of the trajectory of particle motion and are more conveniently interpreted than the traditional seismograms in rectangular coordinates. In the coordinate frames considered above the specified parameters adequately describe any trajectory in space.

The directions of particle motion specified by the coordinates of a spherical frame in the form of azimuth and polar angle records may conveniently be represented on a stereographic projection. Consider a *VSP* seismogram from a blast represented in a spherical coordinate frame and a stereogram (Figure 10). The records of the azimuth and the polar angle calculated with the aid of a computer program [80] are of an intricate form. Rectilinear sections corresponding to linearly-polarized oscillations are observed only within the time interval from 0.223 to 0.230 s. Construction of the stereogram (Figure 10b) involves plotting a sequence of directions $\varphi(t)$, $\omega(t)$. The displacement directions on the stereogram combine to form sets of points which either lie in a plane (11–17, 26–37, 75–87), or assemble in compact groups (18–23, 62–67). The former correspond to plane-polarized, and the latter to linearly-polarized oscillations. The

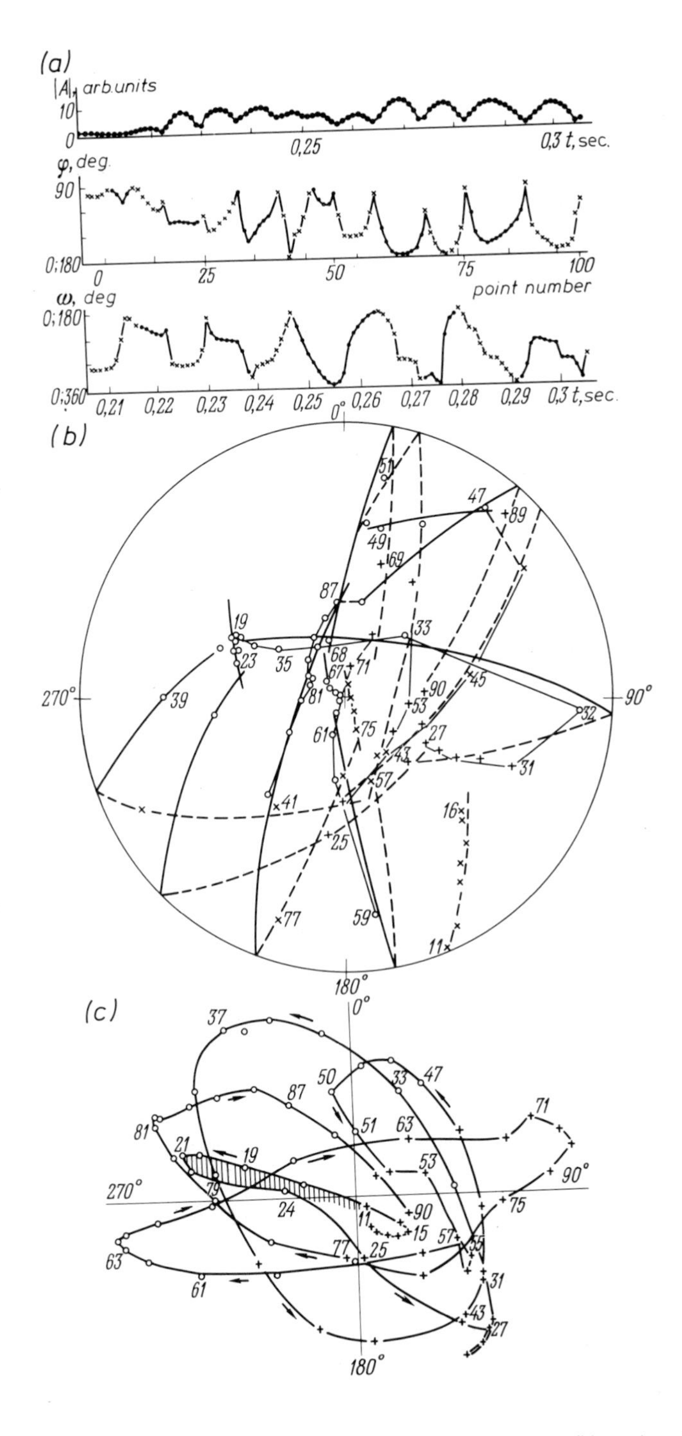

Fig. 10. Seismograms of blasts in spherical coordinates (a), stereogram (b), and evolution (c) ($H = 510$ m, $l = 1$/km).

transition from one polarization plane to another may be smooth or abrupt. The stereogram clearly indicates the direction of particle motion. Starting from the first arrivals, the radius-vector rotates counter clockwise; at point 50 the direction of motion is reversed. All such peculiarities are difficult to notice on a seismogram in spherical coordinates.

However, stereograms only define directions of particle motion and give no indication of amplitudes of oscillations. In order to represent the form of the trajectory of a complexly-polarized oscillation, it is convenient to 'unfold' this trajectory so that it is located in a plane, for example, in a horizontal plane. Two methods are used in practice arbitrarily termed 'evolution of planes' and 'evolution of surfaces'. In both methods the planes of the net are imagined as rotating until they coincide with the horizontal plane of unfolding. In this process all the radius-vectors of a plane-polarized oscillation, moving along minor circles of the stereographic projection, will occupy a definite position in the horizontal plane with their mutual disposition remaining unchanged. Marking the values of the modulus of the vector in terms of the radius-vector, we can plot the evolution of a trajectory of particle motion of any shape. The same method can be employed, if the oscillations have not one, but several polarization planes with arbitrary orientations (Figure 10c). Such a representation presents a sufficiently clear picture of the direction of motion and the shape of the trajectory. Specifically, the section of the seismogram, which in spherical coordinates is identified as belonging to a linearly-polarized oscillation (points 17–24 in Figure 10a), on unfolding (Figure 10c) has the shape of an extended ellipse (shaded area).

In many cases the representation of the trajectory of motion on stereographic projections makes it possible to distinguish from the point of view of polarization type between oscillations whose peculiarities are not evident in spherical coordinates, that is, on the $\varphi(t)$, $\omega(t)$ graphs.

It is worth noting that the evolution of planes may result in ruptures at points corresponding to directions where the orientation of the polarization planes has undergone a sharp change. For this reason in the case of oscillations polarized in space it is convenient to make use of the evolution of surfaces described by the displacement directions. In this case each direction in space is considered in its own meridional plane.

Figure 11 depicts in a stereographic-coordinate frame the direction of particle motion for an oscillation polarized in space. The particle describes an intricate curve. Moving along the curve in the direction of particle motion, we continuously cross one meridional plane after another. For instance, Figure 11a depicts the plane that contains the direction represented by a point on the net.

In the evolution process all meridional planes are imagined as coinciding with the horizontal plane, so that the directions of particle motion lying in those planes move along the minor circles of the net and in the horizontal plane mark the azimuths of directions along which the values of the moduli of the oscillation vector are to be measured.

Thus, one may obtain an evolution of a trajectory of arbitrary shape on the horizontal plane. In this case the shape of such an unfolding may differ somewhat from that obtained with the evolution of planes described above. This is because the angles between the directions in space obtained in the evolution of surfaces may differ from those obtained in the evolution of planes. Figure 11b depicts, for the sake of comparison, two derivatives of the same trajectory obtained by different methods (the evolution of surfaces (line 1) and planes (line 2)).

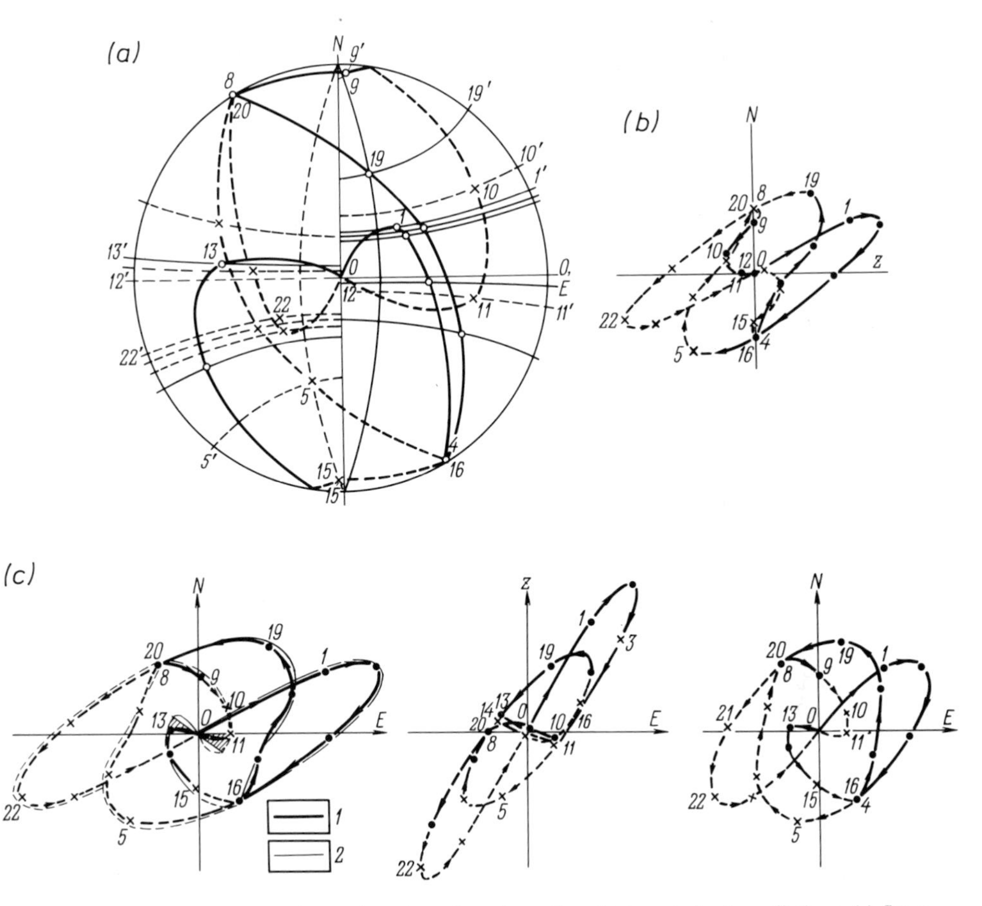

Fig. 11. Methods of representing trajectories of motion of multiple-polarized oscillations. (a) Stereogram of directions of particle motion; (b) evolution in the horizontal plane; (c) projections of trajectory on three mutually-orthogonal planes.

Both curves are quite similar in shape. The greatest difference occurs in sections 11–13, where line 2 must involve a rupture (shown by the shaded area). Hence, the evolution of surfaces makes it possible to represent the trajectory of motion in a horizontal plane without ruptures and with comparatively small distortions of shape.

A trajectory of particle motion may also be represented in the form of projections on three orthogonal planes. Such representation is commonly used in engineering. Figure 11c depicts, for the sake of comparison, the same trajectory (Figure 11a) in orthogonal projections on three mutually perpendicular planes (two vertical and one horizontal). A set of three projections completely determines a trajectory.

Stereographic projections offer unquestionable advantages in that they present a clearer picture.

REPRESENTATION OF PATTERNS OF VARIATION IN DISPLACEMENT DIRECTIONS

In the study of polarization it is not only the trajectory of particle motion at a point that presents considerable interest, but also the patterns of variation of the displacement directions in space or along the line of observations, specifically in the case of *VSP* along vertical profiles.

Let us consider methods of representing the patterns of variation of displacement directions.

All the directions of particle motion in longitudinal waves in the case of homogeneous or axially-symmetrical media lie in the vertical plane, and the variation in direction may be clearly represented in the form of curves, for example, of $\varphi(H)$ for vertical or $\varphi(x)$ for horizontal profiles [36]. For such media, clear-cut relationships have been established in *VSP* between the shape of the curves of directions of particle motion and the velocity characteristics of the medium, and they may be used for interpreting experimental curves to gain information about the medium (see Chapter 6). In the presence of oblique interfaces in the section the variation in directions will no longer be restricted to the vertical, but will include the horizontal plane, as well. The patterns of such variations may be represented by a set of graphs of the angles $\varphi(H)$ and $\omega(H)$ (for a vertical profile). In the absence of sharp velocity interfaces in the section, the $\varphi(H)$ graphs follow smooth curves. Changes in the shape of the $\varphi(H)$ curves may be due to changes in the types of waves or to the intersection of velocity interfaces (see Figure 84). In contrast to the $\varphi(H)$ curves, the shape of the $\omega(H)$ curves depends primarily on the orientation of the interfaces in space.

Line of Visibility. In conditions of asymmetrical media the curves of angles in the vertical and horizontal planes should be considered together. A clear picture of the curves of directions of motion is presented in the stereographic projection. Such a representation of the $\varphi(H)$ and $\omega(H)$ relationships has been termed the 'line of visibility' [31]. On a stereo-net, the line of visibility is the locus of displacement directions (points) for a single wave along a profile (horizontal or vertical; see Chapter 6).

CHAPTER 2

INSTRUMENTATION FOR THREE-COMPONENT OBSERVATIONS USING THE POLARIZATION METHOD

The study of seismic waves involves the analysis of the vector of oscillations and, consequently, requires three-component observations. In the early stages of development of seismic exploration, as with seismology in general, three-component observations were the rule. However, with the introduction of the multi-channel recording of longitudinal reflected waves, practically all observations in seismic exploration were limited to the recording of the vertical component of oscillations. That is why for the last 30 to 40 years the development of the methodology and instrumentation for three-component observations has been practically at a standstill. At the same time specific problems and deficiencies in three-component observations largely discouraged research work based on the use of polarization of seismic waves. Intensification of research aimed at developing the polarization method has primarily involved a substantial improvement in instrumentation and procedures for making three-component observations. Such projects have been carried out in recent years at the Institute of Earth Physics of the Academy of Sciences of the USSR (IEP) in cooperation with KazVIRG and the 'Krasnodarneftegeofizika' Trust.

We shall discuss some special features of instrumentation and procedures used in making measurements. The parameters of the seismic channels and of individual systems of standard instruments making up each channel (seismometer, amplifier, recorder, etc.) and used in solving special problems in seismic exploration and seismology differ in respect of their frequency range, sensitivity (gain), and recording procedure. The choice of parameters merits maximum attention, since they largely determine the efficiency of research.

In this chapter attention will be focused on the classification and appraisal of three-component installations now in use for surface and drill-hole measurements, the method and instruments used to obtain multi-component polarization seismograms needed for polar correlations and the Controllable Directional Reception–I method (CDR–I), the methods and instruments used for obtaining oriented recording in drill-hole observations, and methods for obtaining seismograms for polarization-positional interpretation.

Three-Component Sets

Three-component observations are made with the aid of three mutually-perpendicular seismometers located at a point. The set usually employed was suggested by Golitsyn [59] and consists of a vertical (Z) and two mutually perpendicular horizontal seismometers (X and Y). Symmetrical (homogeneous) three-component sets consist of three mutually-perpendicular seismometers, but in this case they are arranged at an equal angle to the horizontal [24]. Since these installations possess substantial advantages, it is expedient to discuss the main features of the sets of both types in detail.

XYZ SETS

For surface observations when the direction of arrival of the waves is close to the vertical, the *XYZ* sets are more convenient, since each of their seismometers registers waves of a definite type: the vertical seismometer registers longitudinal waves and the horizontal seismometers register transverse waves. In this case the recordings made with the aid of such sets present a clear picture. This is why *XYZ* sets have found wide use in seismology. However, there are deficiencies inherent in installations of this type, and their operation is complicated by difficulties which substantially limit their capabilities.

The main deficiency of the *XYZ* set, which greatly reduces the quality of field recording, is the difficulty of guaranteeing identical frequency response and sensitivity of the vertical and the horizontal seismometers. On permanent seismic stations this is done by changing the parameters of the pendula from time to time and calculating the different characteristics. This rather laborious operation requires not only much time, but also highly skilled personnel. In field operations the conditions under which the installation of seismometers and the observations take place are usually much inferior to those in which static recordings are made. Moreover, in the course of field operations the seismometers are fairly often moved along the traverse, and because of this it is practically impossible to control effectively the great number of seismometers and to guarantee that all of them remain identical. In the absence of rigorous control it is impossible to detect changes in the sensitivity of the channels, which may be quite substantial. This distorts the picture of wave polarization, thereby complicating not only the study of the waves, but also their separation and identification on the seismograms.

When operating *XYZ* sets, it is also impossible to control the polarity of the vertical and horizontal channels directly on each of the seismic records, which is quite essential in field operations. For this reason large errors in the study of the polarization of oscillations (up to 180°) cannot be excluded.

The horizontal instruments of an *XYZ* set are usually oriented along the main points of the compass with the result that the registered horizontal components actually become random components, because all directions of oscillation sources are possible (e.g. when earthquakes are recorded). Having to operate two types of seismometers (horizontal and vertical) under field conditions also complicates the observation process. The factors cited above and primarily the need for control of channel identity make it expedient to employ symmetrical sets. This is even more important for observations beneath the Earth's surface, because in contrast to surface observations the directions of arrival of the waves are in this case not only non-vertical, but vary both along the profile when waves from one shot-point are recorded and also at any point on the profile when charges are exploded at different points.

SYMMETRICAL (HOMOGENEOUS) SETS

Such sets consist of three seismometers whose maximum sensitivity axes make an angle of 35°20′ with the horizontal, the azimuthal difference between neighbouring instruments being 120° (Figure 12a) [24, 36]. In this arrangement the axes of all the seismometers are mutually perpendicular. The seismometers are mounted in special cases of various designs. For surface observations, it is convenient to arrange the instruments at a single

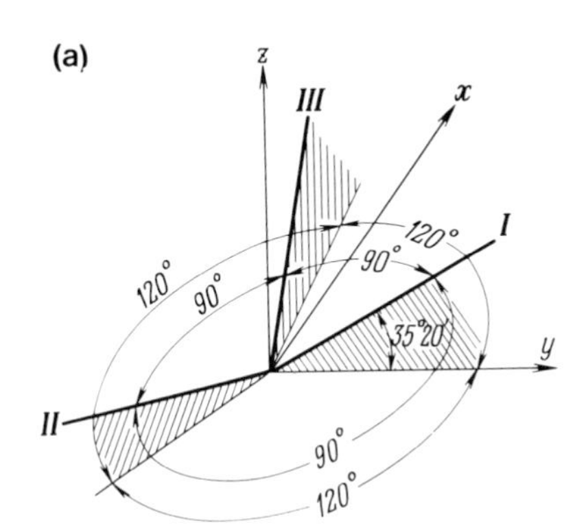

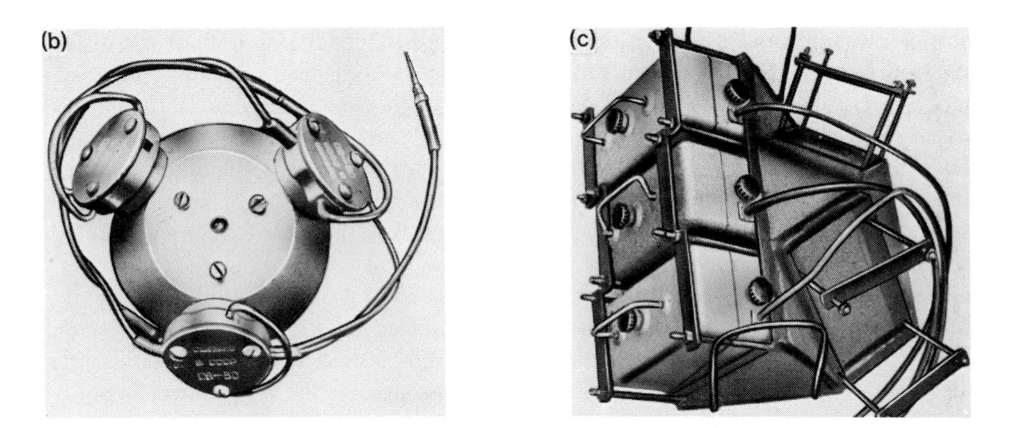

Fig. 12. Three-component symmetrical clusters. (a) diagram of component (I–III) arrangement in a symmetrical cluster; (b) cluster for use in surface seismic prospecting, (c) cluster for use in regional studies (seismometers positioned for identity control).

level. Cases are in use in which seismometers are rigidly mounted in the operating position. Figure 12b shows such an arrangement for seismic exploration. For regional investigations, one may conveniently use a case, which makes it possible either to arrange the seismometers at a common azimuth to control their identity or at different azimuths to make observations (Figure 12c).

Recently Shnirman at IEP has designed a three-component symmetrical field set of seismometers combined in a single cluster on a common mounting with common magnetic and arresting systems. The system is designed especially for polarization observations. A special feature of the seismometers, in addition to the stability and identity of the channels, is that their sensitivity in directions perpendicular to the maximum sensitivity axis is zero. For conventional observations this is a matter of little importance, and so the sensitivity of standard vertical seismometers to vertical displacements may for some types of such instruments be as high as 10%. In observations employing the polarization method the correct shape of the directional diagram is of prime importance.

For observations in drill-holes, inclined seismometers are mounted in a case of cylindrial shape with oblique holes for the seismometers, which are arranged one on top of the other at a common azimuth (for control) or at different azimuths (for observations). We have employed symmetrical sets for different kinds of seismic investigations: from regional investigations based on the recording of distant earthquakes (seismometers with a natural frequency of 1 Hz are used for this purpose) to seismic exploration for ores (here seismometers with a natural frequency of 30–40 Hz are employed).

INSTRUMENTATION CONTROL IN THREE-COMPONENT OBSERVATIONS

The study of wave polarization requires a qualitative analysis of the trajectories of particle motion in the medium, and this is impossible without rigorous control of channel identity. Such control is essential, no matter what subsequent methods of data-processing (analog or digital) are employed.

The main advantage of symmetrical sets over *XYZ* sets is that with the former it is possible to control the identity of the channels under field conditions, both on the surface and underground. The general control of seismometers and channels of a three-component set on the surface is achieved by recording a signal under conditions when all the inclined seismometers are arranged in a common azimuth, as is usually done in seismic exploration. The only difference is that in this case the seismometer axes are not vertical, but inclined at an angle of $35°20'$ to the horizon. However, the rearrangement of seismometers to make recordings in order to control their identity takes time and may only be undertaken occasionally.

With a symmetrical set, high-quality operational control of channel identity directly during observations is possible. Seismometers installed at different azimuths on the surface can be controlled by recording signals from distant sources. In this case the first wave will arrive at the surface from a direction close to the vertical and should be identically recorded by all the seismometers in the cluster. In a drill-hole, the identity of the seismometers in a set can be controlled by recording the direction of the longitudinal wave excited by an explosion or by a shock close to the entrance of the drill-hole. If there are no steeply-dipping interfaces in the section, the direct wave will arrive at the seismometers from a direction close to the vertical and should be identically recorded by all the seismometers in the set.

In principle, two types of control may be used: electrical and seismic. Let us consider them separately.

The electrical method of control operates as follows: actuated by an electric signal (impulse or stationary) applied to working or special excitation coils, the pendula of the seismometers are made to swing, signals from the working coils being used to control their identity. This method enables the parameters of the instruments to be controlled; however, it requires very accurate tuning, in particular, a good correspondence between the signals of the working and the excitation coils. In symmetrical sets such a correspondence can be achieved by comparing signals obtained during the process of electric control with seismic signals recorded by seismometers with parallel sensitivity axes (seismometers oriented in the same direction). A drawback to the electrical method of control is that it is performed only occasionally. Beside, this method cannot be used to control the accuracy of installation of the instruments at the point of observation, that is, their mutual perpendicularity. In order to avoid such errors, it is expedient under field conditions to use, instead of individual seismometers, three-component clusters with seismometers mounted in strictly-defined positions (either in the *XYZ* or symmetrical arrangement), so that their mutual positions remains unchanged.

The seismic method of continuous control has been developed at IEP. It provides for the continuous control of the identity of the channels of a three-component set. The essence of the method is that by the summation of signals from a three-component set, a fourth signal not coinciding with any of the three signal components of the set is recorded.

This recording may at any instant of time be compared with the recording of the same component obtained directly through observation. Using a special analyser [27, 124], one can, from the record of a three-component set of any type, obtain a component of oscillations oriented in an arbitrary direction in space. For instance, one can conveniently use for the purpose of control a component making an equal angle with the axes of all the seismometers of a three-component set. For the symmetrical arrangement this will be the vertical component and for the *XYZ* arrangement the component, which makes an angle of about 55° with all the axes of the three-component set. All channels of the three-component set are involved in the production of this component. Provided the channels of the three-component set are identical, the recording of the control component will be identical with the direct recording of the same component obtained under field conditions. This method makes it possible to control the entire circuit (including the seismometers their mutual arrangement at the point of observation, field recording instruments and reproduction apparatus) and to formalize such control by automatically comparing the records of the same component of oscillations. It should be noted that, in order to control the identity of the channels of a symmetrical set, one may conveniently use the recording of the traditional fourth vertical component.

Let us use examples to illustrate the method of control described above. The traces I–III in Figure 13a correspond to the records of a symmetrical three-component set. The trace Z_Σ was obtained by adding up the signals from the three-component set, the trace Z_d being the direct recording. The fact that both records are practically identical is proof of the identity of all the three channels of the three-component symmetrical set in shape and sensitivity. Figure 13b depicts examples of continuous control in the process of observations performed with the 'Zemlya' and 'Taiga' stations.

The comparison of the *Z*-components obtained by direct recording and by adding up the signals from a three-component set is sometimes conveniently performed by simple subtraction. Figure 13b depicts both these components in opposite phases, the last trace being their difference. The ratio $(Z_d - Z_\Sigma) / (Z_d + Z_\Sigma)$ serves as a measure of identity.

The control proposed above may be performed directly during the process of recording under field conditions or in the course of subsequent processing, if reproducible records have been made in the field. The advantages of the method proposed are evident, the main one being the possibility of continuous control of instrumentation directly during the process of observation. This improves the quality of data obtained during observations and, accordingly, the efficiency of seismic investigations. A drawback to the method is that a fourth component of oscillations has to be obtained under field conditions. However, the operational experience gained shows that the accuracy and simplicity of control fully compensate for the trouble of recording an additional component.

Especially effective is the combination of both control methods (electric and seismic),

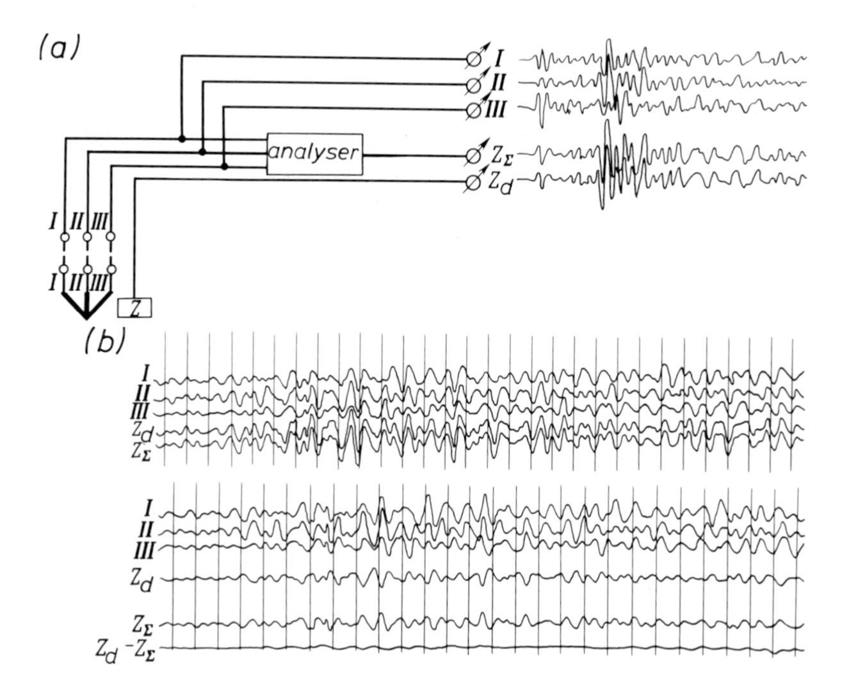

Fig. 13. Block-diagram (a) and seismogram (b) of continuous instrumentation control in three-component observations.

which not only provides for a rigorous assessment of the quality of data obtained in the field, thus enabling unreliable data to be rejected, but, very importantly, enables necessary corrections to be introduced during the course of processing. Obviously, the methods of control described above are essential, when the operational stability of the instruments is poor. With high-quality instruments there is no longer any need for continuous control with all the ensuing complications. The important point is that such methods of control make it possible to obtain satisfactory data commensurate with the aims of the investigation even with instruments of inadequate stability.

Making Multi-Component Polar Seismograms

Polar correlation [24, 36, 51, 53] is achieved with the aid of a multi-component polar seismogram obtained at a single point from seismometers with axes differently oriented in space. Initially, multi-channel azimuthal installations [24] were used to obtain polar seismograms of surface observations. Extensive development of such investigations was hampered by the need to employ a large number of channels at each point. In several instances multi-channel sets have also been used for observations in comparatively shallow drill-holes. However, because of technical difficulties in operating them at small depths and because of the practical impossibility of using them at great depths, it appeared expedient to develop methods of producing multi-channel seismograms with the aid of the pattern technique of the type I [27]. This technique is based on the summation of signals recorded by the seismometers of a three-component set and enables the recording of any oscillation component or sum of components with a specified orientation in space to be obtained. Such a record is equivalent to the recording obtained with a multi-component set.

Polar seismograms are conventionally subdivided into two groups according to their purpose (for polar coorelation or CDR–I). The difference between them lies in the amount of detail and in the orientation of the oscillation components in space. The oscillation components on seismograms of the first group are fixed and evenly distributed in space. We shall use the term 'outline' for such seismograms. They enable the type of polarization of all the recorded waves and the parameters of polarization to be determined. All the components on the seismograms of the second group are conveniently arranged in a single plane (or in its vicinity). The orientation of this plane is determined by the conditions of superposition of waves at a single point. In contrast to outline seismograms, we shall term these planar seismograms. Polar seismograms of both groups are obtained with the aid of special analysers of polar correlation (PC) and CDR–I.

THE PC ANALYSER

We believe that for polar correlation it will suffice to obtain a 23-component outline seismogram with the components uniformly distributed in space. For ease of correlation, the axes of fixed components are arranged along conical surfaces with the cone generatrices making different angles with the horizon. The arrangement of the components on an outline seismogram is shown in stereographic projection in Figure 7. The number of components in the horizontal plane may, for reasons of symmetry, be limited to six (see Figure 7, 1–6), the azimuthal difference between neighbouring components being 30°. Ten components (7–16) are located on a conical surface whose generatrix makes an angle $\psi = 30°$ with the horizontal; the azimuthal difference is 36°. Six components (see Figure 7, 17–22) are located on a conical surface with $\psi = 60°$. The last component (Figure 7, case 23) is the Z-component.

This set includes the components of the XYZ arrangements (cases 1, 4, 23). It is worth noting that the number of components of a polar seismogram together with their location in space is determined by the specific aims of the investigation, and that in this respect the scheme described above may be regarded as neither unique, nor universal.

Multi-component outline seismograms are obtained with the aid of an analyser (Figure 14a). The summation of the signals is accomplished by galvanometers in the recording oscillograph. The galvanometer of each channel receives through the respective resistor ($R1'$, $R2'$, $R3'$ – $R1^{23}$, $R2^{23}$, $R3^{23}$) an amplified input signal from the three-component set. The required phase of the signals applied to the resistors and their nominal value are determined from the calculated summation coefficients. The latter are proportional to the cosines of the angles in space between the directions of the sensitivity axes of the actual seismometers in the installation and those of the components of the polar seimogram. Because of attenuation of input signals during the process of summation, the analyser unit includes six amplifiers (AA1–AA6), which compensate for attenuation and amplify separately in-phase and opposite phase signals from the three-component set. The amplifiers have been designed at IEP and are transistorized (Figure 14b). The first three stages are of the common emitter type. The fourth stage is an emitter follower.

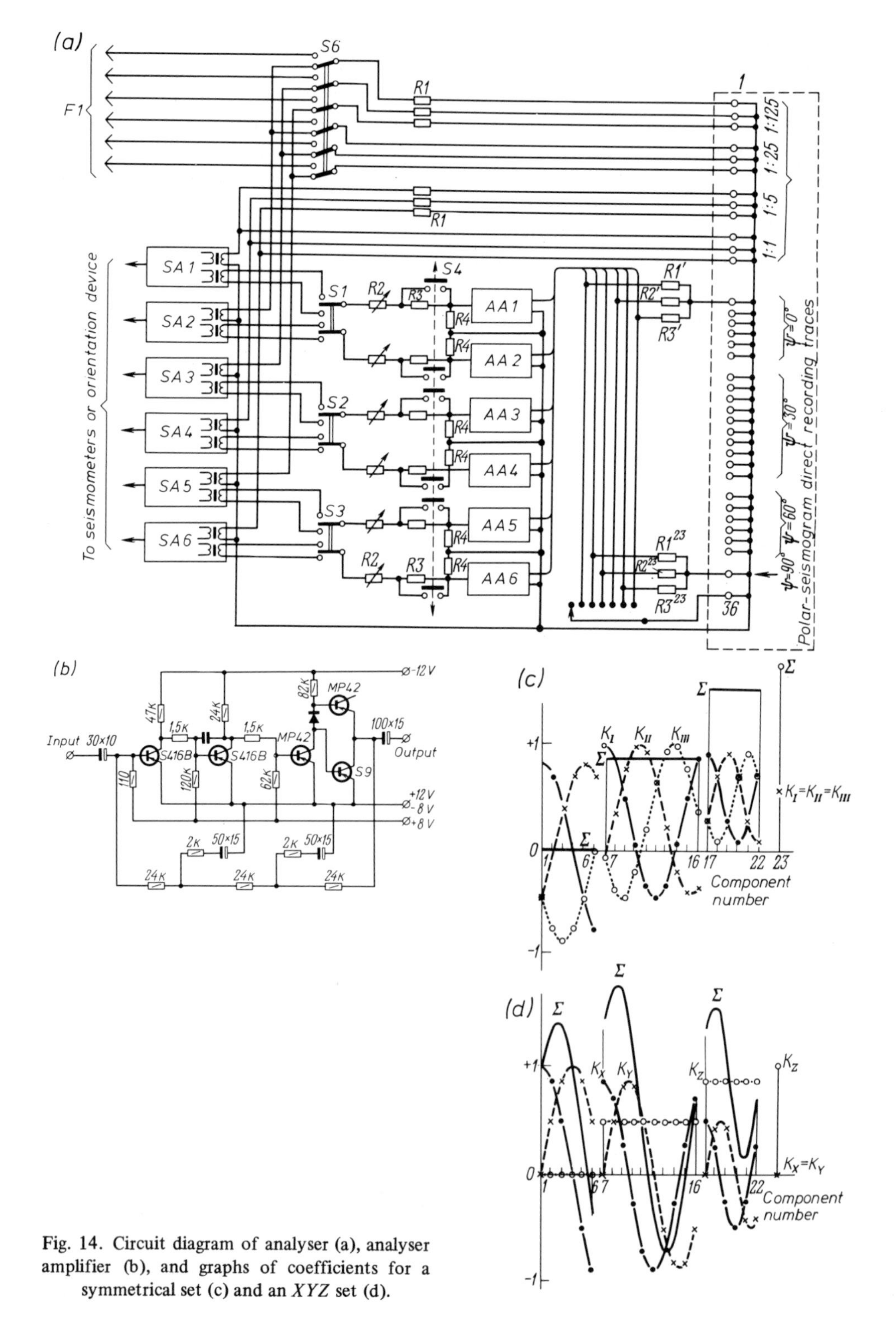

Fig. 14. Circuit diagram of analyser (a), analyser amplifier (b), and graphs of coefficients for a symmetrical set (c) and an XYZ set (d).

The amplifier features strong negative feedback, which improves its linearity and stability. The amplification factor of 500 remains practically unchanged in a temperature range from −40° to +40°C, being little dependent on voltage-supply variations. It draws a power of 50 mW from the two power supplies. The analyser is connected to the output of the station's amplifiers (SA1–SA6). To maintain an undistorted dynamic range, the station's amplifiers operate without any time-dependent adjustment (automatic gain control (AGC), exponential gain control, etc.). The AGC circuit is disconnected, and the AGC transformer is used as an additional output for connection to the analyser. To increase dynamic range, the signals from actual seismometers of the three-component set are recorded at four amplification levels. The signal from each actual seismometer is simultaneously amplified by two amplifiers connected in-parallel. The outputs of each amplifier are connected to two galvanometers, one directly and the other through the resistor R1. The signal from the additional amplifier outputs travels to the switches S1–S3 and then to the resistor network R2–R4. Opposite-phase signals are drawn from the resistors R4 and are fed to the inputs of the six amplifiers of the analyser. The analyser circuit described above is able to produce a 23-component polar seismogram from the signals of both the *XYZ* and the symmetrical three-component sets. The graphs of summation coefficients required to obtain the components are presented in Figure 14c, d, the values of the resistors being shown in Table I.

TABLE I
Coefficients of signal components of seismometers in a symmetrical cluster and corresponding resistance values in the voltage divider of an analyser.

Component number	K_I	K_{II}	K_{III}	Designation in diagram (Fig. 15)	Resistance K ohm	Component number	K_I	K_{II}	K_{III}	Designation in diagram (Fig. 15)	Resistance K ohm
Symmetrical set											
1	0.8192			R1	24.4	6	−0.7071				28.3
		−0.4067			49.1			0.7071		R15	28.3
			−0.4067		49.1						
2	0.707				28.3	7	0.9962				20.1
			−0.7071	R5	28.3			−0.0523			382
									0.0523		382
3	0.4067				49.1	8	0.8660				23.1
		0.4067			49.1			0.3584		R20	55.8
			−0.8192		24.4				−0.3584		55.8
4		0.7071			28.3	9	0.500				40
			−0.7071	R10	28.3			0.7660			26.1
									−0.4226		47.3
5	−0.4226				47.3	10	0.0698			R25	286
		0.8192			24.4			0.9781			20.4
			−0.4067		49.1				−0.1908		104.7

(Table 1 – cont.)

Component number	K_I	K_{II}	K_{III}	Designation in diagram (Fig. 15)	Resistance K ohm	Component number	K_I	K_{II}	K_{III}	Designation in diagram (Fig. 15)	Resistance K ohm
Symmetrical set											
11	–0.3090				64.7	18	0.7071				28.3
		0.9336			21.4			0.7071		R50	28.3
			0.2079	R30	96.2				0.0872		228
12	–0.4226				47.3	19	0.2924				68.4
		0.6428			31.1			0.9063			22
			0.6428		31.1				0.2924		68.4
13	–0.3090				64.7	20	0.0872			R55	230
		0.2250		R35	89			0.7071			28.3
			0.9336		21.4				0.7071		28.3
14	0.0698				286	21	0.2924				68.4
		–0.1736	0.9816		115			0.2924			68.4
					20.4				0.9063	R60	22
15	0.500			R40	40	22	0.7071				28.3
		–0.3746			53.3			0.0872			228
			0.7660		26.1				0.7071		28.3
16	0.8660				23.1	23	0.5736				34.9
		–0.3584			55.8			0.5736			34.9
			0.3746	R45	53.3				0.5736	R65	34.9
17	0.9063				22						
		0.2924			68.4						
			0.2924		68.4						
XYZ set											
1	1.0			R′1	20.0	5	–0.500				40.0
								0.866			23.1
2	0.866				23.1	6	–0.866				23.1
		0.500			40.0			0.500		R′10	40.0
3	0.500				40.0	7	0.866				23.1
		0.866		R′5	23.1				0.500		40.0
4		1.0			20.0	8	0.7071				28.3
								0.500			40.0
									0.500	R′15	40.0

(Table 1 – cont.)

Component number	K_I	K_{II}	K_{III}	Designation in diagram (Fig. 15)	Resistance K ohm	Component number	K_I	K_{II}	K_{III}	Designation in diagram (Fig. 15)	Resistance K ohm
					XYZ set						
9	0.2756				72.7	17	0.500				40.0
		0.8192	0.500		24.4				0.866	R′40	23.1
					40.0						
10	−0.275				72.7	18	0.2588				77.2
		0.8192		R′20	24.4			0.4226			47.3
			0.500		40.0				0.866		23.1
11	0.7071				28.3	19	−0.2588				77.2
		0.500			40.0			0.4226		R′45	47.3
			0.500		40.0				0.866		23.1
12	−0.866			R′25	23.1	20	−0.500				40.0
			0.500		40.0				0.866		23.1
13	−0.7071				28.3	21	−0.2588				77.2
		−0.500			40.0			−0.4226		R′50	47.3
			0.500		40.0				0.866		23.1
14	−0.2756			R′30	72.7	22	0.2588				77.2
		−0.8192			24.4			−0.4226			47.3
			0.500		40.0				0.866		23.1
15	0.2756				72.7	23			1.3	R′55	20.0
		−0.8192			24.4						
			0.500	R′35	40.0						
16	0.7071				28.3						
		−0.500			40.0						
			0.500		40.0						

Recently analysers have been designed for use during seismic exploration and at seismological stations ('Zemlya', 'Cherepakha', and 'Taiga'). Their design is similar to that of the one described above in most respects except in respect of the circuit for the production of opposite-phase signals. Most stations employ phase inverters for this purpose. In individual cases, depending on the recording equipment being used, one has to amplify the signal of each seismometer individually, as in the case of the first analyser prototype described in [27]. Figure 15a depicts the circuit of the low-frequency analyser used with the 'Taiga', 'Cherepakha', and 'Zemlya' stations, which produces a multi-component polar seismogram from the recordings of sets of both types (*XYZ* and symmetrical). The amplifier-inverters (Figure 15b) are balanced amplifiers incorporating transistors T1 and T2. There are emitter followers (T3 and T4) in the output circuit

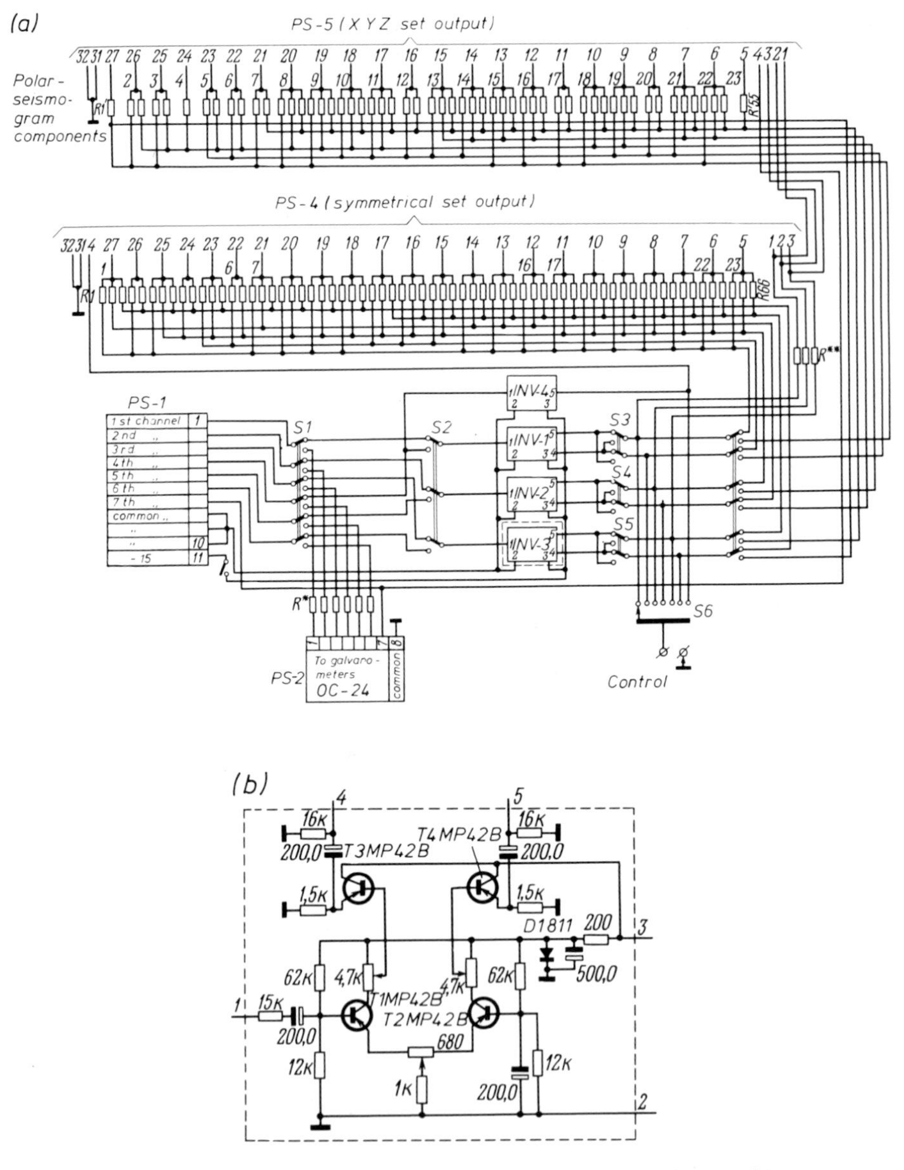

Fig. 15. Circuit diagrams of low-frequency analyser (a) and inverter (b).

of each amplifier to match it to the load and to exclude coupling between the channels. The experience gained with inverters has shown that they have good stability and may conveniently be used to obtain opposite-phase signals.

The analyser is tuned visually with the aid of gain controls. Initially, the station's amplifiers (SA1–SA6) are adjusted for equal amplification, as indicated by the galvanometers (1–12). Next the potentiometer R2 is operated to adjust the six amplifiers of the analyser (AA1–AA6). They are adjusted with the aid of the control galvanometer 36, which is alternately connected to the output of any one of the six amplifiers by means of the switch S5. The operation of the analyser is controlled by recording signals from an audio oscillator applied to the analyser's inputs connected in parallel. This is equivalent to oscillations of a three-component set along its vertical axis. In this case the amplitudes of all the horizontal components of a polar seismogram will be zero. Whether the circuit has been tuned or not may be judged correctly from the ratio of amplitudes of recordings of the components of the polar diagram. If we assume the amplitudes of the recordings of the channels of a three-component set to be unity, we would expect the amplitudes of the signals of the seismometers located on each of the conical surfaces to be identical and equal to $A_{\psi=0} = 0$ (galvanometers 13–18), $A_{\psi=30°} = 0.866$ (galvanometers 19–28), $A_{\psi=60°} = 1.52$ (galvanometers 29–34), $A_{\psi=90°} = 1.73$ (galvanometer 35). For equal channel-sensitivities, these ratios should remain strictly constant. To test whether the summation coefficients have been selected correctly, it suffices to make successive recordings of the audio oscillator's signal with the aid of each of the three analyser channels. The amplitudes of the recordings will correspond to the summation coefficients K_I, K_{II}, and K_{III}. Operational experience has shown that, owing to the stability of the analyser's amplifiers and the use of permament voltage-dividers, the analyser itself requires no adjustment during field operation.

CDR–I ANALYSERS

The analysers discussed above feature one voltage-divider, which enables a set of oscillation components rigidly fixed in space to be obtained. A special feature of such analysers is that the oscillation components obtained with their aid are uniformly distributed in space. This does not ensure the required efficiency of directional reception and precludes the free choice of spatial oscillation components, which would enable the use of the parts of the directional diagram with maximum selectivity. For maximum sensitivity, the axis of the directional diagram should coincide in direction with the displacement in a wave presumed to be an unwanted wave, whereas the area around the plane of zero displacement for the unwanted wave where an optimum analysis of the useful wave is possible (with the maximum signal-to-noise ratio) should be analysed in detail. For such detailed analysis of the area of maximum selectivity, the components of the polar diagram should best be arranged in the plane of zero displacement perpendicular to the maximum-sensitivity axis of the diagram or in its vicinity, for example, along two conical surfaces and below this plane. The generatrices of these conical surfaces make an angle of 10° with the plane.

The CDR–I analyser (Figure 16a) is conventionally subdivided into two channels A and B. Channel A serves to obtain outline polar diagrams and Channel B planar polar seismograms in the plane of zero displacements of the unwanted wave. Both channels together provide planar polar seismograms in the plane of zero displacements of unwanted waves polarized in different directions, the orientation of such seismograms in space being arbitrary.

Channel A includes amplifier-inverters 1, which serve to amplify the input signal and produce two output signals of equal amplitude, one of which is in-phase with the input signal, the phase of the second being opposite. The voltage dividers 2 divide the signals coming from the amplifier-inverters into 21 signals in proportion to the coefficients

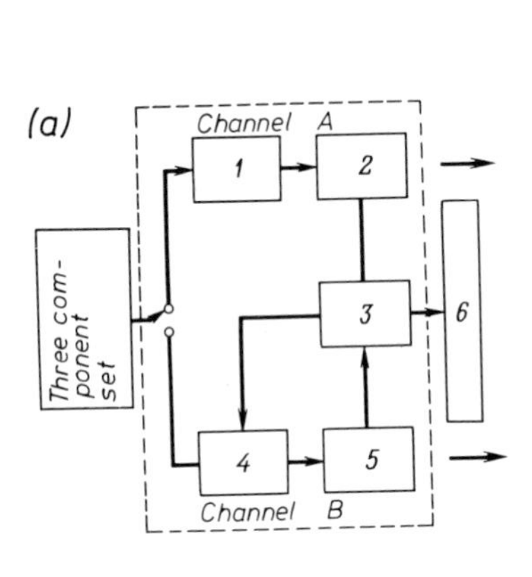

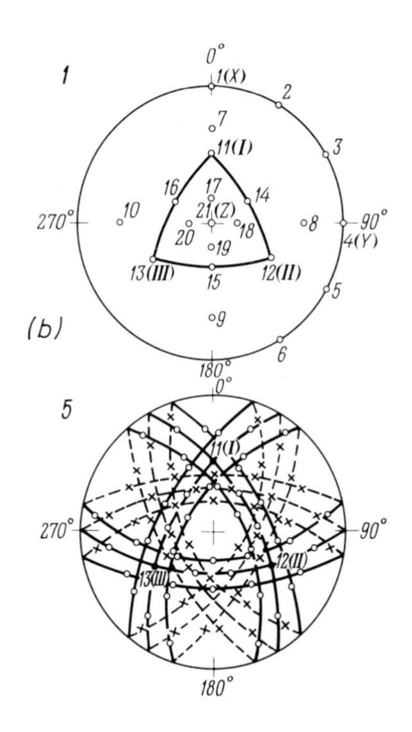

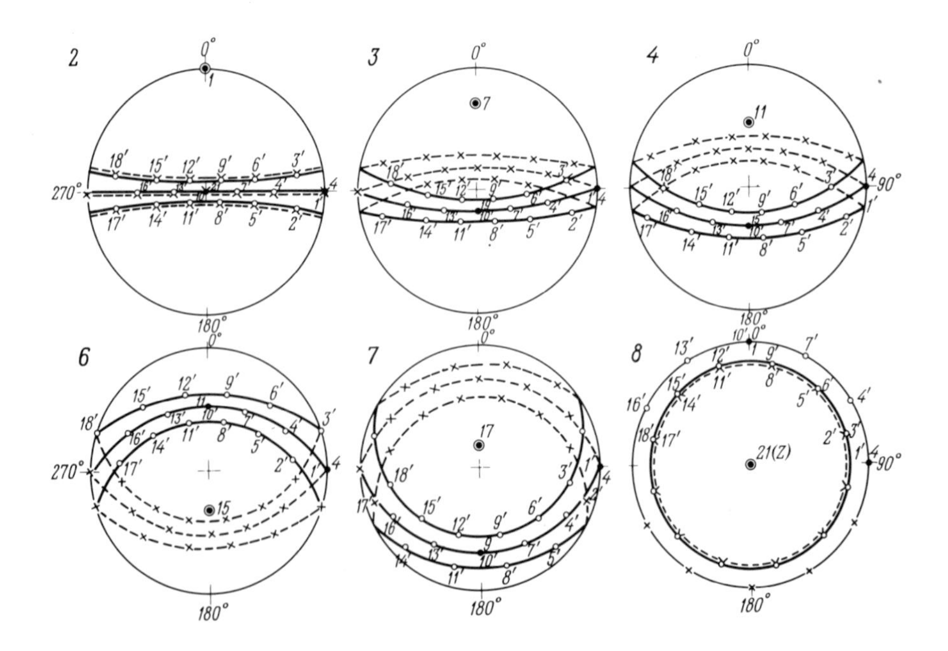

Fig. 16. Block diagram of selective polarization analyser (SPA) for CDR–I (a), and diagrams of arrangement of components of planar seismograms in space.

of the components of the outline polar seismogram. Commutation circuit 3 common to both channels serves to transmit the signals to the appropriate channels of the analyser. Channel *B* has amplifier-inverters 4 and voltage dividers 5 similar to those of Channel *A* the only difference being that the dividers serve to produce 18 signals proportional to the summation coefficients for the components located in the plane of zero displacements of unwanted waves. The analyser signals are recorded by a multi-channel oscillograph on photographic paper or on magnetic tape 6.

An outline polar seismogram obtained at the output of Channel *A* (Figure 16b, 1) includes 21 components uniformly distributed in space. The outline seismogram consists of six (1–6) horizontal components ($\psi = 0$) with the difference in azimuth between neighbouring components equal to $\Delta\omega = 30°$, four components (7–10), with $\psi = 20°$ and $\Delta\omega = 90°$, three components with $\psi = 35°$ and $\Delta\omega = 120°$, coinciding with the axes of the symmetrical set I, II, and III, (11–13), three components (14–16) with $\psi = 55°$ and $\Delta\omega = 120°$ located in the planes of the symmetrical installation, four components (17–20) with $\psi = 70°$ and $\Delta\omega = 90°$, and a vertical component (21) with $\psi = 90°$.

The planar polar seismograms obtained at the output of Channel *B*, as well as in the course of the combined operation of Channels *A* and *B*, represent sets of six components (1′, 4′, 7′, 10′, 13′, and 16′) located in planes orthogonal to the components of the outline polar seismogram (1–21), and six components arranged along conical surfaces, the generatrices of which make angles of 10° with the orthogonal planes above (components 3′, 6′, 9′, 12′, 15′, and 18′) and below (components 2′, 5′, 8′, 11′, 14′, and 17′) each plane. The arrangment of planar polar seismograms perpendicular to axes I, II, and III of a three-component set obtained with Channel *B* is depicted in Figure 16b, 5.

An appropriate connection of Channel *A* outputs to the Channel *B* input has made it possible, with the aid of one Channel *B* divider, to obtain a set of planar polar seismograms with a relative arrangement of the components similar to that described above, but with an arbitrary orientation in space.

The relative arrangment of the components of planar polar seismograms is determined by the inclination angles of the components of the outline polar seismogram. If the angles of inclination are equal, the relative arrangement of the components is independent of the azimuths of the components of the outline seismogram.

Figure 16b, 2–4, 6–8 are schematic representations of the arrangement of the components of planar polar seismograms for the components of an outline seismogram making different angles with the horizontal. The components of planar seismograms for all horizontal components of an outline seismogram (Figure 16b, 1) are arranged in the vertical plane and in its vicinity, the orientation of this plane being determined by the azimuth of the component of the outline seismogram (Figure 16b, 2). The arrangement of the components of planar polar seismograms for different components of an outline diagram is depicted in Figure 16b. The circuit diagram of the CDR–I analyser is shown in Figure 17. The voltage dividers A (resistors R1A–R46B) are stacks of precision resistors connected in an appropriate fashion. The analyser can operate in four modes. Let us consider each mode separately.

Mode 1 – the production of an outline seismogram (see Figure 16, Channel *A*). The input signals are fed to the inputs of inverter-amplifiers INV1A + INV4A via switch

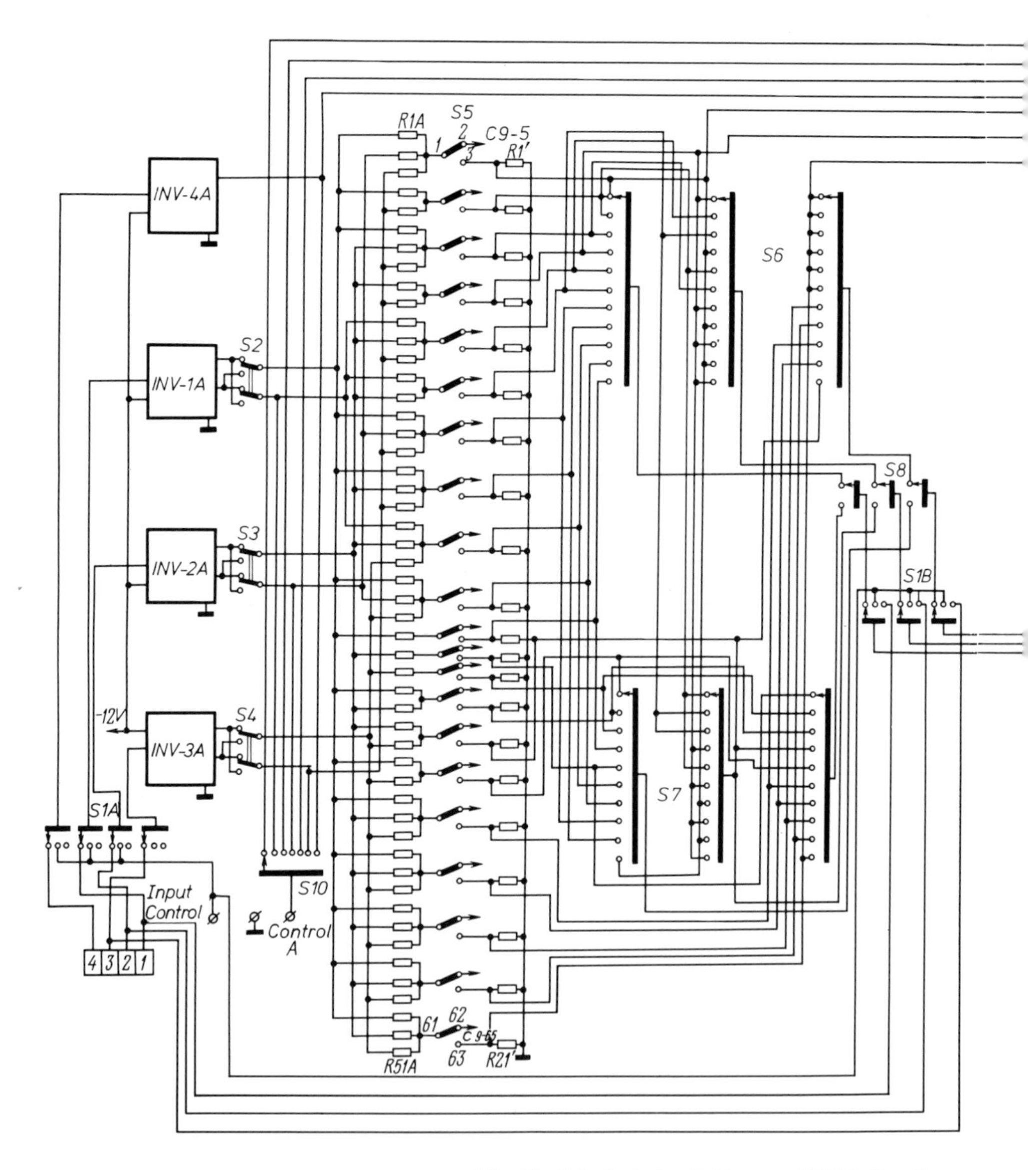

Fig. 17. Principal circuit diagram of CDR–I analyser.

S1A, the input of INV4A receiving the signal from the vertical seismometer that controls the correct operation of the analyser. Switches S2, S3, and S4 are provided, if necessary, correcting the phase of the input signal. In-phase and opposite phase (with respect to the input) signals from the outputs of the inverter-amplifiers are fed to the divider of Channel *A*, whose output consists of 21 signals proportional to the components of

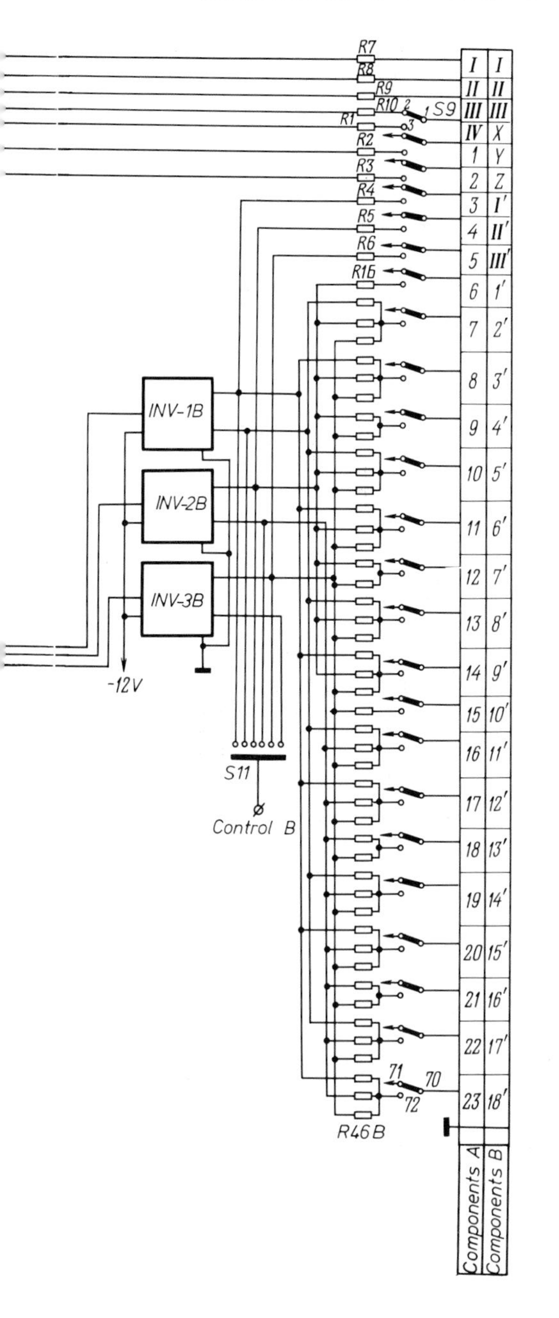

an outline polar seismogram. These signals are fed via switches S5 and S9 to the analyser output.

Mode 2 – the production of fixed planar polar seismograms (Channel *B*). The input signals are fed via switch S1B to the inputs of the inverter-amplifiers INV1B–INV3B.

From the inverter outputs the signals are fed to divider B, which produces 18 output signals proportional to the components of a planar polar seismogram, these signals being transmitted via switch S9 to the analyser output.

Mode 3 – the production of arbitrarily oriented planar seismograms (Channel *A–B*). The input signals are fed via switch S1A to the inverter-amplifiers INV1A–INV4A and then to divider A, as in the case of Mode 1. Switch S5 feeds these signals to switches S6 and S7, which select the appropriate components of an outline seismogram to be introduced into Channel *B*. Signals thus selected are fed via S8 and S1B to the input of INV1B–INV3B and subsequently to divider *B*. Signals from the outputs of divider *B* (18 components of a planar polar seismogram) are fed via switch S9 to the analyser output.

Mode 4 – control and tuning. A control signal used for tuning inverter-amplifiers is fed via switches S1A and S1B to the inverter inputs of both channels. Switches S10 and S11 terminating at control hubs are provided for reading out signals from inverted outputs. Resistors R1′–R21′ serve as load resistors for Channel *A* when the channels operate in series and R1–R10 are adjustment resistors.

An example of the use of the CDR–I analyser is considered in Chapter 3. It should be pointed out that the solution of certain problems aimed at improving the efficiency of CDR–I requires extremely accurate orientation of the planar polar seismograms. This refers specifically to the study and separation of converted transmitted waves recorded under conditions of superposition of intense longitudinal waves. Under such conditions it may prove inadequate to obtain a set of planar seismograms fixed in space, and a precise orientation of the planar polar seismogram will become necessary. To this end it will be convenient to connect the analyser producing planar polarization seismograms with a position sensor, which orients the signals of a three-component set in accordance with the orientation of the planar polar seismogram.

Instrumentation and Methods for Producing Oriented Records

In the study of trajectories of particle motion it is expedient to link them to space coordinates. In surface observations this is a rather simple process. At each observation point the sets are oriented either along the traverse (mainly when explosions are being recorded) or on the points of a compass, when the direction to the source is not known beforehand (e.g. when earthquakes are being recorded). In drill-hole observations, correlation involves great difficulties [47, 49]. Moreover, during the processing of surface and drill-hole three-component observation data, it becomes necessary to re-orient the records. Let us consider methods of obtaining oriented records.

For the production of oriented three-component records, drill-hole tools with specified orientation have been developed. A tool that utilizes the Earth's magnetic field [4, 129] has been tested for open drill-holes, and another that utilizes the gyroscopic effect [22] for cased drill-holes. Orientation of the cluster in the tool [4] is performed with the aid of a cam used in inclinometers only in the plane of the hole's axis. In this case the orientation of the seismometers in space is performed in accordance with inclinometer data obtained previously during the process of drilling the hole. All such tools have not yet passed the experimental stage. The tools featuring specified orientation have the following drawbacks.

1. orientation is performed only with respect to the azimuth, and corrections related to the inclination of hole's axis cannot be introduced. At the same time in many cases and especially in ore-rich areas, the holes deviate greatly from the vertical.
2. It is difficult to provide rigid contact between the three-component cluster and the walls of a hole, because the cluster has to be rotated in the tool. This may result in parasitic vibrations severely distorting the records of useful signals.
3. The design of the drill-hole tool is complicated by the necessity to house in it tracking and driving gear that rotates the cassette.
4. The size of the drill-hole tool has to be appreciably increased.

To overcome these difficulties we have proposed and developed a method of producing oriented records based on CDR–I [116, 117, 120]. This method has made it possible to discard the orientation of the cluster in the drill-hole in favour of determining its actual position, and to obtain the oriented records themselves on the surface with the aid of a special electronic orientation-device.

DETERMINATION OF CLUSTER ORIENTATION IN A DRILL-HOLE

In order to determine the orientation, it suffices to find the direction of the axis of the cluster and the azimuth of the axis of one of the seismometers (e.g. the first, ω_I).

The direction of the axis of the cluster in the tool pressed against the wall of the drill-hole may be assumed to coincide with that of the hole's axis defined by two angles: the azimuth ω_0 and the angle with the vertical φ_0. These two angles are usually measured with the aid of an inclinometer. To determine the cluster's orientation in uncased drill-holes, IEP in collaboration with Kaz VIRG has developed instrumentation based on the use of the Earth's magnetic field [125]. The instrumentation consists of cluster position sensors (CPS), a three-component cluster itself and a surface metering-post. Sets of two types have been developed: CPS–I and CPS–II. They have different angle-metering circuits and employ different orientation procedures.

CPS–I. A cassette with the seismometers SM1–SM3, a relay switch RS1 (circuit diagram in Figure 18a), and three position sensors are housed in a borehole tool. A compass from an inclinometer with a contact device on the magnetic pointer and a circular rheocord R1 under it is used as the azimuth sensor. The compass is suspended on two mutually-perpendicular circular cardan joints, which guarantee that the compass retains a horizontal position at all inclination angles. There is a solenoid S1 under the rheocord of the compass whose core is joined to the axis of the magnetic pointer, and hence, in the absence of current in solenoid S1, the pointer is free and points in the N–S direction. When the current is switched on, the pointer's axis together with the solenoid's core move downwards and in so doing press the contacts against the rheocord, fixing the position of the magnetic pointer. Underneath it, there are two mutually perpendicular inclination angle sensors R2 and R3, similar to those used in inclinometers. The tool is connected to the surface post by means of a seven-wire logging cable. To carry out a measurement, wires 6 and 7 are connected to the power supply, the contacts of all the rheocords are pressed against the rheocords, and the resistances are in turn measured at the surface with the aid of a bridge. After the measurements have been completed the

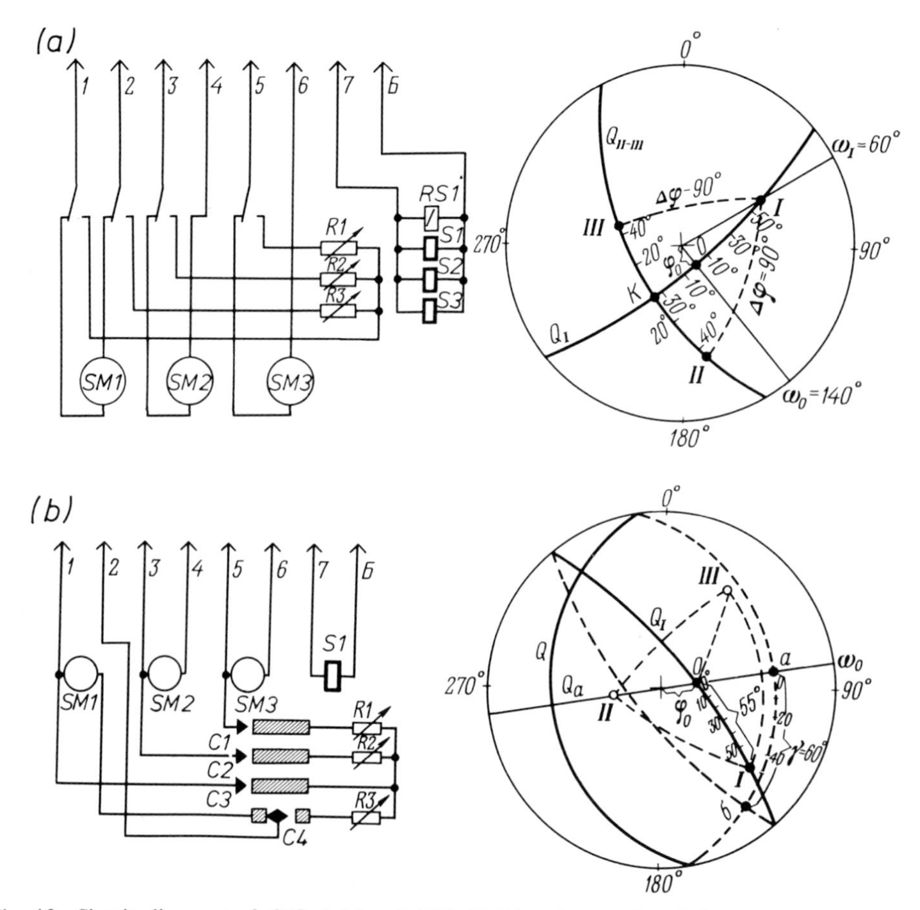

Fig. 18. Circuit diagrams of CPS–I (a) and CPS–II (b) and examples of determination of actual position of a set in a borehole.

solenoid coils and the relays are disconnected from the power supply, and the appropriate wires of the cable are connected to the seismometers. The CPSs are calibrated on a special table used for calibrating inclinometers. The next stage is the plotting of the rheocord resistances against the tool's orientation graphs.

To determine the orientation of the cluster from the readings of CPS–I, the measured resistances are translated into the angles φ_0, ω_0, and ω_I, which enable the actual position of the cluster in space to be determined. Such plots are conveniently made on a stereo net. To this end the initial data (Figure 18a) [(the direction of the axis of the drill-hole (point 0) described by the values of φ_0 and ω_0 and the azimuth of the axis of the first seismometer in the cluster (line ω_I)] are plotted on the net. The plots in Figure 18a have been made for the case $\omega_0 = 140°$, $\varphi_0 = 15°$, $\omega_I = 60°$. To find the position of the axis of the first seismometer in the cluster on the line ω_I, we must find the direction (point 1), which makes an angle of 55° with point 0. The newly-found point 1 and point 0 lie in the same plane Q_I. Measuring an angle of 35° in this plane from point 0 in the direction opposite to point 1; we obtain the direction K (point K). The angle between points 1 and K is 90°. To determine the position of axes II and III of the two remaining seismometers, construct the plane Q_{II-III} perpendicular to the axis I. The plane Q_{II-III} will pass through point K. The directions of the axes of seismometers II and III are symmetrical about point K, making with it angles of 45°. Hence, the actual position of the seismometer cluster is fixed by the directions of axes I, II, and III. The mutual orthogonality ($\Delta\varphi = 90°$) of all the axes in the cluster may serve as proof that the plots were made correctly.

CPS–II. The circuit diagram of CPS–II and the stereogram are depicted in Figure 18b. To determine the position of the cluster with the aid of CPS–II, a third angle γ of the cluster's rotation about its axis has been measured in addition to the two angles fixing the direction of the axis of the drill-hole (φ_0, ω_0). To this end an additional circular rheocord has been rigidly attached to the inclinometer housing in the immediate vicinity of the coil containing rheocords R1 and R2, which measure the angles φ_0 and ω_0. In the recording mode, all three seismometers are connected to the cable wires. In the angle-measurement mode, the cable is connected via contacts C1–C3 and via switch S4 to the appropriate rheocords. The rheocord resistance corresponding to the angles φ_0, ω_0 and γ could be measured in turn from the surface.

To determine the orientation of the cluster from the readings of angles φ_0 and ω_0, the position of the cluster's axis is plotted on the stereo-net (point 0 in Figure 18b), and then the plane Q is constructed perpendicular to this axis. The vertical plane Q_a passing through the cluster's axis 0 will intersect the plane Q along the direction a. The angle between the two directions γ is measured in the plane Q from this direction, and the direction b is determined. The first direction (point a) corresponds to the line of intersection between plane Q and the vertical plane Q_a passing through the axis of the drill-hole. The second direction (point b) coincides with the line of intersection between plane Q and plane Q_1, which passes through the axis of the first seismometer I and the axis of the drill-hole (0). To find the position of the axis of seismometer I in the plane Q_1, it suffices to measure an angle of 55° from point 0 on the net in the direction of point b. The positions of the other two seismometers are determined in the same way as in the case of CPS–I. All plots in the example cited in Figure 18b refer to the case $\varphi_0 = 22°$, $\omega_0 = 79°$, and $\gamma = 60°$.

Operating experience with CPS–II is proof of its reliability. Future designs may be based on CPS–II. It is worth noting that there is often no need to measure three angles. This applies to observations in vertical drill-holes characteristic of oilfields or in oblique drill-holes for which inclinometer data have been obtained previously, for instance, during the course of drilling. In such cases it is sufficient to measure only one angle ω_I or γ and to determine the azimuth of the first seismometer in the cluster.

Gyroscopic Position Sensor CPS–G. For observations in cased drill-holes or in drill-holes in areas with strong magnetic anomalies where it is impossible to utilize the Earth's magnetic field gyroscopic position sensors have been developed. The sensors measure only one angle, namely, the azimuth of the axis of one of the seismometers in a three-component cluster. The principal unit of the gyroscopic sensor CPS–G is a gyroscope including the gyromotor, the arrester, and the circular measuring rheocord. The latter has been made specially to replace the sectional rheocord of a standard gyroscope.

The circular rheocord is connected with the vertical coil of the gyroscope housing and with brushes fixed with respect to the tool's housing and, therefore, also with respect to the cassette containing the seismometers. The difference between various sensor models is mainly in the type and dimensions of gyroscopes employed, as well as in their arresting systems.

Sensor CPS–G–I employs a 60-mm-diameter gyroscope, and accordingly it is intended mainly for oil wells. There is only one position with respect to the tool housing in which the gyroscope is arrested. This enables the azimuth of the axis of the cluster to be determined by successive (from point to point) translation of the cluster's rotation angle with respect to the gyroscope axis. At the top of the drill-hole, the tool is oriented in the specified direction (towards the blast point or to the North). The gyroscope axis will be oriented in the prescribed direction (point 0 in Figure 19a). The tool is lowered after the gyromotor has been switched on and the gyroscope has been freed, with the gyroscope axis retaining the specified direction. If the tool's rotation angle about the specified direction at the first observation point (position 1), which may generally be identified with the azimuth of the first seismometer's axis ω_{I}, is γ_1, the axis of the gyroscope after it had been switched off and arrested will assume position ω_{I}. This is the position in which the blast waves are received (position 2). At the next point the operations are repeated in the same order, but this time ω_{I} will be the initial position (position 3), and the gyroscope axis will assume position 4.

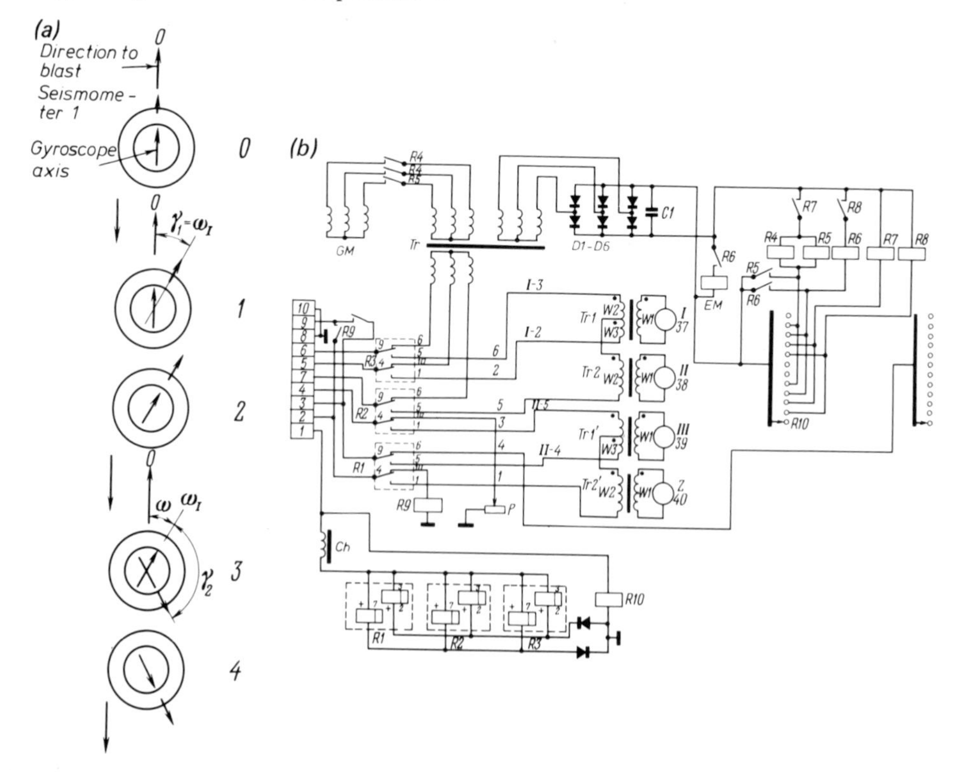

Fig. 19. Gyroscopic position sensor (CPS–G–I). (a) orientation diagram, (b) circuit diagram of drill-hole tool.

The drill-hole section of CPS–I includes the control circuit that provides for three modes of operation of the tool: 'clamping', 'orientation', and 'blast recording', and performs appropriate switching of the gyromotor circuits, the gyroscope, and the clamping gear, as well as a power supply consisting of a three-phase transformer and rectifier (Figure 19a). The sensor is manipulated, and the power supply and sensor operation are controlled from a surface dashboard.

CPS–G–II has been developed for small-diameter drill-holes. The gyromotor in CPS–G–II is always arrested in the same position, and the method of successive reading of azimuths described above is employed.

CPS–G–III, employing a device enabling the gyromotor to be arrested in any position, has been designed by Isaenko of the Byelorussian Scientific Research Institute for Exploration Geology (Bel NIGRI). The device makes it possible to determine the azimuth at every point directly from the readings of the angles. Moreover, such an arresting system improves the reliability of the instrument under difficult field conditions.

The position sensors are housed in the drill-hole tools and are rigidly joined to the three-component clusters. We shall not dwell on other drill-hole tool designs, because they are not specifically peculiar to three-component observations.

ORIENTED RECORDS

The actual position of the cluster thus determined enables signals oriented with the aid of an electronic orientation device to be obtained. There are two possible orientation procedures (direct and stage-by-stage).

Direct Orientation consists of direct translation of signals from the seismometers of a cluster arbitrarily oriented in space into signals of the seismometers of a cluster oriented in the specified direction by summing the former in appropriate proportion. The proportions are determined by correction factors proportional to the cosines of the angles between the axes of the actual (I, II, and III) and the oriented (I_a – III_a) seismometers. Nine angles: $\mathrm{I}_a - \mathrm{I} = \gamma_1$; $\mathrm{I}_a - \mathrm{II} = \gamma_2$; $\mathrm{I}_a - \mathrm{III} = \gamma_3$; $\mathrm{II}_a - \mathrm{I} = \gamma_4$; $\mathrm{II}_a - \mathrm{II} = \gamma_5$; $\mathrm{II}_a - \mathrm{III} = \gamma_6$; $\mathrm{III}_a - \mathrm{I} = \gamma_7$; $\mathrm{III}_a - \mathrm{II} = \gamma_8$; $\mathrm{III}_a - \mathrm{III} = \gamma_9$ are measured on the stereonet (Figure 20a).

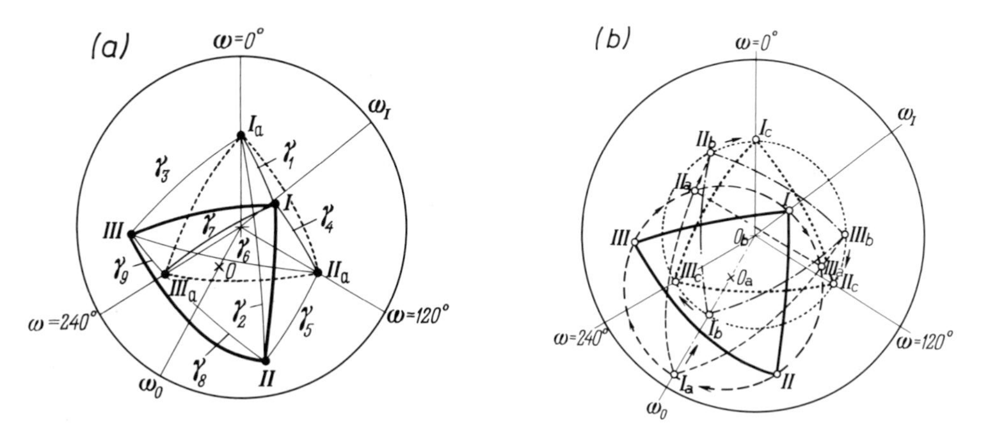

Fig. 20. Production of oriented signals. (a) direct orientation; (b) step-by-step orientation.

The values of $\cos \gamma_i$ correspond to the values of the correction factors K sought. Resistances R1–R9 are computed from the formula $R_i = 7.5/k\Omega/K$ or are read off a nomogram. In this case $7.5/k\Omega$ is taken as unity. Direct orientation greatly simplifies the circuitry, but complicates the operations, necessitating the computation of correction factors.

Stage-by-stage Orientation, in contrast to direct orientation in which each fictitious seismometer is oriented independently, permits the orientation of a cluster as a whole (Figure 20b). Let the actual position of the cluster I, II, III in space be such that the direction of the cluster's axis (the drill-hole axis) is determined by the inclination angle φ_0 and the azimuth ω_0 (point 0_a) with the azimuth of the cluster's first seismometer being ω_I. In our case $\omega_\text{I} = 53°$. The first stage (operation *a*) consists of making the azimuth of the first seismometer ω_I coincide with the azimuth of the cluster's axis ω_0. As a result the axes of the seismometers of the actual cluster I, II, III will occupy the positions of those of the fictitious cluster *a*: I_a, II_a, III_a. The second stage (operation *b*) involves the vertical positioning of the axis of the fictitious cluster (point 0_a moves to the position 0_b). This results in the orientation of the axes of all three seismometers of the fictitious cluster *a* in the respective directions of the fictitious cluster *b* (I_b, II_b, III_b), so that they will occupy positions on the surface of a cone whose generatrix makes an angle of 35° with the horizontal. The third stage (operation *c*) consists of the clockwise rotation of fictitious cluster *b* about the vertical axis until the azimuth of the first seismometer I_b coincides with the meridian ($\omega = 0°$) or in any specified direction. This completes the orientation of the fictitious cluster *c*. In our case the directions of the fictitious seismometers I_c, II_c, and III_c are specified by the following coordinates: $\omega_{\text{I}_c} = 0°$; $\omega_{\text{II}_c} = 120°$; $\omega_{\text{III}_c} = 240°$, and $\varphi_{\text{I}_c} = \varphi_{\text{II}_c} = \varphi_{\text{III}_c} = 35°$.

Stage-by-stage orientation is technically convenient, because it permits the use of data obtained directly from CPS (φ_0, ω_0, ω_I) and obviates the necessity to perform angle measurement and correction-factor determinations under field conditions. Direct orientation has been used both in observations with CPS–I and CPS–II. The procedure of stage-by-stage orientation may serve as a basis for the design of an easily-operated device which would automatically produce orientation corrections from parameters read off CPS.

The Electronic Orientation Device (Figure 21a) has two modes of operation: one for measuring CPS and the other for orientation. In the measuring mode the switch SW1 is closed, the pilot light L1 lights up, and the current flowing to the drill-hole tool over wires 7 and B makes it assume the position corresponding to the measurement of the angle sensors. Simultaneously relay Rel.1 is switched on, and the cable wires 1, 2, 3, and 5 are connected to switch S1. The CPS rheocords, by means of the switch S1, are in turn connected to terminals C and D, so that rheocord readings can be taken with the aid of universal bridges. After the measurements have been completed, SW1 is switched off, and the contacts of Rel.1 assume another extreme position, connecting the seismometers of the drill-hole tool via sockets F1 and F2 to the outputs of orientation-device amplifiers. The measurement data obtained serve to calculate the factors (resistances) of the variable resistors R1–R9. To introduce the factors obtained into the orientation device, the resistors R1–R9 are in turn connected into the circuit by the switch S2.

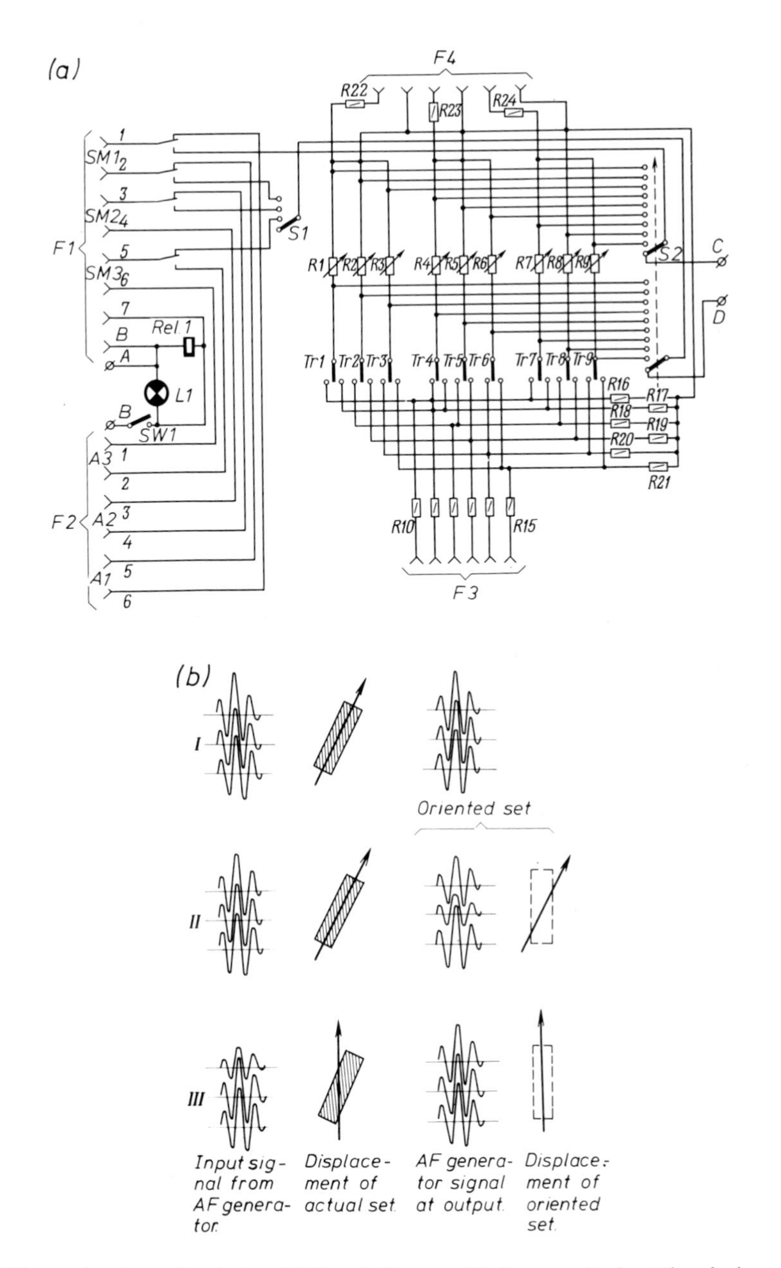

Fig. 21. Electronic orientation device. (a) Circuit diagram; (b) diagram of orientation device operation control.

A set of decade resistance dividers can conveniently be used for this purpose. Next, the resistors are connected to the appropriate signal phase in accordance with the signs of the factors. As a result, a signal with an amplitude proportional to the oriented signals of the three-component cluster appears across resistors R22–R24, which serve as load resistors. Signals from F4 are transmitted to the oriented recording galvanometers and simultaneously to the analyser inputs.

The orientation device requires only an infrequent control, since all its elements that participate in the actual orientation process are passive elements. The control of the orientation device in operation usually reduces to the control of equality of gain of its channels. To adjust gain, a signal from an audio generator is applied to the amplifier inputs connected in-parallel, and slide gain controls are adjusted to obtain equal amplitudes on galvanometers 1–3.

To gain clearer insight into the orientation process, let us consider the operation of the system for the production of oriented polar seismograms as a whole. Suppose that the actual cluster in the drill-hole is inclined. The application of identical signals to the channel inputs, for instance, an audio-generator signal to the inputs connected in-parallel, with the orientation device in the 'off' position, is equivalent to an axial oscillation of a symmetrical three-component cluster (Figure 21b,I). In this case, at the output of the system we obtain identical records from all three channels. With the orientation device in the 'on' position the intensities of the records from the three channels will be different (Figure 21b,II). This is because the orientation device has moved the cluster's axis to the vertical position, so that now the directions of particle motion, simulated by an identical input signal, do not coincide with the cluster's axis. If now, with the orientation device switched on, generator signals with an amplitude corresponding to the vertical oscillation of the cluster are applied, identical records will be obtained at the channel outputs, because the fictitious oriented cluster will be vertical, this corresponding to vertical particle displacements (Figure 21b,III).

The above considerations can be used to control the operation of instrumentation. For example, with the orientation device switched off and with an identical signal applied to the channel inputs, the output signals will be identical, and the amplitudes of the records of polar seismogram components must correspond to the orientation of the components, that is, the horizontal components must vanish, and the amplitudes of components located on the same conical surface must be equal. If, on the other hand, the input signals are not equal, the output signals will also not be equal, and the amplitudes of the components of the polar seismogram will no longer correspond to their orientation in space. This will be apparent already from the fact that the amplitudes on the record of horizontal components of a polar seismogram (channels 1–6) are not zero. With the orientation device switched on and with the signals at the inputs of different channels adjusted to conform to a definite proportion between individual channels (achieved as a result of orientation executed in accordance with orientation input signals), the output amplitudes will again be equal, and the amplitudes of the components of the polar seismogram will correspond to their orientation in space. With equal input signals the amplitudes at the outputs of the channels are not equal, and the amplitudes of the polar seismogram components do not correspond to their orientation.

Fixed- and Tracking-Components Block

The polarization-positional correlation of waves is executed with the aid of summary tracking-component seismograms. In the case of analog processing, these components are chosen by the interpreter from summary seismograms of fixed or tracking components obtained in the course of processing of magnetic films. To compile a summary seismogram of fixed components, appropriate components at every point are chosen and in turn recorded anew. A method of producing summary seismograms has been developed for this purpose, and a block of fixed and tracking components [126] in the form of an attachment to the magnetic station has been fabricated (Figure 22). Two types of blocks are in use – one for working without and another for working with a control seismometer. Their circuit diagrams differ little in principle. Let us consider the first variant of the circuit.

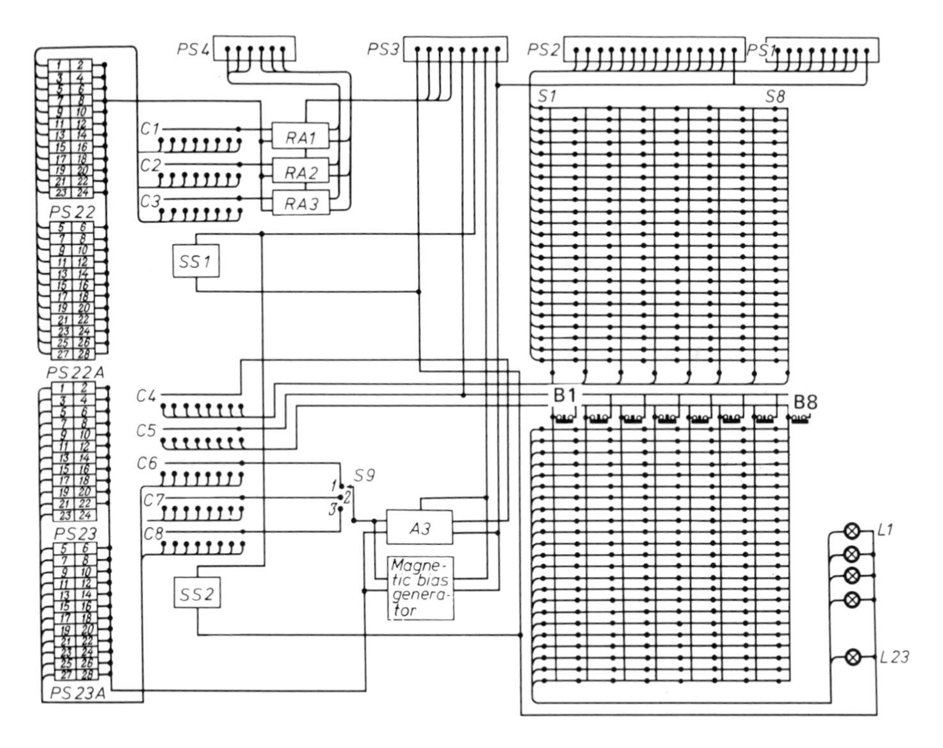

Fig. 22. Circuit diagram of the block of fixed and tracking components.

The pick-up heads of the station's drum are connected via sockets PS22 and PS22A to the contacts of stepping switch SS1 and then to input 3 of reproducing amplifiers RA1–RA3. The connection makes it possible to read signals off only three heads at a time. After each revolution, the control signal from the drum's cams moves the stepping relay to another position with the result that the reproducing heads are in turn, synchronously with the revolutions of the drum, connected to the reproducing amplifiers, so that during the first revolution the signals are read off the first channel triad, and during the second revolution off the second, etc. The new recording proceeds automatically, with the cycle being completed after eight revolutions of the drum, after which the stepping switches return to their initial positions.

The amplified signals are fed via sockets PS4 to the analyser where the 23-component polar diagram is formed. These signals are fed via sockets PS1 and PS2 to switches S1–S8 of the fixed- and tracking-components block. The corresponding contacts of the upper plates of the switches are connected to form parallel stacks which in turn are connected to the analyser output, so that the order of each switch step corresponds to the number of the analyser channel. The slip contacts of switches S1–S8 are connected to contact group C4 of stepping switch SS2. The slip contact of this group is connected to the input of the recording amplifier. The recording amplifier's input via the switch S9 is connected with the sliding contacts of the C6–C8 group. The contacts of these groups are connected via sockets PS23 and PS23A to the 24 pick-up heads of drum D2 (see Fig. 23).

Thus, by selecting the necessary analyser components with the aid of switches S1–S8 one can automatically transfer the original seismogram from the left to the right drum.

By means of successive transfer cycles, it is possible to obtain on the drum's film all 24 traces of the selected fixed or tracking components. The second plates of switches S1–S8 serve for manual and automatic control of correct component selection. The contacts of the second plates of switches S1–S8 are connected in-parallel, and the stacks are connected to appropriate control lights L1–L23. Pressing in turn the buttons B1–B8, one may check whether the components have been selected correctly from the glow of pilot lights. The slide contacts of the switch plates are connected with the fifth contact group of the stepping switch SS, and as a result a pilot lamp, showing the component of the polar seismogram recorded at the moment, lights up every time the drum makes a revolution.

To simplify connection, the block of fixed and tracking components is provided with its own magnetic bias generator. The circuits of all trace-reproducing amplifiers, as well as the recording amplifiers, are similar to that of the analyser's internal amplifiers. The block operating with a program drawn up by the interpreter is able to produce a summary seismogram of tracking components.

Equipment for Three-Component Observations and Data Processing Used in Seismic Prospecting

During the development of the polarization method, equipment for three-component field observations and for data-processing has been devised [79]. Instrumentation to be used for the polarization method has to satisfy more exacting specifications than that used for conventional observations. This first of all applies to the identity of the channels both from the point of view of the shape of their frequency response and their sensitivity. To this end special methods of continuous control of channels as a whole and separate amplifier-recorder channels have been designed. Let us consider the equipment used for observations and for data-processing separately. Note that equipment designed for drill-hole observations is the most complex.

FIELD EQUIPMENT FOR VSP USING THE POLARIZATION METHOD (PM–VSP)

The field set is based on the standard seismic station 'Poisk–I–48–MOV–OV' provided with additional special blocks. The simplified block-diagram of the field station is depicted in Figure 23a. The main blocks are: drill-hole tool 1, surface tool control block 2, non-oriented recording amplifiers 3, orientation device 4, magnetic head switch 6, oriented amplifiers 8, and analyser 7. When the drill-hole tool is raised or lowered, it is connected

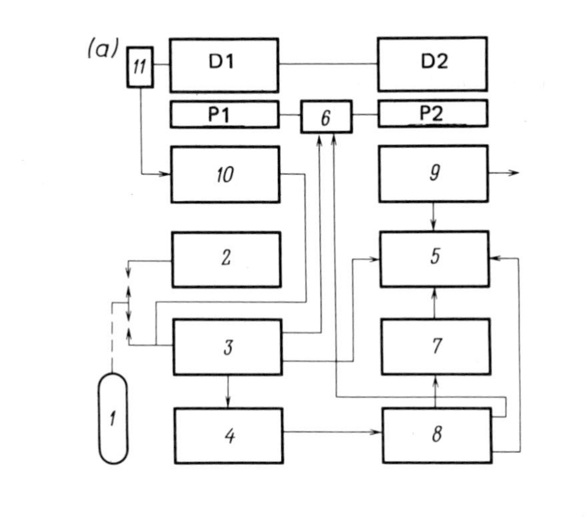

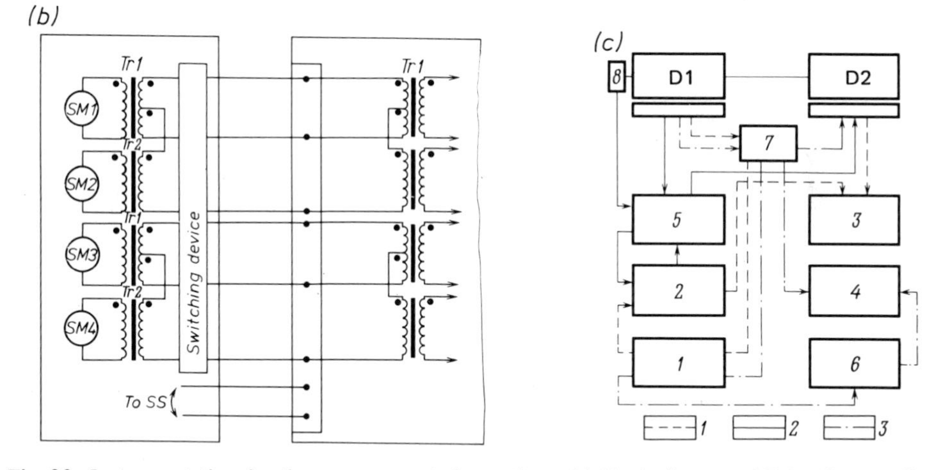

Fig. 23. Instrumentation for three-component observations. (a) block diagram of PM–VSP recording station; (b) wiring diagram of the drill-hole tool; (c) block diagram of data-processing station for: (1) producing polar seismograms, (2) recording fixed and tracking components on magnetic film, (3) orienting field records.

by means of a seven-wire cable to the tool-control block, and when blasts are recorded to the non-oriented signal amplifier of station 3. The drill-hole tool contains: a three-component seismometer cluster, a control seismometer, a cluster position sensor (magnetic or gyroscopic) and a clamping device.

Eight symmetrical transformers, four of them housed in the tool, are used to transmit signals from the four seismometers via the seven-wire cable. The drill-hole tool connection diagram is depicted in Figure 23b. Block 2 serves to control the clamping device and the position sensor. After the tool has been prepared for operation, the cable wires are connected to the station for blast recording.

The non-oriented recording amplifiers at the station's input have three non-oriented signal outputs connected to the following blocks: the orientation device, the magnetic head switchboard, and the oscillograph.

Orientation Device. We have used the output winding of the AGC transformer of each amplifier as an additional output for the orientation device (all observations are made without a time-variable gain control). Correction factors are introduced into orientation block 4 corresponding to the actual position of the cluster in the drill-hole.

Magnetic Head Switchboard. The magnetic head switchboard serves to record non-oriented signals on magnetic film (drum D1). It connects in turn the amplifier outputs to the appropriate pick-up heads P1 and P2. We have used two switching variants. In the first, three-component cluster signals from eight blasts (24 channels) have been recorded. The second employs five tracks for each blast to record the fourth control seismometer and to mark the instant of the blast, so that only five explosions can be recorded on the non-oriented recording film (the drum D1).

Oscillograph. This is designed for photo-oscillographic recording of non-oriented signals from a three-component cluster on two gain levels with a 5 : 1 difference in gain and a signal from the fourth control seismometer.

The outputs from the orientation device are connected to the inputs of oriented signal amplifiers 8 which also have three outputs connected to the following blocks: (a) the analyser to which an additional output similar to that of non-oriented recording amplifiers is connected; (b) the magnetic head switchboard for recording signals on the drum D2; and (c) oscillograph 5 for recording oriented signals. The 23 components of a polar seismogram from the analyser output are transmitted for direct recording on an oscillograph.

The amplifier-recorder channels are controlled by a signal from generator 10 automatically applied to the amplifier inputs after each explosion with the aid of the free cam 11 of the magnetic recorder. The switching circuit provides for disconnection of the seismometers from the amplifiers during the time the control signal from the generator is recorded. The list of recorded signals, apart from signals from the drill-hole tool and from the seismic station, also includes signals from surface seismometers, which facilitate control of excitation conditions and combination of observations along the vertical and horizontal profiles. The signals from seven such control seismometers are recorded on a photo-oscillograph with the aid of amplifiers 9.

To sum up, the equipment is capable of producing the following data under field conditions: magnetograms of non-oriented signals from drill-hole seismometers (a three-component cluster and a control seismometer); magnetograms of oriented signals from a three-component cluster and the instant of the blast; photo-oscillograms containing a non-oriented recording on two amplification levels, an oriented record, a polar seismogram and the records of control channels.

EQUIPMENT FOR DATA-PROCESSING

An analog-processing station has also been assembled on the basis of the standard 'Poisk–I–48–MOV–OV' station additionally equipped with an analyser, an orientation device, and a block of fixed and tracking components 5. The block diagram of the processing station is presented in Figure 23c. The processing station is capable of executing the following operations.

1. Provision of polar seismograms from a field magnetic film (drum D1). The signals from a three-component drill-hole cluster are transmitted to amplifiers 1 and then to the analyser 2. The 23 signals from the analyser output are recorded by oscillograph 3. When required, the polar seismogram may be recorded on magnetic film (drum D2) with the aid of amplifiers 4.

2. Selection and recording on a magnetic film of fixed or tracking components according to a program drawn up by the interpreter. The summary magnetograms are reproduced on an oscillograph for subsequent position correlation.

3. Orientation and re-orientation of field records. Such an operation may be necessary, if orientation has not been performed during the course of field observations, or if it is deemed necessary to introduce corrections into the orientation. To this end the non-oriented signals are read off the magnetic film (drum D1) and after amplification transmitted to orientation device 6, amplified by amplifiers 4 and recorded on magnetic film (drum D2).

The equipment for surface observations using the polarization method has to satisfy special requirements. The foremost among them relates to the increased number of channels. This requirement stems, on the one hand, from the necessity to record not just one, but a minimum of three oscillation components at each point, and on the other, from the extension of two- and three-dimensional observation systems. Here we have in mind equipment with up to several hundred channels. The major trend in this direction is the design of powerful electronic systems to increase the channel density immediately following the seismometers. Moreover, in order to study the trajectories of particle motion, it is necessary to obtain undistorted records in a wide dynamic range. Only digital recording can satisfy the requirements formulated above.

There are also additional requirements that the seismometers must satisfy. First of all, they must be assembled in a three-component cluster, that is, they must be designed as a single instrument in one housing and with a common lead. We may recall that symmetrical sets have appreciable advantages over the traditional *XYZ* sets. The period of natural vibrations of the seismometers must, of course, be such as to guarantee the frequency range of recording needed for the solution of a particular problem. An additional point is that for quantitative interpretation of data with the aid of the polarization method the directional diagram of the seismometers must be regular in shape and have zero sensitivity to displacements at right angles to the displacement of the pendulum.

There are good prospects for sub-LVL observations with the aid of the polarization method. Special seismometers are needed for such observations equipped with devices capable of lowering and orienting the seismometers in uncased drill-holes, drilled to penetrate the entire LVL, together with gear to extract them from the boreholes after recording. Operational models of such sub-LVL seismometers have already been produced and described (see Chapter 6). A technology for sub-zone observations has also been developed, which does not substantially increase operational costs.

Although the bulk of the data-processing should be performed on a digital computer, at some stages analog processing can be very helpful. Accordingly, analog-processing equipment should be modernized mainly in the direction of increased speed and processing potentialities.

CHAPTER 3

POLAR CORRELATION OF SEISMIC WAVES

Polar correlation (PC) facilitates the study of seismic waves from the point of view of their polarization. Polar correlation serves to determine the parameters of polarization of seismic waves in resolving the following problems: identification of simple and complex interferential waves; determination of their type, etc. Polar correlation is tantamount to the analysis of peculiarities in oscillation records depending on the orientation in space of their components, and is performed with the aid of a polar seismogram. The latter is a collection of records of oscillation components, each having a definite orientation in space. Figure 24 is a diagrammatic representation of a conical polar set-up with

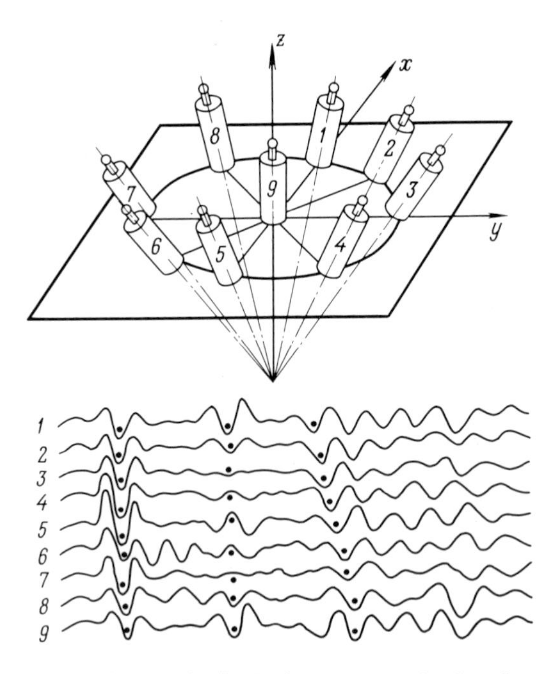

Fig. 24. Conical polar cluster of inclined seismometers and polar seismogram [141].

inclined seismometers and a corresponding seismogram of waves with different polarization directions. Polar correlation is a further development of correlation principles in the sphere of observations at a point, which serve as the basis and promote the progress and efficiency of all seismic methods. This determines the advantages and the potentialities of polar correlation as compared with uncorrelated three-component recording. To make the situation clear, we may draw an analogy with positional observations. A three-component record is in this respect similar to a positional seismogram obtained during observations at distant points in which phase correlation is impossible. A polar seismogram is similar to a positional one on which phase correlation of waves is ensured. This makes it possible, during the course of analysis, to separate regular waves with stable polarization parameters from irregular unwanted background waves with unstable polarization, just as in positional correlation, waves producing records with shapes stable along the traverse are separated. In addition, polar correlation makes possible the visual determination of wave types and polarization parameters from polarization seismograms.

The visual (qualitative) methods of polar correlation consist of the analysis of the shapes of records and lineups on polar seismograms. Visual methods are usually adequate for separating simple oscillations on the seismograms, and in certain cases also, for qualitative determination of the position of the polarization plane of plane-polarized oscillations.

Polar correlation performed on seismograms first of all shows up the recording intervals of simple linearly-polarized waves and plane-polarized waves, as well as recording regions of waves with more complex polarization. An important factor in the analysis of the wave field in polar correlation is the splitting-up of a complex oscillation record into various space components. This can be achieved, if the waves arrive at the seismometer with a certain time-lag relatively to one another. In many cases the oscillation components may be separated visually.

Polar correlation also enables the nature of polarization of the dominant waves to be determined, and an optimum component and vector to be found for each wave on the polar seismogram.

It should be pointed out that, despite the fact that a rather complete analysis of particle motion trajectories, including polar correlation, can be performed at present on a digital computer with the aid of ADPS 'Polarizatziya' (see Chapter 4), visual analysis of polar seismograms is very important both at the data-production and data-processing stages. The methods of three-dimensional analysis and plotting discussed below facilitate the evaluation of data quality, timely planning, and field operation control, and help to understand the computer-processed polarization parameter data and interpret them.

Derivation of Equations for Line-Ups and Amplitudes of Elliptically-Polarized Oscillations

Two characteristics determine the polar seismogram of every wave: the phase and the amplitude (dynamic). The phase characteristic determines the dependence of the wave's phase-shift β on the directions in space of the components of the polar seismogram, defined by their azimuths ω_i and angles with the horizontal ψ_i. The shape of the polar seismogram's line-up, a term we shall, by analogy with the position seismogram, use to describe the line that joins the extremal wave phases, depends only on the wave's polarization type. The expression

$$\beta = \beta(\omega_i, \psi_i)$$

is the equation for the line-up.

The wave's amplitude charactersitic describes the dependence of the extremal amplitudes A recorded on the polar seismogram on the orientation in space of its components.

The expression $A = A(\omega_i, \psi_i)$ is the equation for the wave's amplitudes on the polar seismogram.

The equations for the line-up and amplitudes provide a complete description of the wave records on the polar seismograms. In analysing these equations, one is able to formulate the characteristics that permit the separation of the main types of seismic waves according to the nature of their polarization. We shall deduce the equations of polar correlation for an elliptically-polarized oscillation, because linear polarization is a particular case of elliptical polarization.

EQUATION FOR THE LINE-UP

Consider two linearly-polarized oscillations with different amplitudes A'_0 and B'_0 and initial phases β'_1 and β'_2, but with equal periods T, with arbitrary displacement directions ω'_1, φ'_1 for the oscillation $A'(t)$ and ω'_2, φ'_2 for the oscillation $B'(t)$.

For harmonic oscillations $A'(t)$ and $B'(t)$ we have the formulae

$$A'(t) = A'_0 \sin\left(\frac{2\pi}{T} + \beta'_1\right),$$
$$B'(t) = B'_0 \sin\left(\frac{2\pi}{T} + \beta'_2\right). \tag{19}$$

Find the projections (components) of each of the interfering oscillations on one direction and sum the components obtained. The projections of each of the component oscillations on the direction described by angles ω and ψ may be expressed by the equation

$$A'_{\omega\psi} = A'_0 \cos\Delta\varphi'_1 \sin\left(\frac{2\pi}{T} + \beta'_1\right),$$
$$B'_{\omega\psi} = B'_0 \cos\Delta\varphi'_2 \sin\left(\frac{2\pi}{T} + \beta'_2\right), \tag{20}$$

where $\Delta\varphi'_i$ is the angle in space between the displacement vector and the direction of the polar-seismogram component.

The analytical expression for $\cos\Delta\varphi'_i$, where $\Delta\varphi'_i$ is the angle between two straight lines, is

$$\cos\Delta\varphi'_i = \frac{l_1 l_2 + m_1 m_2 + n_1 n_2}{\sqrt{l_1^2 + m_1^2 + n_1^2}\,\sqrt{l_2^2 + m_2^2 + n_2^2}}$$

where

$$l_1 = \cos\varphi'_i \sin\omega'_i; \qquad l_2 = \sin\psi \sin\omega;$$
$$m_1 = \cos\varphi'_i \cos\omega'_i; \qquad m_2 = \sin\psi \cos\omega;$$
$$n_1 = \sin\varphi'_i; \qquad n_2 = \cos\psi,$$

Hence

$$\cos\Delta\varphi'_i = \sin\psi \cos\varphi'_i + \cos\psi \sin\varphi'_i \cos(\omega - \omega'_i). \tag{21}$$

The projections of oscillations $A'(t)$ and $B'(t)$ on the direction ω, ψ will take the form

$$A'_{\omega\psi} = A'_0 [\sin\psi \cos\varphi'_1 + \cos\psi \sin\varphi'_1 \cos(\omega - \omega'_1)] \sin\left(\frac{2\pi}{T} t + \beta'_1\right),$$
$$B'_{\omega\psi} = B'_0 [\sin\psi \cos\varphi'_2 + \cos\psi \sin\varphi'_2 \cos(\omega - \omega'_2)] \sin\left(\frac{2\pi}{T} t + \beta'_2\right). \tag{22}$$

Hence the projections of the displacement components on the direction ω, ψ are proportional to $\cos\Delta\varphi'_i$, that is, all the amplitudes on a polar seismogram vary according to the cosine law. Each of the expressions (2) may be regarded as an equation of a theoretical seismogram corresponding to a displacement in the ω, ψ direction, with the amplitudes of records on these seismograms being determined by the expressions:

$$A'_{\omega\psi} = A'_0 \sin\psi \cos\varphi'_1 [1 + \cot\psi \tan\varphi'_1 \cos(\omega - \omega'_1)] \sin\left(\frac{2\pi}{T} t + \beta'_1\right),$$
$$B'_{\omega\psi} = B'_0 \sin\psi \cos\varphi'_2 [1 + \cot\psi \tan\varphi'_2 \cos(\omega - \omega'_2)] \sin\left(\frac{2\pi}{T} t + \beta'_2\right). \tag{23}$$

If one fixes the values of ψ and ω, that is, imagines a seismometer whose maximum sensitivity axis points in the direction determined by these angles, then Equations (23) will each describe one of the interfering oscillations recorded by this seismometer.

We use the conventional method of summation of two harmonic oscillations with equal directions and with equal periods, but with different amplitudes and initial phases to sum the components of both oscillations.

To this end, we will substitute the values of amplitudes of the components from Equations (23) in Formula (3) and obtain, after several transformations for the phase shift of the summary oscillation, the following expression:

$$\tan\beta' = \frac{\cot\psi \left[\cos\omega \left(\frac{A'_0}{B'_0} \sin\beta'_1 \sin\varphi'_1 \cos\omega'_1 + \sin\beta'_2 \sin\varphi'_2 \cos\omega'_2\right) + \right.}{\cot\psi \left[\cos\omega \left(\frac{A'_0}{B'_0} \cos\beta'_1 \sin\varphi'_1 \cos\omega'_1 + \cos\beta'_2 \sin\varphi'_2 \cos\omega'_2\right) + \right.}$$
$$\frac{\left. + \sin\omega \left(\frac{A'_0}{B'_0} \sin\beta'_1 \sin\varphi'_1 \sin\omega'_1 + \sin\beta'_2 \sin\varphi'_2 \sin\omega'_2\right)\right] +}{\left. + \sin\omega \left(\frac{A'_0}{B'_0} \cos\beta'_1 \sin\varphi'_1 \sin\omega'_1 + \cos\beta'_2 \sin\varphi'_2 \sin\omega'_2\right)\right] +}$$
$$\frac{+ \frac{A'_0}{B'_0} \sin\beta'_1 \cos\varphi'_1 + \sin\beta'_2 \cos\varphi'_2}{+ \frac{A'_0}{B'_0} \cos\beta'_1 \cos\varphi'_1 + \cos\beta'_2 \cos\varphi'_2}. \tag{24}$$

This expression is the equation for the line-up of the summary oscillation on the polarization seismogram. The terms in round brackets in Equation (24) depend only on the parameters of the original oscillations. The coefficients in front of the round brackets determine the dependence of the phase shift on the azimuth of the polar seismogram component, and the coefficients in front of the square brackets depend on the component's inclination angle.

The interference of the original harmonic oscillations with arbitrary directions φ_1', ω_1' and φ_2', ω_2' and phases β_1 and β_2 may be reduced to the interference of two mutually perpendicular harmonic oscillations

$$A(t) = A_0 \sin\left(\frac{2\pi}{T}t + \beta_1\right),$$

$$B(t) = B_0 \sin\left(\frac{2\pi}{T}t + \beta_2\right)$$

with directions $\varphi_1 \omega_1$ and $\varphi_2 \omega_2$ for which the phase shift $\beta_1 - \beta_2 = 90°$, that is, to elliptically-polarized oscillations. Putting $\beta_1 = 0$ we obtain $\beta_2 = 90°$ and can re-write Equation (24) in the form

$$\tan\beta = \frac{m \cot\psi \sin\varphi_2 \cos(\omega - \omega_2) + \cos\varphi_2}{\cot\psi \sin\varphi_1 \cos(\omega - \omega_1) + \cos\varphi_1}. \tag{25}$$

Let us take as the origin of the azimuth scale the azimuth of the first oscillation $\omega_1 = 0$, and allow for the condition of orthogonality of the directions φ_1, ω_1 and φ_2, ω_2:

$$\sin\varphi_1 \sin\varphi_2 \cos\omega_2 + \cos\varphi_1 \cos\varphi_2 = 0. \tag{26}$$

Then the equation for the line-up of elliptically-polarized oscillations will, after appropriate transformations, assume the form

$$\tan\beta = \frac{m[\cot\psi \sin\varphi_2 \cos(\omega - \omega_2) + \cos\varphi_2]\,(\tan^2\varphi_2 \cos^2\omega_2 + 1)^{1/2}}{\tan\varphi_2 \cos\omega_2 - \cot\psi \cos\omega}, \tag{27}$$

where $m = B_0/A_0 = b/a$ is the elliptical form-factor.

EQUATION OF AMPLITUDES

The equation for the amplitudes of a wave on a polar seismogram $C'_{\omega\psi}$ may be obtained by substitution of the amplitude values from Expressions (20) in Formula (2):

$$(C'_{\omega\psi})^2 = (A_0' \cos\Delta\varphi_1')^2 + (B_0' \cos\Delta\varphi_2')^2 + 2A_0'B_0' \cos(\omega - \omega') \times \\ \times \cos(\omega - \omega_2') \cos(\beta_1' - \beta_2'). \tag{28}$$

After substitution of values of $\cos\Delta\varphi_1'$ and $\cos\Delta\varphi_2'$ from Equation (21), Equation (28) will take the form

$$(C'_{\omega\psi})^2 = \sin^2\psi\,\{(A_0' \cos\Delta\varphi_1')^2 \cos^2\varphi_1' [1 + \cot\psi \tan\varphi_1' \cos(\omega - \omega_1')]^2 + \\ + (B_0')^2 \cos^2\varphi_2' [1 + \cot\psi \tan\varphi_2' \cos(\omega - \omega_2')] + \\ + 2A_0'B_0' \cos\varphi_1' \cos\varphi_2' [1 + \cot\psi \tan\varphi_1' \cos(\omega - \omega_1')] \times \\ \times [1 + \cot\psi \tan\varphi_2' \cos(\omega - \omega_2')] \cos(\beta_1' - \beta_2')\}. \tag{29}$$

Substitute the interference of mutually-perpendicular oscillations with a phase shift of $\beta_1 - \beta_2 = 90°$ for the interference of the original oscillations, as we have done in deducing the equation for the line-up. In the same simplifying assumptions, $\omega_1 = 0 = \beta_1$ and $\beta_2 = 90°$, we obtain an equation for the amplitudes of elliptically-polarized oscillations

$$C^2_{\omega\psi} = A_0^2(\sin\psi \cos\varphi_1 + \cos\psi \sin\varphi_1 \cos\omega)^2 + \\ + B_0^2 [\sin\psi \cos\varphi_2 + \cos\psi \sin\varphi_2 \cos(\omega - \omega_2)]^2, \tag{30}$$

where A_0 and B_0 are the polarization ellipse's semi-axes.

Taking into account the Orthogonality Condition (26), we may write Equation (30) in the form

$$C^2_{\omega\psi} = A_0^2 \frac{[\sin\psi \tan\varphi_2 \cos\omega_2 - \cos\psi \cos\omega]^2}{1 + \tan^2\varphi_2 \cos^2\omega_2} + \\ + B_0^2 [\sin\psi \cos\varphi_2 + \cos\psi \sin\varphi_2 \cos(\omega - \omega_2)]^2. \tag{31}$$

Hence, Equations (27) and (31) are equations for the line-up and polar-seismogram amplitudes for elliptically-polarized waves. Polar correlation equations for all types of seismic waves may easily be obtained from Formulae (27) and (31).

Analysis of Equations for Line-Ups and Amplitudes of Linearly-Polarized Waves

Let us analyse the equation for the line-up and amplitudes for different seismic waves. We shall start our analysis with linearly-polarized waves, because they are of special interest on account of the fact that simple three-dimensional waves are linearly-polarized waves.

The equations for the line-up and amplitudes of a linearly-polarized oscillation on a polar seismogram can be obtained from the general equations for the line-up (24) and the amplitudes (28) of a polar seismogram.

EQUATIONS FOR LINE-UP AND AMPLITUDES

The equation for the line-up may be obtained from Formula (24). Putting $A_0' = 0$, we have $\tan\beta' = \tan\beta_2'$ and $\beta_1' = \beta_2'$; that is, the phase shifts of all the components on the polar seismogram are equal to the phase of the second oscillation β_2'. For $\beta_0' = 0$, $\tan\beta' = \tan\beta_1'$, that is, the phase shifts are equal to the phase of the first oscillation $\beta' = \beta_1'$. The equation for the amplitudes of a linearly-polarized oscillation whose displacement vector points in the direction ω_1', φ_1' may be obtained from (28), if we put $B_0' = 0$. Then $C'_{\omega\varphi} = A_0' \cos\Delta\varphi_1'$.

Substituting herein the value of $\cos\Delta\varphi_1'$ from (21), we obtain

$$C'_{\omega\psi} = A_0' [\sin\psi \cos\varphi_1' + \cos\psi \sin\varphi_1' \cos(\omega - \omega_1')],$$
$$C'_{\omega\psi} = A_0' \sin\psi \cos\varphi_1' [1 + \cot\psi \tan\varphi_1' \cos(\omega - \omega_1')].$$

For ψ = const., the amplitude on the seismogram is independent of the azimuth of the seismometer axis.

The following conclusions may be drawn from the equations obtained above.

1. On a polar seismogram the line-up is always vertical.
2. Since the relative phase shift $\Delta\beta$ is zero on all polar-seismogram components, there may be a 180° phase inversion on individual seismogram components.
3. The recorded form of a linearly-polarized oscillation is identical on the records of all the polar-seismogram components, the amplitudes of the recordings varying according to the cosine law.

Such phase and amplitude characteristics of the recoridngs may serve as tests for separating linearly-polarized waves on polar seismograms.

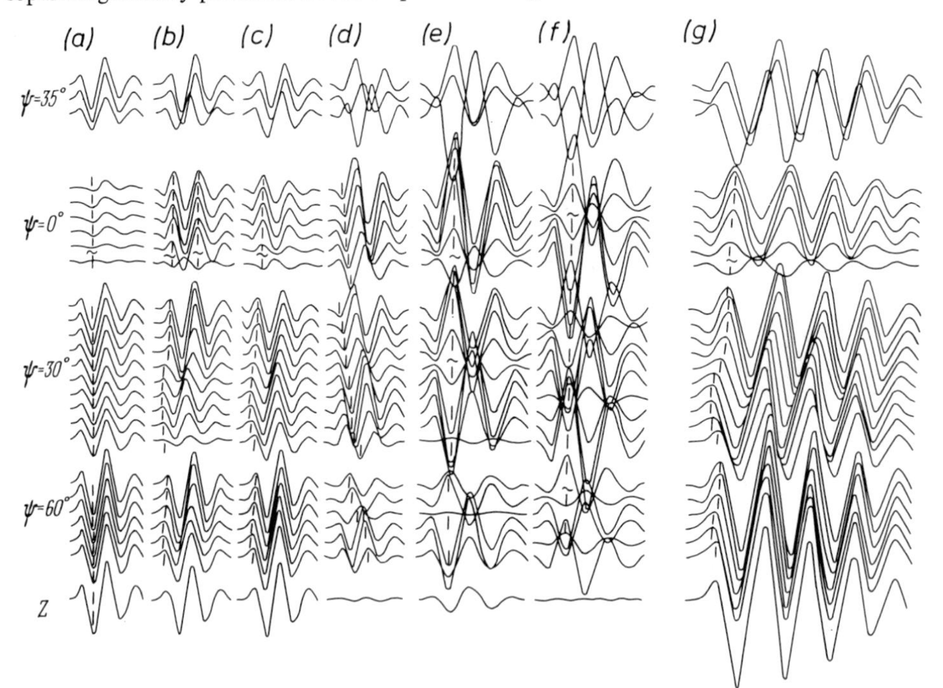

Fig. 25. Theoretical polar seismograms. (a) Linearly-polarized oscillation ($\varphi_P = 10°$, $\omega_P = 45°$); (b, c) elliptically-polarized oscillation in a vertical plane with an inclined ellipse (b) ($\varphi_P = \varphi_{SV} = 45°$, $\omega_P = 45°$, $\omega_{SV} = 225°$) and a vertical ellipse (c) ($\varphi_P = 10°$, $\omega_P = 45°$, $\varphi_{SV} = 80°$, $\omega_{SV} = 225°$); (d) elliptically-polarized oscillation in a horizontal plane ($\varphi_{SV} = 80°$, $\omega_{SV} = 225°$, $\varphi_{SH} = 90°$, $\omega_{SH} = 135°$); (e, f) linearly-polarized oscillations for *SV* (e) and *SH* (f) waves ($\varphi_{SH} = 80°$, $\omega_{SH} = 225°$, $\varphi_{SH} = 90°$, $\omega_{SH} = 135°$); (g) elliptically-polarized Rayleigh wave oscillation.

Figure 25 is a representation of theoretical polar seismograms calculated for waves with different types of polarization. The initial pulses are assumed to be Berlage pulses

$$A = at^{-b} e^{-c\omega t} \sin \omega t,$$
$$B = at^{b} e^{-c\omega t} \sin (\omega t + \beta),$$

where $a = 1; b = 3; c = -0.5; \beta = 90°$.

Let us consider seismograms of linearly-polarized waves. The longitudinal *P* wave (Figure 25a) is linearly-polarized with displacements directed almost along the vertical (φ_P = 10°, ω_P = 45°). The *S* waves are also linearly-polarized, but their displacement directions are close to horizontal. The displacement directions of the *SV* wave (Figure 25e) are determined by the angles φ_{SV} = 80°, ω_{SV} = 225°, and the *SH* wave (Figure 25f), φ_{SH} = 90°, ω_{SH} = 135°. The seismograms clearly show all the characteristics that serve to separate linearly-polarized oscillations (vertical line-ups with a 180° phase inversion (Figure 25e, f) and without it (Figure 25a), the conservation of the form of the recording, and the cosine law of amplitude variation on the seismogram. The relative amplitudes are functions only of the inclination angle of the polar seismogram's components.

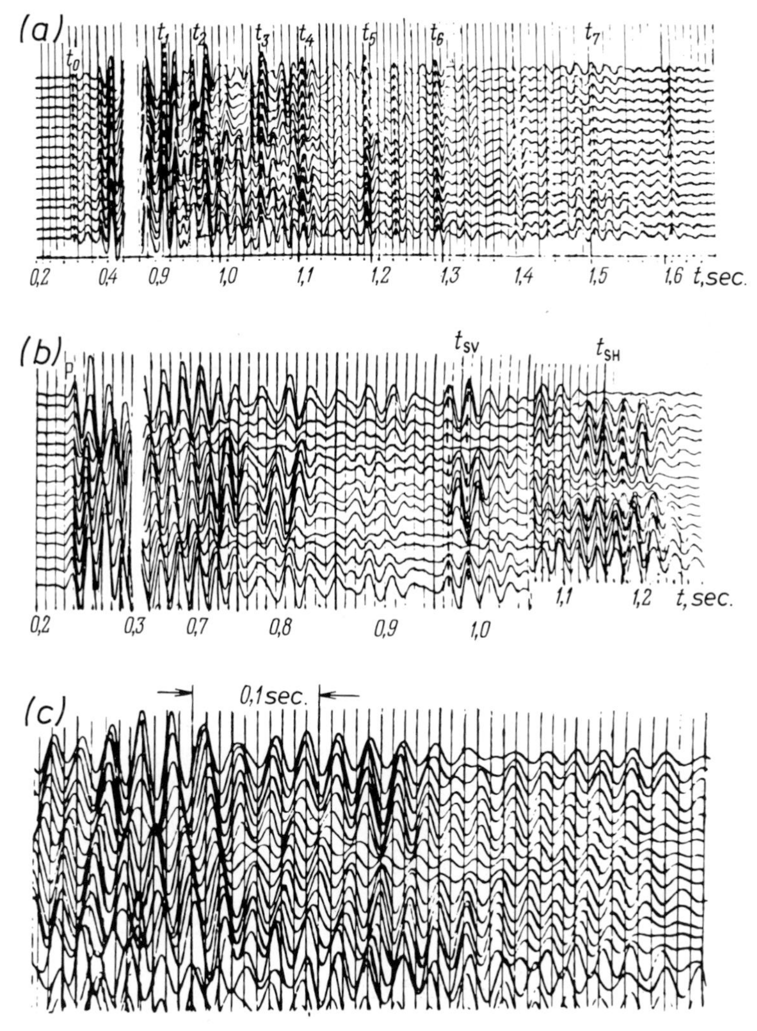

Fig. 26. Observed polar seismograms (polar set, ψ = 40°) of waves with different types of polarization. (a) Linearly-polarized *P*-wave oscillations; (b) linearly-polarized *S*-wave oscillations (*SV* on the right, *SH* on the left); (c) multiply-polarized oscillations.

The observed polarization seismograms of linearly-polarized oscillations in longitudinal refracted (t_0) and reflected ($t_1 - t_7$) waves are presented in Figure 26a, those of transverse t_{SV} and t_{SH} waves, in Figure 26b, and of multi-polarized oscillations, in Figure 26c. Examples of polar seismograms of linearly-polarized oscillations corresponding to different sets are to be seen in Figure 27.

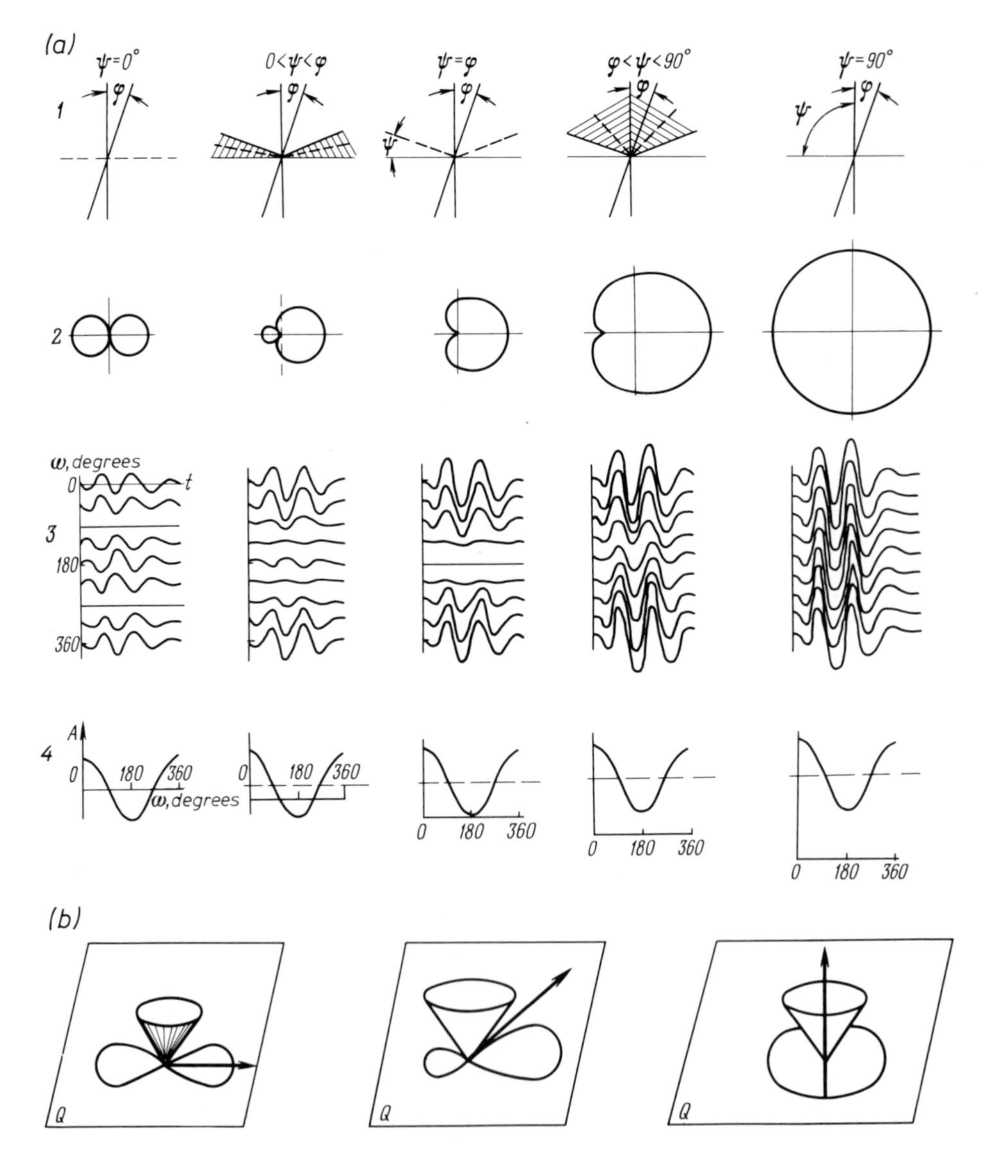

Fig. 27. Directional diagram and theoretical seismograms of polar sets. (a) for sets with different seismometer inclination angles in a fixed displacement direction; (b) for different displacement directions for a fixed seismometer inclination angle: (1) diagram of mutual arrangement of seismometer axes and displacement direction, (2) directional diagrams in polar coordinates, (3) theoretical polar seismograms, (4) directional diagrams in rectangular coordinates.

The variations of the relative recorded amplitudes of linearly-polarized oscillations as a function of the azimuth of the maximum sensitivity axis of the seismometers may conveniently be represented in the form of graphs.

DIRECTIONAL DIAGRAMS OF POLAR SETS

We shall apply the term directional diagram of a polar set to the dependence of the relative amplitude of a recording on the orientation of the seismometer's axis in space. In contrast to the directional diagram of a seismometer, the displacement direction in the diagram of a polar set is fixed, and the orientation of the seismometer axes is variable. The shape of the diagram depends on the direction of the displacement vector, and because of that, the term 'directional diagram of a set' is a conventional one, and its meaning is conditional on the determination of the direction of the displacement vector. The directional diagrams of polar sets are very important for polar correlation, and so we shall consider such diagrams for sets of various types in more detail.

Horizontal Set. The directional diagram for a horizontal set ($\psi = 0$, Figure 27a) is independent of the direction of the displacement vector and consists of two tangent circumferences, corresponding to the equation of amplitudes of a polar seismogram with horizontal components. The maximum amplitude will be recorded on the horizontal component coinciding in azimuth with that of the displacement vector. The recorded amplitude of a component whose axis is at right angles to the displacement direction will be zero. For transverse waves in a homogeneous and isotropic medium, this component lies in the plane tangent to the wave front. The records of the components on both sides of the zero-amplitude component will display 180° phase inversions. The amplitudes on the seismogram vary according to the cosine law, and there are regions of small and zero amplitudes. Hence, for a given set its directional diagram determines the form of the polar seismogram.

Conical Sets. The recorded amplitude of an inclined component depends on the direction of the displacement vector and on the direction of the component of the polar set. Let us consider the variation in the form of the directional diagram of a set with the variation of the component's inclination angle from 0° to 90°, the direction of the displacement vector remaining constant (angle φ).

Figure 27a1 depicts the fields of variation of angle ψ (shaded area) and the directional diagrams (Figure 27a2). For $\psi = 0$ (horizontal set), the directional diagram consists of two tangent circumstances. For $0 < \psi < \varphi$ the directional diagram is no longer symmetrical. Phase inversions and regions of small and zero amplitudes are typical of a polar seismogram. The amplitudes of components that differ by 180° are no longer equal. For $\psi = \varphi$ the diagram is asymmetrical with one of its parts degenerating into a point. There is no phase inversion on the polar seismogram (all oscillations are in-phase), but there is a region of small amplitudes (including zero amplitudes). The zero amplitude corresponds to the component perpendicular to the displacement vector. For $\varphi < \psi < 90°$ the diagram is asymmetrical. All oscillations on the polar diagram are in-phase, and there is no phase inversion. The differences between the amplitudes on the seismograms are less pronounced. The region of small amplitudes exists only for ψ values that are close to φ.

The differences between the amplitudes of records of seismometers whose azimuth differ by 180° decrease with increase in the angle ψ. For $\psi = 90°$, all components become vertical, and the diagram assumes the form of a circumference. All the records on the polar diagram are identical and correspond to the record of the Z component.

In rectangular coordinates the directional diagram of a conical set is a cosinusoid whose parameters are determined by the oscillation vector. The shape of the diagram does not change as the inclination angle of the seismometers varies, the change being limited only to its position with respect to the abcissa, and hence the directional diagrams may be represented by one cosinuisoid (Figure 27a4). Let us consider this point in more detail. For $\psi = 0$, the axis of the cosinusoid coincides with the abcissa of the rectangular coordinate frame. Displacing the abcissa parallel to the axis of the cosinusoid by a distance less than its amplitude, we may obtain a family of directional diagrams for the range of angles $0 < \psi < \varphi$. For $\psi = \varphi$, the abcissa is tangent to the cosinusoid's extrema. Moving the abcissa still farther away, we may obtain diagrams corresponding to the case $\varphi < \psi < 90°$. Finally, a cosinusoid with its abcissa at infinity corresponds to a diagram which, in polar coordinates, has the form of a circumference.

Hence, the form of directional diagrams of sets and therefore the character of the polar seismogram for the same displacement vector, also are conditioned by orientation of the set's components.

Similar directional diagrams may be obtained for one set (with a fixed seismometer inclination angle) for a variable diaplacement direction. For example, for a horizontal displacement, the directional diagram is symmetrical and consists of two tangent circumferences (Figure 27b2). In the case of an inclined displacement, the diagram is asymmetrical about the displacement, the directional diagram for a vertical displacement being a circumference.

Analysis of Equations for Line-Ups and Amplitudes of Elliptically-Polarized Waves

The general form of equations for line-ups and amplitudes of elliptically-polarized waves is given by Expressions (25) and (31). Let us consider these equations. The analysis of the equation for the line-up of a polar seismogram presents the greatest interest.

CLASSES OF LINE-UPS

Let us analyse the extrema of a curve corresponding to the line-up (25), regarding the phase shift β as a function of the azimuth ω of the polar seismogram's component. To this end, we may write the expression of the derivative $\partial\beta/\partial\omega$, which after trigonometric transformations takes the form:

$$\frac{\partial\beta}{\partial\omega} = \frac{m \cot\psi\,[\cot\psi \sin\varphi_1 \sin\varphi_2 \sin(\omega_2 - \omega_1) +}{[\cot\psi \sin\varphi_1 \cos(\omega - \omega_1) + \cos\varphi_1]^2 +}$$

$$\frac{+ \cos\varphi_2 \sin\varphi_1 \sin(\omega - \omega_1) - \cos\varphi_1 \sin\varphi_2 \sin(\omega - \omega_2)}{+ m^2\,[\cot\psi \sin\varphi_2 \cos(\omega - \omega_2) + \cos\varphi_2]^2}\,.$$

Equating the derivative to zero and putting $\omega_1 = 0$, we obtain with the aid of elementary simplifications the expression

$$a \sin \omega + b \cos \omega = c,$$

where a, b, and c are trigonometric parameters independent of ω:

$$a = \cos \omega_2 ; \quad b = \cos^2 \varphi_1 \sin \omega_2 ; \quad c = \cot \psi \sin \varphi_1 \cos \varphi_1 \sin \omega_2 .$$

The solution of this equation satisfies the following inequality:

$$c^2 \leqslant a^2 + b^2 .$$

Substituting the values of a, b, and c in this inequality and expressing ω_2 from Condition (26) in terms of φ_1 and φ_2, we may reduce the inequality to the following form:

$$\cos 2\varphi_1 + \cos 2\varphi_2 \geqslant -2 \sin^2 \psi .$$

Transforming the latter inequality, we obtain

$$\cos^2 \varphi_1 - \sin^2 \varphi_1 + \cos^2 \varphi_2 - \sin^2 \varphi_2 \geqslant -2 \sin^2 \psi$$

or

$$\cos^2 \varphi_1 + \cos^2 \varphi_2 \geqslant \cos^2 \psi . \quad (32)$$

If the normal to the polarization plane makes an angle φ_n with the vertical, the following equality, which determines the mutual orthogonality of the three directions in space, must be satisfied

$$\cos^2 \varphi_1 + \cos^2 \varphi_2 + \cos^2 \varphi_n = 1,$$

whence

$$\cos^2 \varphi_1 + \cos^2 \varphi_2 = \sin^2 \varphi_n . \quad (33)$$

Substituting Expression (33) into Inequality (32), we obtain

$$\varphi_n \geqslant 90^\circ - \psi . \quad (34)$$

Inequality (34) determines the condition of existence of extrema on the line-up of a polar seismogram.

It follows from (34) that line-ups may be divided into two classes according to their shape: line-ups with and without extrema. Extrema exist in the case, when the angle between the vertical and the normal to the polarization plane φ_n exceeds the angle between the vertical and the polar seismogram's components ($90^\circ - \psi$). This condition may be visualized as follows. If the normal to the polarization plane n_2 lies inside the cone formed by the axes of inclined seismometers of the set, the line-up has no extremum, and vice versa, if the normal n_1 lies outside the cone, the line-up has an extremum (Figure 28a). Hence, a polar set in the form of a conical surface divides the half-space into two regions: internal – the region free of extrema, and external – the region of extrema. A change in the inclination angle ψ of the seismometers changes the ratio of the two regions: the greater the angle ψ the greater the region where the extrema exist.

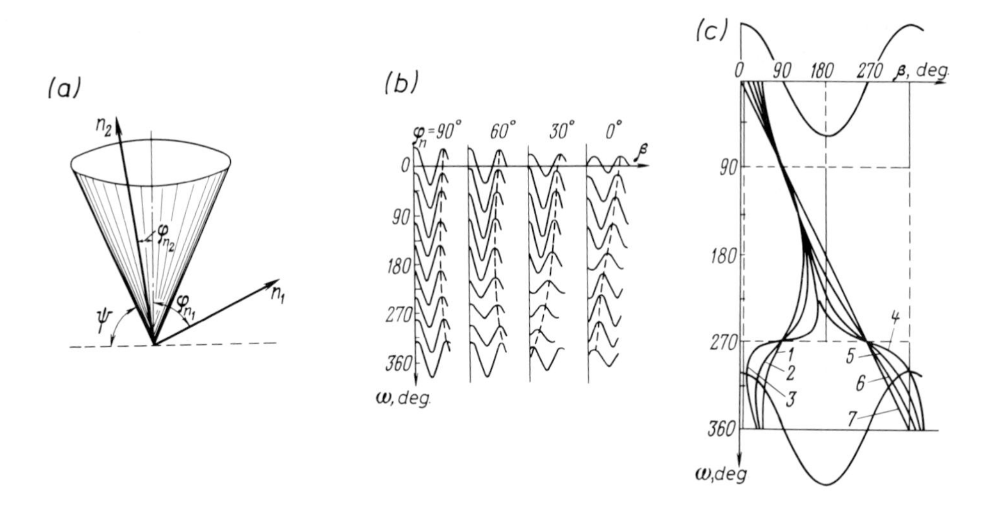

Fig. 28. Classes of line-ups. (a) Regions of existence and absence of extrema; (b) theoretical polar seismograms for a conical cluster (ψ = 45°); (c) family of line-ups (ψ = 45°) for different angles φ_n: 1–90°, 2–75°, 3–60°, 4–45°, 5–30°, 6–15°, 7–0°.

The problem of the presence or absence of extrema on the line-up is very important in polar correlation. This is because, during the course of continuous phase correlation, phase shifts as large as a full period are observed for a closed conical set, correlation along an axis without extrema resulting in a shift equal to a period, there being no shift in case of a line-up with an extremum, so that in the last case we obtain the same phase. Examples follow of experimental and theoretical polar seismograms to illustrate the point.

Let us consider theoretical polar seismograms corresponding to a conical set with a seismometer inclination angle of ψ = 45° for a variable angle φ_n (Figure 28b). With φ_n equal to 90° and 60°, the normal to the polarization plane lies outside the cone of the set, and the line-ups have extrema. With φ_n equal to 0° and 30°, there are no longer any extrema on the line-ups, and when the correlation cycle is completed there is a phase shift of 360°, that is, of one period.

Families of line-ups in the case of circular polarization (ω = 1) for different angles of inclination of the normal with the polarization plane (φ_n varies from 0° to 90°), but for an identical azimuth ω_n = 90°, are depicted in Figure 28c for ψ = 45°. There are two families of line-ups in Figure 28c. The first includes line-ups 1, 2, and 3 corresponding to the angles $\varphi_n \geqslant 90° - \psi$. They are characterized by extrema. When the angle φ_n is increased (φ_n enters the cone of the set), the completion of a full correlation cycle results in a shift by a full period. For the sake of clarity, Figure 28c depicts two sections of the sinusoid. Hence, the condition $\varphi_n \geqslant 90° - \psi$ does not hold for the curves of the second family 4–7, and there are no extrema on the line-ups. With φ_n = 0, the polarization plane is horizontal, and the line-up is an inclined straight line.

The curvature of the line-ups increases as the angle φ_n increases, and the line-ups deviate increasingly from a straight line. For the first family, the line in question is the vertical axis, and for the second, it is an inclined straight axis (line 7). Thus, the presence or absence of an extremum on the line-up is determined solely by the position in space of the polarization plane, the shape of the line-up depending on other polarization parameters, as well, such as the ratio of the semi-axes of the ellipse and its orientation in the polarization plane. A polar seismogram of a wave polarized in the vertical plane is depicted in Figure 25bc.

Let us consider the most common cases of orientation of the polarization plane: vertical and horizontal.

VERTICAL POLARIZATION PLANE

Oscillations polarized in the vertical plane are quite common, because the interference of various types of longitudinal waves being propagated in the vertical plane results in an interferential oscillation polarized in this plane. Such polarization is encountered in the study of converted waves. Rayleigh surface waves are also known to be polarized in the vertical plane.

Equation of Line-Ups. Since a normal to the vertical polarization plane makes an angle $\varphi_n = 90^\circ$ with the vertical, the condition for the existence of an extremum (34) is satisfied for all conical sets irrespective of the value of the angle ψ. Hence, the line-ups of waves polarized in the vertical plane are characterized by an extremum.

In the case of a vertical polarization plane, ω_1 and ω_2 are either equal of differ by 180°. In that case it follows from the general equation of the line-up of an elliptically-polarized oscillation (27) that

$$\tan \beta = \frac{m(\cot \psi \tan \varphi_2 \cos \omega \pm 1)}{\tan \varphi_2 \pm \cot \psi \cos \omega}, \tag{35}$$

(where the *plus* sign refers to $\omega_2 = 0$ and the *minus* sign to $\omega_2 = 180^\circ$).

As will be seen from the equation, the shape of the line-up depends substantially on many parameters, which characterize both the oscillation (ω, φ_2) and the polar set (ψ, ω).

Of special interest may be the case of polar correlation of a wave polarized in the vertical plane and recorded by a horizontal set $\psi = 0$. This is because up to now such waves have been mainly recorded by *XYZ* sets.

In this case the equation of the line-up may be obtained from Equation (35) for $\psi = 0$:

$$\tan \beta = m \tan \varphi_2 .$$

Hence, the phase shift on a polar seismogram depends only on the oscillation parameters and is constant for all the components of a set with $\psi = 0$, provided m = const. and φ_2 = const. It follows from the considerations presented above that for $\psi = 0$ the line-ups never have an extremum, since the normals are always inside the set, no matter how great the angles φ_n are. However, in the case of a vertical polarization plane $\varphi_n = 90^\circ$, the records on all seismograms for $\psi \neq 0$, including the case of very small ψ, will display extrema. Thus, in the case of $\psi = 0$ and $\varphi_n = 90^\circ$, there are two contradictory conditions valid at a time, one favouring an extremum, the other acting against it. This is the clue to the ambiguity of correlation, which may formally be executed with a double phase inversion about the vertical line-up or with two breaks and with a full period phase shift (Figure 28c, line 8). Hence, in the case of elliptically-polarized oscillations in the vertical plane, a polar seismogram of a horizontal set is similar to that of a linearly-polarized oscillation, and a vertical line-up with a 180° phase inversion is peculiar to it. This means that formally polar correlation in this case can be executed in the same way as in the case of a linearly-polarized oscillation. Such a situation is realized when the seismometers are arranged in a plane perpendicular to the plane of polarization. However, in contrast to a linearly-polarized oscillation, the line-ups of an oscillation polarized elliptically in the vertical plane have an extremum when $\psi \neq 0$. The above mentioned peculiarities of polar seismograms of waves polarized in the vertical plane are clearly evident from the example of a theoretical seismogram for the interference of P and SV waves, which is frequently observed in the initial part of the seismograms. The interference results in an elliptically-polarized oscillation in the vertical plane with an inclined polarization ellipse (see Figure 25b).

As an example of an observed polar seismogram of a wave polarized in the vertical plane, let us consider a polar seismogram of a transverse S wave (Figure 29). In this case with $\psi = 0$, there is a line-up with a 180° phase inversion. With $\psi = 30°$ and $\psi = 60°$, the line-ups have extrema.

In the case when the axis of the polarization ellipse is vertical (vertical ellipse), a case frequently encountered on the seismograms, the line-up equation takes the form

$$\tan \beta = m \cot \psi \cos \omega, \qquad \varphi_1 = 0, \quad \varphi_2 = 90°;$$

$$\tan \beta = -\frac{m}{\cot \psi \cos \omega}, \qquad \varphi_1 = 90°, \quad \varphi_2 = 0.$$

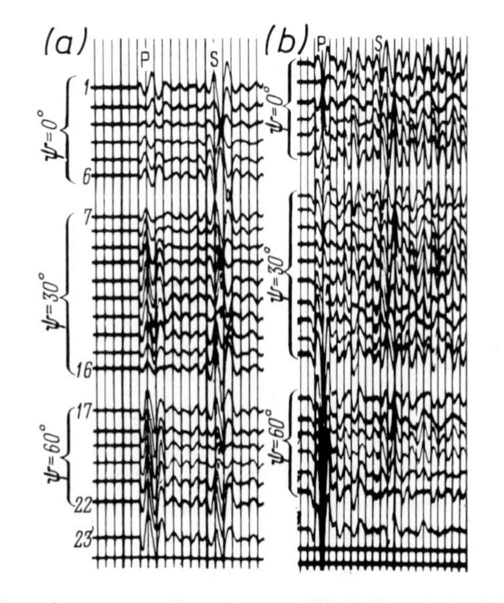

Fig. 29. Observed polar seismograms of an S wave elliptically-polarized in the vertical (a) and horizontal (b) planes.

For $\psi = 0$

$$\tan \beta = 0, \qquad \beta = 0, \qquad \varphi_1 = 0;$$
$$\tan \beta = \infty, \qquad \beta = 90^\circ, \qquad \varphi_2 = 0.$$

that is, only one component of the vertical ellipse is recorded by a horizontal set.

For $\psi = 90^\circ$ (the vertical Z component)

$$\tan \beta = \infty, \qquad \beta = 90^\circ, \qquad \varphi_1 = 0,$$
$$\tan \beta = 0, \qquad \beta = 0, \qquad \varphi_2 = 0.$$

Hence, the phase shift between the vertical and horizontal components is 90°. This serves as an additional criterion for identifying the vertical polarization ellipse on polar seismograms, alongside such properties of records of elliptically-polarized oscillations as the conservation of shape of the record on all components, and the cosine law of amplitude variation. A polar seismogram of a Rayleigh wave may also serve as an example of a polar seismogram of an oscillation polarized in a vertical ellipse (Figure 25g).

Let us consider the variation in shape of a line-up with variation in the ratio of the semi-axes of the polarization ellipse. Theoretical polar seismograms (Figure 30) have been calculated for a conical set $\psi = 45^\circ$ and for several fixed values of m in the range from zero to infinity. The shape of the polarization ellipses considered above is depicted in Figure 30b. The last trace on each theoretical seismogram corresponds to the recording of the vertical seismometer and fixes the initial phase of the oscillation. In the case of a linearly-polarized oscillation with a vertical displacement $m = 0$ (or $B_0 = 0$), a vertical line-up may be traced on the seismograms, and all the amplitudes on it are equal. As m is increased, the line-ups become increasingly curved. In the limiting case, when $m = \infty$, that is, $A_0 = 0$, the oscillation again becomes linearly-polarized, but this time in the horizontal plane. The phase and the direction of this oscillation differ from those of the linearly-polarized oscillation for $m = 0$ by 90°. In this case as we trace the same extremum on the seismogram, we change the phase by 180°, as indicated by the dashed line on the last seismogram. The line-up has discontinuities and consists of separate vertical segments with a phase shift of 180° between them (part of the line-up with an inverted phase is shown by the dashed line). The recorded amplitude of the vertical seismometer decreases with an increase in m, vanishing for $m = \infty$, since in this case the displacement is horizontal. The family of line-ups corresponding to the above-mentioned ratios of the semi-axes of the polarization ellipse is depicted in Figure 30c. All the line-ups have extrema and intersect at two points corresponding to the components whose azimuths coincide with that of the normal to the polarization plane ω_n. The recorded amplitudes on the components are zero. Dashed lines indicate the parts of the line-ups corresponding to oscillations with phases inverted by 180°.

It will be seen from the analysis of theoretical seismograms and line-ups that it is important for polar correlation to study phase shifts on the records of low-amplitude channels, where the curvature of the line-up is at its greatest. Phase shifts on these channels enable the type of polarization and the form-factor of the ellipse to be determined. For this purpose multi-component polar seismograms must be available on which the difference between the directions of adjacent components is relatively small.

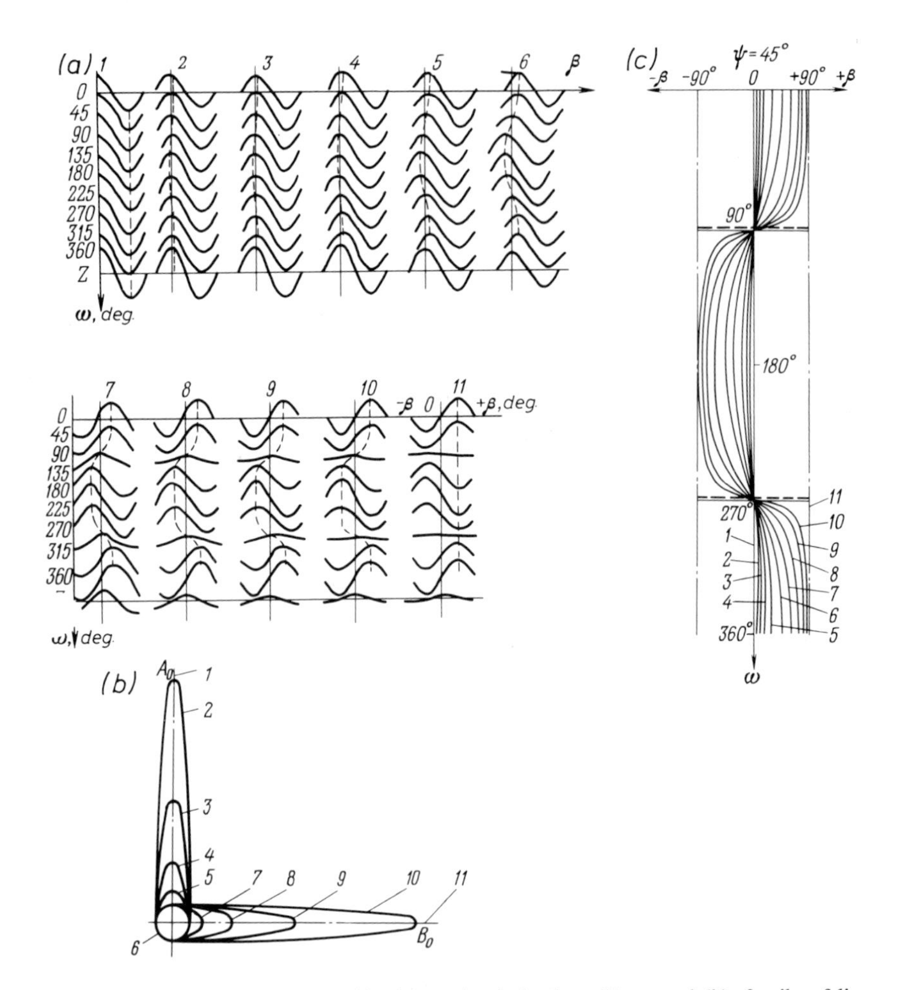

Fig. 30. Theoretical polar seismograms (a), shape of polarization ellipses and (b), family of line-ups (c). 1 – $m = 0$; 2 – $m = 0.6$; 3 – $m = 0.12$; 4 – $m = 0.25$; 5 – $m = 0.5$; 6 – $m = 1$; 7 – $m = 2$; 8 – $m = 4$; 9 – $m = 8$; 10 – $m = 16$; 11 – $m = \infty$.

Equation of Amplitudes. For a wave polarized in the vertical plane, the equation of amplitudes may be obtained from Equation (30) with $\omega_1 = 0$, $\omega_2 = 0$ or $\omega_1 - \omega_2 = 180°$:

$$C^2_{\omega\psi} = A_0^2 \cos \Delta\varphi_1 + B_0^2 \cos \Delta\varphi_2 ,$$

$$C^2_{\omega\psi} = A_0^2(\sin \psi \cos \varphi_1 + \cos \psi \sin \varphi_1 \cos \omega)^2 + B_0^2(\sin \psi \cos \varphi_2 + \cos \psi \sin \varphi_2 \cos \omega)^2$$

where φ_1 is at right angles to φ_2 ; φ_1 and $\varphi_2 \neq 0$.

For a polarization ellipse with a vertical axis, and for $\varphi_1 = 90°$, $\varphi_2 = 0$

$$C^2_{\omega\psi} = A_0^2 \cos^2 \psi \cos^2 \omega + B_0^2 \sin^2 \psi ,$$

and for $\varphi_1 = 0, \varphi_2 = 90°$

$$C^2_{\omega\psi} = A_0^2 \sin^2 \psi + B_0^2 \cos^2 \psi \cos^2 \omega.$$

In the simplest case of circular polarization ($m = 1$)

$$C_{\omega\psi} = A_0(1 - \cos^2 \psi \sin^2 \omega)^{1/2}$$

HORIZONTAL POLARIZATION PLANE

Oscillations, polarized in the horizontal plane, present great interest in the study of transverse waves.

Equation of Line-Ups. In the case of a horizontal polarization plane ($\varphi_1 = \varphi_2 = 90°$ and $\omega_2 = 90°$), the line-up equation assumes the form:

$$\tan \beta = m \tan \omega.$$

Hence, if the polarization plane is horizontal, the phase angle does not depend on the inclination angle of the polar set's components ψ, but on the shape of the polarization ellipse (on the ratio of the semi-axes).

In contrast to the vertical polarization plane, in the case of the horizontal plane, the form of the line-up on polar seismograms of different sets is conserved. Since the normal to the horizontal polarization plane is always vertical ($\varphi_n = 0$), that is, it is always located inside the cone irrespective of the value of ψ, the line-ups of waves polarized in the horizontal plane on the polar seismograms display no extrema. An example of a polar seismogram of an oscillation polarized in the horizontal plane is depicted in Figure 25d. The polar seismogram has been computed for the case of interference of *SV* and *SH* waves, which usually takes place in anisotropic media. The oscillation parameters are $\varphi_{SV} = 80°$, $\omega_{SV} = 225°$, $\varphi_{SH} = 90°$, $\omega_{SH} = 135°$.

Figure 29b depicts a seismogram for an *S* wave the first phases of which are polarized in the horizontal plane (inclined line-ups on all sets, and minimum intensity of the *Z*-component recording), with subsequent phases polarized in the plane, which makes an angle of about 30° with the horizon (minimum amplitude on the 22nd channel).

Thus, irrespective of the shape of the polarization ellipse of an oscillation polarized in the horizontal plane, the line-ups on polarization seismograms of different sets are characterized by the absence of extrema and by a phase shift of a period accompanying the completion of a correlation cycle.

Equation of Amplitudes. For an elliptically-polarized oscillation, the equation of amplitudes may be obtained from Equation (31) with $\varphi_1 = \varphi_2 = 90°$ and $\omega_2 = 90°$:

$$\begin{aligned} C_{\omega\psi} &= \sqrt{A_0^2 \cos^2 \omega \cos^2 \psi + B_0^2 \sin^2 \omega \cos^2 \psi} \\ &= \cos \psi \sqrt{A_0^2 \cos^2 \omega + B_0^2 \sin^2 \omega}\ , \end{aligned}$$

that is, the amplitudes on the polar seismogram vary with ψ in accordance with the cosine law. The limits of amplitude variation are from $A_0 \cos \psi$ to $B_0 \cos \psi$. To separate oscillations polarized elliptically in the horizontal plane, one should, apart from applying phase criteria (shape of the line-up), also pay attention to the conservation of the form of the recording on all the components of the polar set.

PLANES OF EQUAL PHASE SHIFTS (EQUIPHASE PLANES)

Let us consider the patterns of variation of phase shifts as a function of the orientation of oscillation components of waves polarized elliptically in a plane arbitrarily oriented in space. In particular, we shall be interested in the case when a vertical line-up is observed on the polar seismogram of an elliptically-polarized wave. We may demonstrate that in this case the components of the polar set will lie in one plane.

The shape of the line-up of an elliptically-polarized wave is determined by equation (25). It follows from this equation that

$$\begin{aligned} 0 = {} & (m \tan \beta \sin \varphi_1 \sin \omega_1 - \sin \varphi_2 \sin \omega_2) \sin \omega + \\ & + (m \tan \beta \sin \varphi_1 \cos \omega_1 - \sin \varphi_2 \cos \omega_2) \cos \omega + \\ & + (\cos \varphi_2 - m \tan \beta \cos \varphi_1) \tan \psi. \end{aligned} \tag{36}$$

Equation (36) is an equation of a plane and determines an equiphase plane (a plane in which the relative phase shift between all recorded components is constant, that is, $\Delta\beta = 0$). There is a definite plane of equal phase shifts corresponding to each value of β. The position of this plane in space is determined by the direction of the normal to it (by the azimuth ω_{ep} and the angle with the vertical φ_{ep}). The values of ω_{ep} and φ_{ep} may be determined from the formulae

$$\tan \omega_{ep} = \frac{m \tan \beta \sin \varphi_1 \sin \omega_1 - \sin \varphi_2 \sin \omega_2}{m \tan \beta \sin \varphi_1 \cos \omega_1 - \sin \varphi_2 \cos \omega_2}, \tag{37}$$

$$\begin{aligned} \tan \varphi_{ep} = {} & [(m \tan \beta \sin \varphi_1 \cos \omega_1 - \sin \varphi_2 \cos \omega_2)^2 + \\ & + (m \tan \beta \sin \varphi_1 \sin \omega_1 - \sin \varphi_2 \sin \omega_2)^2]^{1/2} \times \\ & \times (m \tan \beta \cos \varphi_1 - \cos \varphi_2)^{-1} \end{aligned} \tag{38}$$

Let us consider the relative position of the polarization and the equal phase-shift planes. To this end, we shall first write the equation for the polarization plane. Let the polarization plane be defined by two components of a polarized oscillation, each of which is determined by the angles φ_1, ω_1 and φ_2, ω_2, where φ_1 and φ_2 are the angles that the first and the second components make with the vertical, and ω_1 and ω_2 are the azimuths of these components. We shall use a rectangular-coordinate frame, placing its origin at the point of intersection of the two directions. We shall obtain the equation of the polarization plane in the form

$$\begin{aligned} 0 = {} & (\sin \varphi_1 \cos \varphi_2 \cos \omega_1 - \cos \varphi_1 \sin \varphi_2 \cos \omega_2) X + \\ & + (\cos \varphi_1 \sin \varphi_2 \sin \omega_2 - \cos \varphi_2 \sin \varphi_1 \sin \omega_1) Y + \\ & + [\sin \varphi_1 \sin \varphi_2 \sin (\omega_1 - \omega_2)] Z \end{aligned} \tag{39}$$

Let us find the relative position of the polarization plane and the equiphase plane. The angle γ between two planes is determined by the formula

$$\cos\gamma = \frac{A_1A_2 + B_1B_2 + C_1C_2}{(A_1^2 + B_1^2 + C_1^2)^{1/2}\,(A_2^2 + B_2^2 + C_2^2)^{1/2}}, \tag{40}$$

where A, B, and C are coefficients determined by the following expressions:

for the polarization plane from Equation (39),

$$\begin{aligned} A_1 &= \sin\varphi_1\,\cos\varphi_2\,\cos\omega_1 - \cos\varphi_1\,\sin\varphi_2\,\cos\omega_2, \\ B_1 &= \cos\varphi_1\,\sin\varphi_2\,\sin\omega_2 - \cos\varphi_2\,\sin\varphi_1\,\sin\omega_1, \\ C_1 &= \sin\varphi_1\,\sin\varphi_2\,\sin(\omega_1 - \omega_2); \end{aligned}$$

for the equiphase plane from Equation (36),

$$\begin{aligned} A_2 &= m\tan\beta\sin\varphi_1\,\sin\omega_1 - \sin\varphi_2\,\sin\omega_2, \\ B_2 &= m\tan\beta\sin\varphi_2\,\cos\omega_1 - \sin\varphi_2\,\cos\omega_2, \\ C_2 &= m\tan\beta\cos\varphi_1 - \cos\varphi_2. \end{aligned}$$

Substituting those coefficients into Expression (40) and carrying out appropriate simplifications, we obtain $\gamma = 90^\circ$. Consequently, all equiphase planes are orthogonal to the polarization plane. Hence, a vertical line-up on a polar seismogram may correspond not only to a linearly-polarized wave, but also to an elliptically-polarized wave. In the latter case the components of the polar set must be located in a plane orthogonal to the polarization plane, that is, in an equiphase plane. The normals to all equiphase planes lie in the wave's polarization plane, and all equiphase planes intersect along a normal to the polarization plane.

The following conclusions may be drawn from Equations (37) and (38): (1) for an equiphase plane with zero phase shift ($\beta = 0$) the direction of the normal is determined by the coordinates $\omega_{ep} = \omega_2$ and $\varphi_{ep} = \varphi_2$; (2) for an equiphase plane with $\beta = 90^\circ$, the direction of the normal is determined by the coordinates $\omega_{ep} = \omega_1$ and $\varphi_{ep} = \varphi_1$. Hence, the normals to the equiphase planes with $\beta = 0^\circ$ and $\beta = 90^\circ$ coincide in direction with those of the axes of the polarization ellipse. In both cases $\beta = 0^\circ$ and $\beta = 90^\circ$, the orientation of the equiphase planes is independent of the shape of the polarization ellipse and is determined exclusively by the ellipse's position in this plane. In all intermediate cases $0 < \beta < 90^\circ$, the position of the polarization plane depends also on the ellipse's form m.

Let us consider the position of the equiphase planes for some particular cases of polarization. Figure 31a depicts a family of equiphase planes for an elliptically-polarized wave with $m = 1$ (circular polarization). The equiphase planes with values $\beta = 0^\circ$ and $\beta = 90^\circ$ pass through the directions of the polarization ellipse's axes (A_0, B_0).

A change in the polarization parameters (the shape of the ellipse or the orientation in space of the polarization plane) changes the position of the equiphase planes in space. Figure 31b depicts the position in space of the equiphase plane ($\beta = 30^\circ$) for an ellipse with $m = 0.5$, when the inclination angle of the polarization plane is changed. Figure 31c depicts the dependence of the position of the equiphase planes ($\beta = 30^\circ$) on the shape of the polarization ellipse for a fixed orientation in space of the polarization plane.

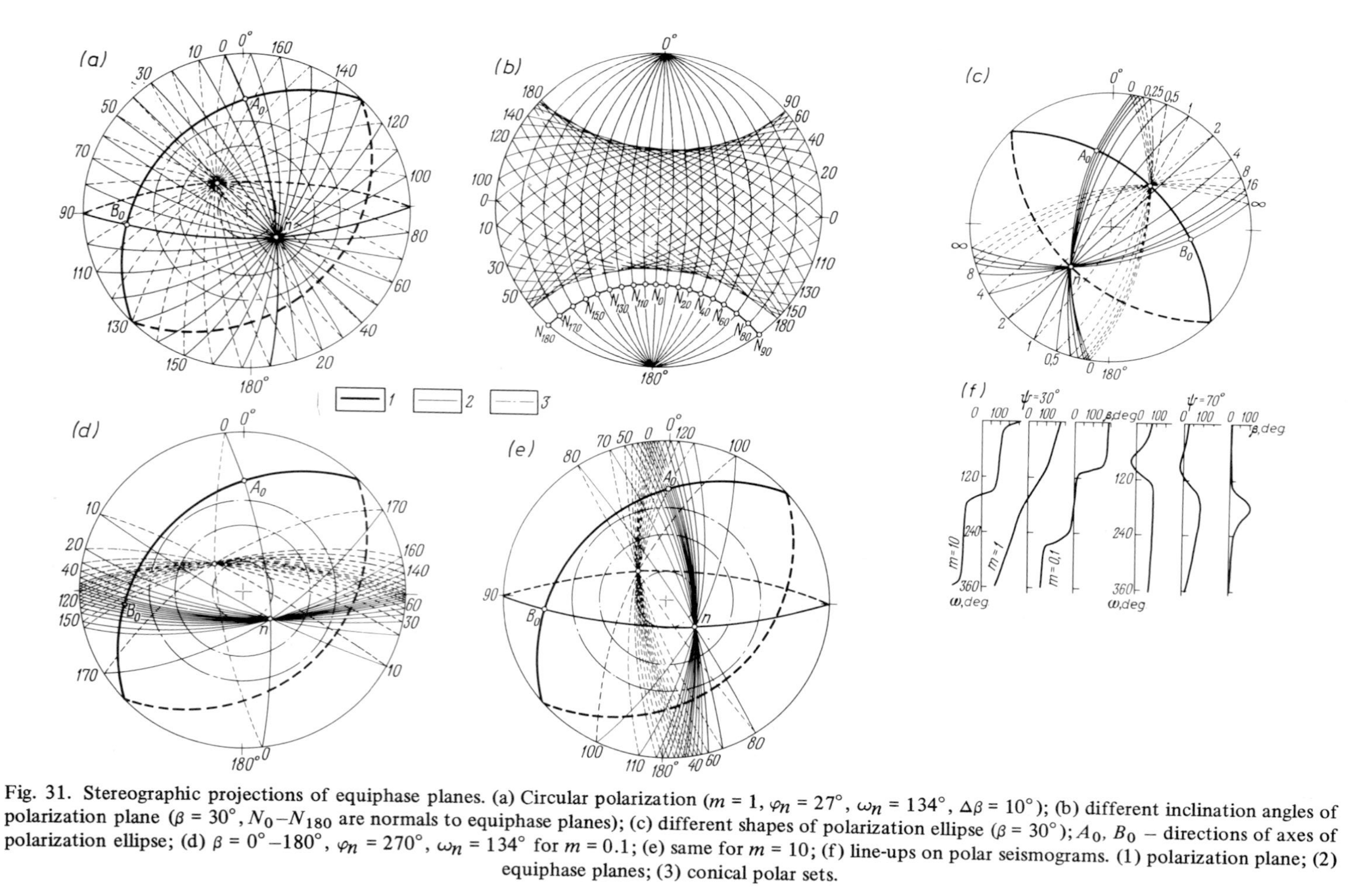

Fig. 31. Stereographic projections of equiphase planes. (a) Circular polarization ($m = 1$, $\varphi_n = 27°$, $\omega_n = 134°$, $\Delta\beta = 10°$); (b) different inclination angles of polarization plane ($\beta = 30°$, $N_0 - N_{180}$ are normals to equiphase planes); (c) different shapes of polarization ellipse ($\beta = 30°$); A_0, B_0 – directions of axes of polarization ellipse; (d) $\beta = 0°-180°$, $\varphi_n = 270°$, $\omega_n = 134°$ for $m = 0.1$; (e) same for $m = 10$; (f) line-ups on polar seismograms. (1) polarization plane; (2) equiphase planes; (3) conical polar sets.

It will be seen from Figure 31c that with $m = 0$ or $m = \infty$, linearly-polarized oscillations may be observed in the direction A_0 and B_0. In those cases, all the components are recorded in an equal phase. It would be interesting to compare the families of equiphase planes for waves with the same polarization plane, but with polarization ellipses of different shapes. Figures 31d and e depicts the families of equiphase planes corresponding to the same directions of the polarization ellipse's axes as in Figure 31a. However, in contrast to Figure 31a, which dealt with circular polarization ($m = 1$), Figures 31d and e deal with elliptical polarization ($m = 0.1$ and $m = 10$). A comparison of the plots demonstrates that the orientation of equiphase planes with β equal to 0° or 90° is the same in all cases, because the directions of the polarization ellipse's a axes have not been changed. However, in contrast to Figure 31a, where the equiphase planes are uniformly distributed in space, in the other two cases there is no such uniformity.

We may compare the equiphase axes of polar seismograms for the cases of polarization discussed above. Making use of stereograms for equiphase planes (Figure 31), we may easily plot the line-ups of polar seismograms with an arbitrary arrangement of components. To this end it suffices to plot a line on the stereogram corresponding to the set, and then to read off the values of the phase shift as functions of the azimuth of the inclination angle of an oscillation component. Let us consider for the above cases the line-ups of polar seismograms for sets with inclination angles of components equal to 30° and 70° (Figure 31f). There are no extrema on the line-ups for the set with $\psi = 30°$, because the normal to the polarization plane n makes an angle of 27° with the vertical and therefore is located inside the cone of the polarization set. Regions of sharp phase variations are clearly visible on the line-ups for $\psi = 30°$: with $m = 10$, these regions coincide with the azimuths of the 0°–20° and 145°–180° components; with $m = 0.1$ they coincide with the azimuths of the 100°–120° and 240°–260° components. With $m = 1$, the phase-shift variation is more uniform. A characteristic feature of the line-ups corresponding to a polar set with $\psi = 70°$ is the presence of extrema, because the normal to the polarization plane is located outside the cone of the polar set.

Plotting Polar Seismograms

The equations presented above enable the line-ups and amplitudes of polar seismograms for the principal types of seismic waves to be calculated. However, in many cases such calculations are lengthy and laborious. This applies in the first instance to multiply-polarized oscillations, which are the result of interference of oscillations with different periods. It frequently turns out much easier to use graphoanalytical methods employing stereographic projections [26, 86, 109] for plotting seismograms. Such methods of plotting polar seismograms place no limitation on the number of interfering oscillations with different amplitude and phase parameters and periods. Let us consider the method of plotting seismograms for harmonic oscillations. It should, however, be pointed out that such graphs may be plotted for oscillations of any form, including pulsed types as well. This will bring about no changes in the plotting methods, and we shall assume the oscillation form to be sinusoidal only for the sake of simplicity.

In contrast to the analytical methods of plotting seismograms in which the angles between the set's components with the coordinates ω, ψ and the directions of particle motion in the waves ω_i, ψ_i are calculated analytically, the specified directions in the graphoanalytical method are represented on the stereographic projection, and the operations of calculating the angle between them (which are rather laborious) are replaced by the measurement of the angle on the net. Let us illustrate the use of graphoanalytical methods by plotting polar seismograms for several cases of wave polarization.

LINEARLY-POLARIZED OSCILLATION

Let the linearly-polarized oscillation be specified in the form $A = A_0 \sin [(2\pi/T)t]$, where $A_0 = 2$. The directions of particle motion (specified by the coordinates φ_A, ω_A) and the components of the polar set 1–23, are shown on the stereogram (Figure 32a).

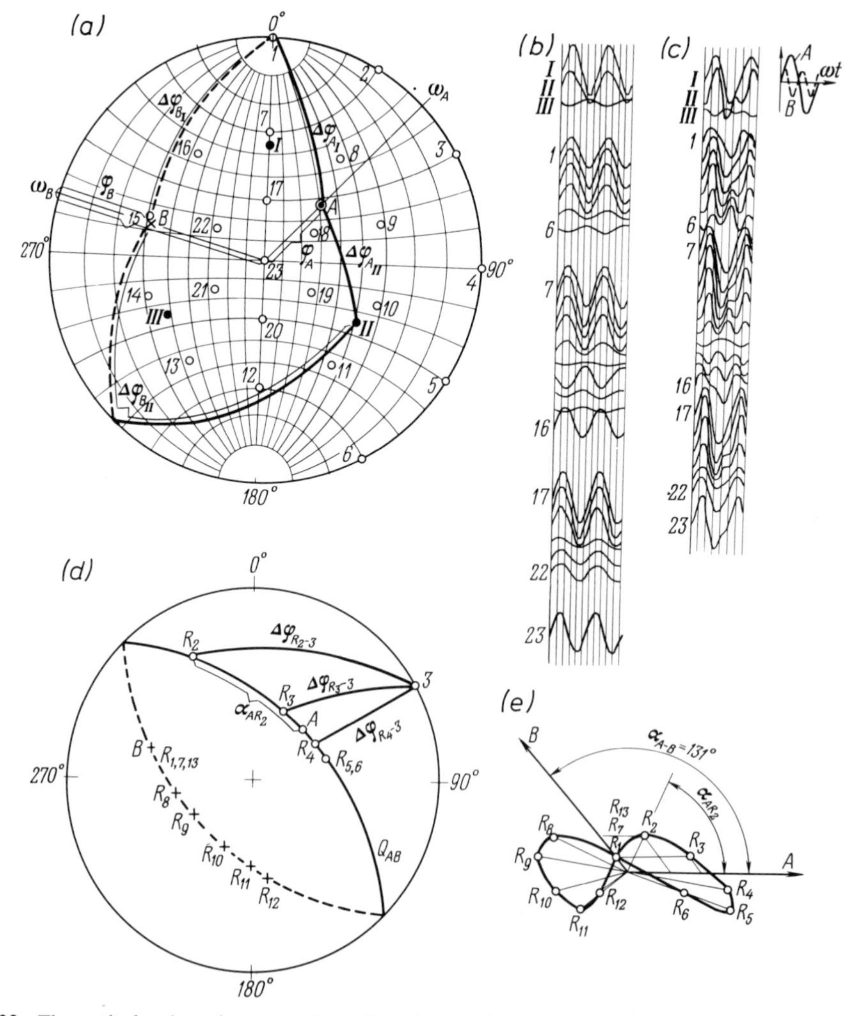

Fig. 32. Theoretical polar seismogram for a linearly-polarized oscillation and for interference of two linearly-polarized oscillations with different periods. (a) Stereogram of directions of oscillatory motions A and B, polar seismogram components (1–23), and a symmetrical set (I, II, III); (b) polar seismogram of a linearly-polarized oscillation; (c) polar seismogram of an interferential oscillation; (d) plots in stereographic projection; (e) geometrical summation of vectors.

Oscillations on any of the set's components 1–23 are projections of the vector of motion on the particular component and are determined by the expression $A_i = A_0 \sin[(2\pi/T)t] \times \times \cos\Delta\varphi_i$, where $\Delta\varphi_i$ is the angle between the direction of particle motion A and the appropriate component of the set. To obtain any component, one need only measure the angles $\Delta\varphi_i$ and calculate A in specified steps of $\Delta(2\pi/T)t$. For example, for component 1 of the polar set, $\Delta\varphi_1 = 60°$ (Figure 32a). Varying $\Delta(2\pi/T)t$ in steps of 30° from 0° to 360° we obtain

$$
\begin{aligned}
A_{0°} &= A_0 \sin 0° \cos 60° = 0\\
A_{30°} &= A_0 \sin 30° \cos 60° = 0.5\\
A_{60°} &= A_0 \sin 60° \cos 60° = 0.87\\
A_{90°} &= A_0 \sin 90° \cos 60° = 1\\
A_{120°} &= A_0 \sin 120° \cos 60° = 0.87 \quad \text{etc.}
\end{aligned}
$$

The oscillation components for all other components of the polar set can be calculated in the same way. A polar seismogram of a linearly-polarized oscillation calculated with the aid of the method described above is depicted in Figure 32b.

The principles of plotting the polar seismograms remains the same in the case of interference of an unlimited number of a linearly-polarized oscillations with different directions of arrival, phase shifts, and intensities.

INTERFERENCE OF TWO LINEARLY-POLARIZED OSCILLATIONS WITH DIFFERENT PERIODS

Let the linearly-polarized oscillations be specified in the form $A = A_0 \sin(2\pi/T)t$, $B = = B_0 \sin[(4\pi/T)t + \beta]$, where $A_0 = 2$; $B_0 = 1$; $\beta = 30°$ (Figure 32c, right-hand side). The directions of particle motion in the interfering oscillations are determined by the coordinates (Figure 32a): $\omega_A = 45°$, $\varphi_A = 40°$, $\omega_B = 285°$, $\varphi_B = 120°$. The projection of the interferential oscillation on the direction of any one of the components 1–23 of the polar set will be determined by the arithmetical sum of the projections of oscillations $A + B$, each of which, in its turn, is calculated as the projection of a linearly-polarized oscillation on the directions 1–23:

$$
A = A_0 \sin\frac{2\pi}{T} t \cos\Delta\varphi_A, \qquad B = B_0 \sin\left(\frac{4\pi}{T} t + \beta\right) \cos\Delta\varphi_B,
$$

where $\Delta\varphi_A$ and $\Delta\varphi_B$ are the angles between the direction of a polar-set component 1–23 and the directions of particle motion in the interfering oscillations A and B.

For component 1, the angles $\Delta\varphi_A$ and $\Delta\varphi_B$ are equal to 63° and 77°, respectively. With $(2\pi/T)t = 0$

$$
A = 0, \qquad B = 1 \cdot \sin 30° \cos\Delta\varphi_B = 0.0112.
$$

For $(2\pi/T)t = 30°$

$$
\begin{aligned}
A &= 2 \sin 30° \cos\Delta\varphi_A = 0.454;\\
B &= 1 \sin 90° \cos\Delta\varphi_B = 0.225.
\end{aligned}
$$

For $(2\pi/T)t = 60°$

$$A = 2 \sin 60° \cos \Delta\varphi_A = 0.795,$$
$$B = 1 \sin 150° \cos \Delta\varphi_B = 0.112, \quad \text{etc.}$$

The next step is calculation of the sum $A + B$ and plotting of one polar-seismogram trace. For another component, the corresponding angles $\Delta\varphi_A$, $\Delta\varphi_B$ are determined, and similar calculations are repeated. The angles $\Delta\varphi_A$ and $\Delta\varphi_B$ may be measured once for each component of the polar set. It will be seen from the seismogram calculated for the case of interference of two linearly-polarized waves (Figure 32c) that the predominant oscillation recorded on the components close to the direction of oscillation A (8, 9, 17, 18) is the low-frequency oscillation A. At the same time high-frequency oscillations B predominate on the components close to oscillation B (14, 15). The explanation is that the angle between the directions of motion in the oscillations is quite close to 90°.

With the graphoanalytical method, a somewhat different procedure for plotting polar seismograms is also possible: first, the radius-vector of the resultant oscillation for a specified value of $(2\pi/T)t$ is plotted on the stereo-net, next the projections of the radius-vector on all the components of the polar set are calculated. This variant is interesting insofar as it makes it possible, in the process of solving the direct problem, to obtain information on the position of the polarization plane and the shape of the trajectory of particle motion in a plane, in the case of a plane-polarized wave, or in space, in the case of an oscillation polarized in space. Let us consider an example in which this procedure is used to plot a seismogram in the case of interference of two linearly-polarized oscillations illustrated in Figure 32c. The directions of motion in oscillations A and B are plotted on the stereo-net according to their coordinates, and the angle between them is measured (in our example α_{A-B} is 131°). The oscillation components A_i, B_i are calculated for each value of $\Delta(2\pi/T)t$, and the vectors **A** and **B** are added up geometrically (Figure 32e). This can be conveniently done graphically on millimetre paper. The magnitude of the resultant, that is, the radius-vector modulus $|R_i|$, and the angle between the radius-vector and one of the oscillations, for instance A, (the angle α_{AR_i}) are determined. For the value $(2\pi/T)t = 0$, $A = 0$, and therefore the modulus and the direction of the radius-vector coincide with those of oscillation B (point R_1 in Figure 32e). With $(2\pi/T)t =$ $= 30°$, $A = 2 \sin 30° = 1$, $B = 1 \sin 90° = 1$, we obtain as a result of geometrical summation of the vectors **A** and **B**, a radius-vector with the modulus $|R_2|$ and the angle α_{AR_2} (Figure 32e). To determine the position of the radius-vecto in space at the specified instant of time, we measure the angle α_{AR_2} from the direction of A towards the direction of B in the plane Q_{AB} (Figure 32d), which contains the direction of motion of both oscillations, and obtain point R_2. The moduli $|R_i|$ and the direction of the radius vectors for other instants of time are determined in the same way (points R_1–R_{13} in Figure 32d). The set of directions R_i thus obtained with the coordinates $\omega_{R_i}\varphi_{R_i}$ will determine the polarization plane. Finally all the radius-vectors are projected on the directions of the polar-set components, this requiring the measurement of the angles $\Delta\varphi_{R-(1-23)}$ between the radius-vector and components 1–23. The amplitude of the total oscillation will be equal to $A_\Sigma = \sum_i |R_i| \cos \Delta\varphi_{R_i-(1-23)}$.

INTERFERENCE OF THREE LINEARLY-POLARIZED OSCILLATIONS

Let us consider the general case, when the oscillations have different intensities and frequencies and do not lie in one plane. Suppose we have three oscillations with the shape and direction of motion as shown in Figure 33a. We find the sum of oscillations A and B using the procedure described above. The shape of the trajectory of the resultant oscillation, which lies in the plane Q_{AB}, is depicted in Figure 33b. Each of the points 0–24 in the figure corresponds to a radius-vector, with the intervals between the radius vectors being $\Delta(2\pi/T)t = 30^\circ$. The plotting procedure involves the determination of the moduli $|R_i|$ and the angles α_{AR_i} for every radius-vector. For example, for point 1 we have $|R_1| = 4.2$, $\alpha_{AR_1} = 24^\circ$. We measure the angle α_{AR_1} from point A towards point B (Figure 33c) in the plane Q_{AB}. We obtain point R_1 – the direction in space of the resultant of the two oscillations A and B at instant 1. We then draw the plane Q_{R_1C} through the point R_1 and C and measure the angle between the directions of R_1 and C in this plane ($\alpha_{R_1C} = 63^\circ$). Next, we sum vectors R_1 and C geometrically, just as was done in the case of oscillations A and B. We find the modulus of the total oscillation $|R|$ and the angle α_{RC}. The radius-vector lies in the same plane Q_{R_1C}. To determine its position in space, we measure the angle α_{RC} from point C and obtain point R. The same procedure is used to find the direction and the moduli of all the other radius-vectors, the set of which determines the trajectory of particle motion (points 0–24 in Figure 33d).

Now we can plot the seismogram of any component, projecting all the radius-vectors on its direction. For instance, to plot the seismogram of component 2, we measure on the net all the angles between the radius-vectors (1–24) and the direction of component 2: $\Delta\varphi_{1-2}$, $\Delta\varphi_{2-2}$, $\Delta\varphi_{3-2}$, etc. (Figure 33d). The polar diagram of all 23 components is depicted in Figure 33e. It will be seen from the stereogram and from the polar seismogram that the particle moves in a complex three-dimentional trajectory first in a clockwise direction, then passes an equilibrium position (point 12), and reverses its direction. The reversal of direction of particle motion is described on the seismogram by the change in the inclination angle of the line-up. Similar plots may be drawn up for any number of interfering oscillations of arbitrary frequency, intensity, and initial phase.

Determination of Parameters of Linearly-Polarized Oscillations

Polar seismograms make it possible to determine the direction of the vector of particle motion in space. Of special interest are the longitudinal waves for which the direction of the vector of particle motion in homogereous media coincides with the direction of wave propagation.[1] We shall not dwell here on methods of interpretation based on the direction of particle motion, but shall discuss in detail methods of determining the direction of particle motion in linearly-polarized waves.

In contrast to polar correlation, which employs mainly the phase parameters of waves, the determination of the direction of the vector of particle motion involves only the use of the relative recorded amplitudes of waves on a polar seismogram. We recall that

[1] It should be remembered that on the Earth's surface, just as on any other interface, the direction of the vector of particle motion need not necessarily coincide with that of the incident ray [114].

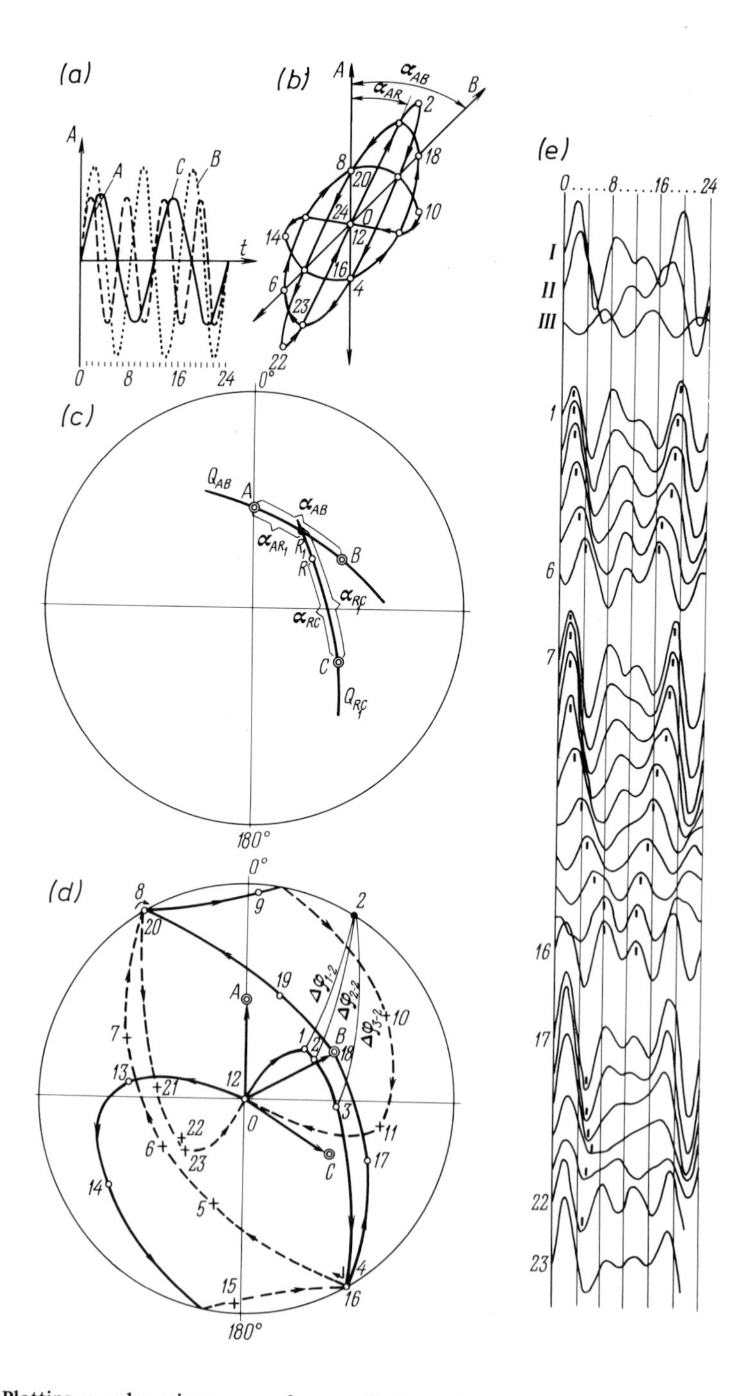

Fig. 33. Plotting a polar seismogram of an oscillation polarized in space. (a) Original oscillations; (b) trajectory of motion of the total oscillation A and B; (c) plotting the trajectory in space; (d) diagram of motion of the total oscillation (stereogram); (e) polar seismogram.

multi-component polar seismograms can be obtained by two methods: from observations using multi-component (8–16) polar sets and by obtaining signal patterns from three-component sets. Whereas polar seismograms obtained with the aid of both methods are equivalent for the purpose of polar correlation, the former method of obtaining polar seismograms permits more accurate evaluation of particle motion. The advantage depends precisely on the number of independent measurements and hence on the number of seismometers in the polar set. For qualitative assessment of particle motion, the seismograms obtained with the aid of both methods are equivalent. Let us consider qualitative and quantitative methods of determining particle-motion trajectories from multi-component and three-component polar seismograms.

QUALITATIVE ASSESSMENT

The directions of particle motion in a linearly-polarized oscillation can be assessed qualitatively visually from a three-component and a polar seismograms.

The sector of displacement direction can be determined from the signs of arrivals on the three-component seismogram. Three mutually-orthogonal planes passing through the seismometer axes divide the entire space into eight sectors. The displacement directions in each sector have different signs on the three-component seismogram. For *XYZ* sets, four sectors occupy the upper half-space, and four sectors occupy the lower half-space. The position of the sectors for a symmetrical set is shown on the graticule (Figure 34a). In contrast to the *XYZ* set, some sectors in this case partially occupy the upper and partially the lower half-spaces. Similar signs on the records of all three of the set's seismometers correspond to the directions enclosed in the pyramid formed by the seismometer axes (shaded areas). For all other directions, the signs of arrivals on the records of one or of two seismometers are different.

The direction of motion is close to that of the polar-seismogram component with maximum amplitude. On the polar seismogram (Figure 34b), the maximum recorded amplitude of the *P* wave is on the 18th channel. The recording on the 23d component is close to it, but somewhat smaller in amplitude. We may assume that qualitatively the direction of particle motion in the *P* wave coincides with that of the 18th component with the coordinates $\omega = 60°$ and $\varphi = 60°$.

The direction of motion may also be qualitatively assessed from the components with zero or minimum amplitudes as the direction normal to the plane of zero displacements. On the seismogram (Figure 34b), the first phase of the *P* wave is practically not recorded (or is recorded with minimum amplitudes) on components 3, 18, 21, and 23. This means that the plane in which these components lie is normal to the direction of particle motion. Indeed, we can easily see on the net with the components marked on it (Figure 34d), that these components lie in one vertical plane Q_0. In this case, the direction of particle motion in the *P* wave will coincide with the direction of the normal to this zero-amplitude plane and will be described by the coordinates $\omega_n = 150°$, $\varphi_n = 90°$. Other qualitative features of the record agree with the directions of motion in the *P* wave thus determined. For example, phase inversion takes place on the records of components 2 and 4 and 13 and 4 arranged on both sides of the zero-displacement plane. It should be pointed out that the nature of motion in the *P* wave shows that it is a complex wave. The next arrival after the first oscillation P_1, with direction of motion close to the horizontal, is

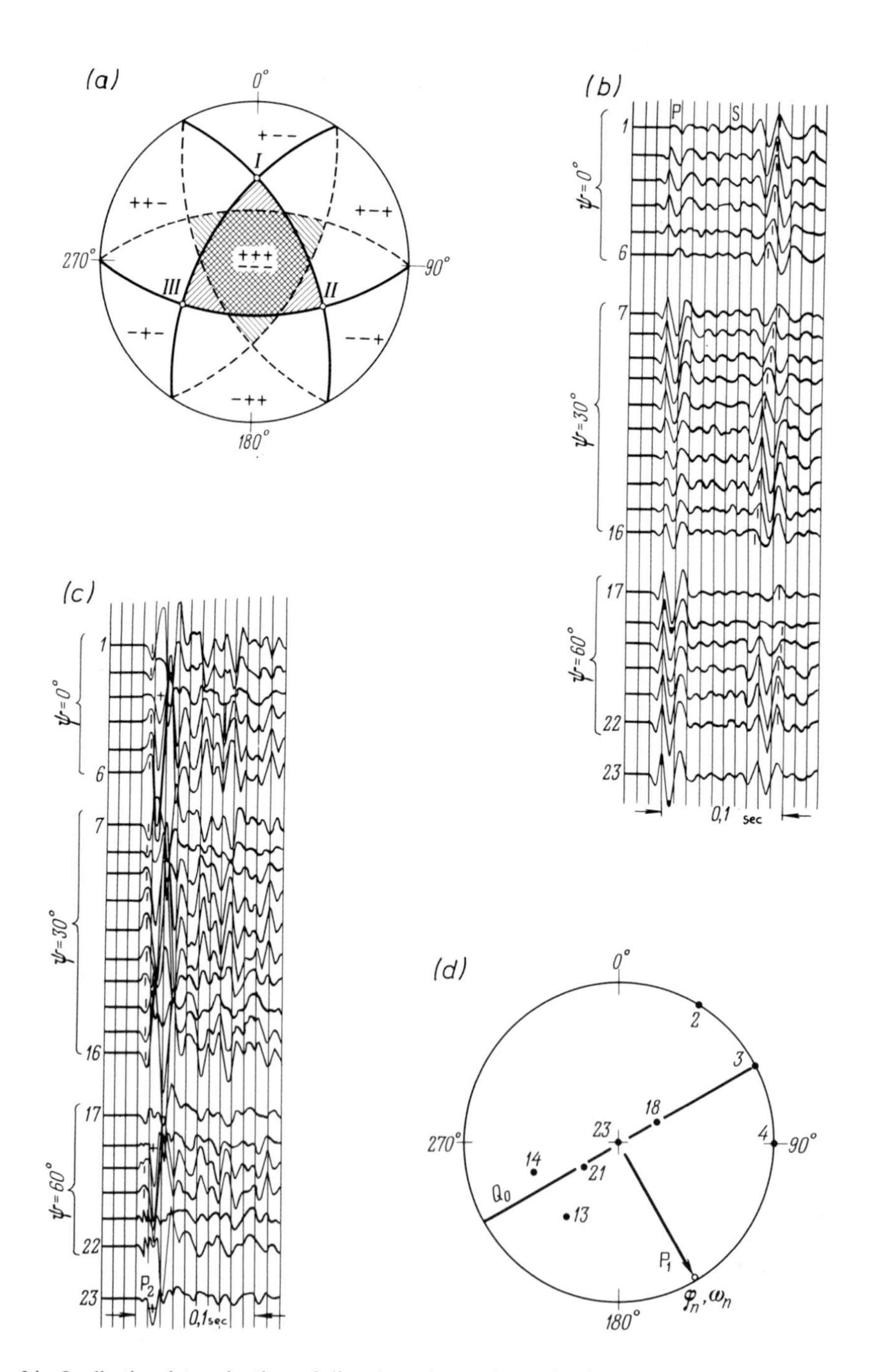

Fig. 34. Qualitative determination of direction of particle motion in a linearly-polarized oscillation. (a) from the signs of arrivals on records of a three-component set (I–III); (b) from the component with the maximum recorded amplitude; (c, d) from the normal to the plane of zero displacements (seismogram and stereogram).

a more intense P_2 wave with a different direction of motion. The arrival of this wave is clearly marked on the components lying in the zero-displacement plane of the record of the P_1 wave, especially on components 18 and 23. In many cases zero amplitudes are not observed on the seismograms, and components displaying phase inversion are used to determine the directions.

QUANTITATIVE DETERMINATION

The directions of motion may also be determined from the values of relative amplitudes read off a polar seismogram. Let the direction of the displacement vector $\mathbf{A}_0$ be specified in space by the angle it makes with the vertical φ and by its azimuth ω. Orient the axis of the rectangular coordinate frame so that its azimuth coincides with that of the displacement vector ($\omega = 0$). The angle in space $\Delta\varphi$ between the direction of the displacement vector and that of the seismometer axis (ψ, ω) will be determined by Expression (21)

$$\cos \Delta\varphi = \cos \omega \cos \psi \sin \varphi + \sin \psi \cos \varphi$$

The maximum amplitude of the record is

$$A_{\omega\psi} = A_0 \cos \Delta\varphi = A_0(\cos \omega \underbrace{\cos \psi \sin \varphi}_{p} + \underbrace{\sin \psi \cos \varphi}_{q}),$$

$$A_{\omega\psi} = A_0(p \cos \omega + q) = A_0 p \cos \omega + A_0 q.$$

Hence, the dependence of the amplitude of the oscillation component $A_{\omega\psi}$ on the azimuth of the set's component ω for a specified φ and for a constant ψ (a conical set) is described by a cosinusoid whose axis is generally displaced with respect to the axis by a distance $A_0 q = A_0 \sin \psi \cos \varphi$ (Figure 35a), and whose extremal values are equal to $A_0(p + q)$ and $A_0(q - p)$, respectively. The relationship obtained may be used to determine the direction of the displacement vector with the aid either of the anlytical or the graphoanalytical method.

Graphoanalytical Method. This method is used to determine the direction of the displacement vector from the recorded amplitudes with the aid of directional diagrams. As demonstrated previously, the shape of the directional diagrams in polar coordinates depends on the inclination angle of the components and the displacement vector. The orientation of the directional diagram, on the other hand, depends on the difference between the azimuths of the displacement vector and the direction assumed as the original.

It is most appropriate to determine the direction of the displacement vector from the directional diagrams in rectangular coordinates by plotting the observed directional diagram and comparing it with the graticule of theoretical directional diagrams. The latter in rectangular coordinates consists of a family of cosinusoids with different amplitudes. To plot an observed directional diagram, one has to mark the amplitude values measured on a polar seismogram on tracing paper using the same scale as that of the graticule with the axis of the observed diagram parallel to that of the cosinusoid. The observed diagram should be laid on top of the theoretical diagram, so that one of the theoretical diagrams on the graticule would best approximate the observed points (see Figure 35a). The theoretical diagram thus selected is drawn on tracing paper.

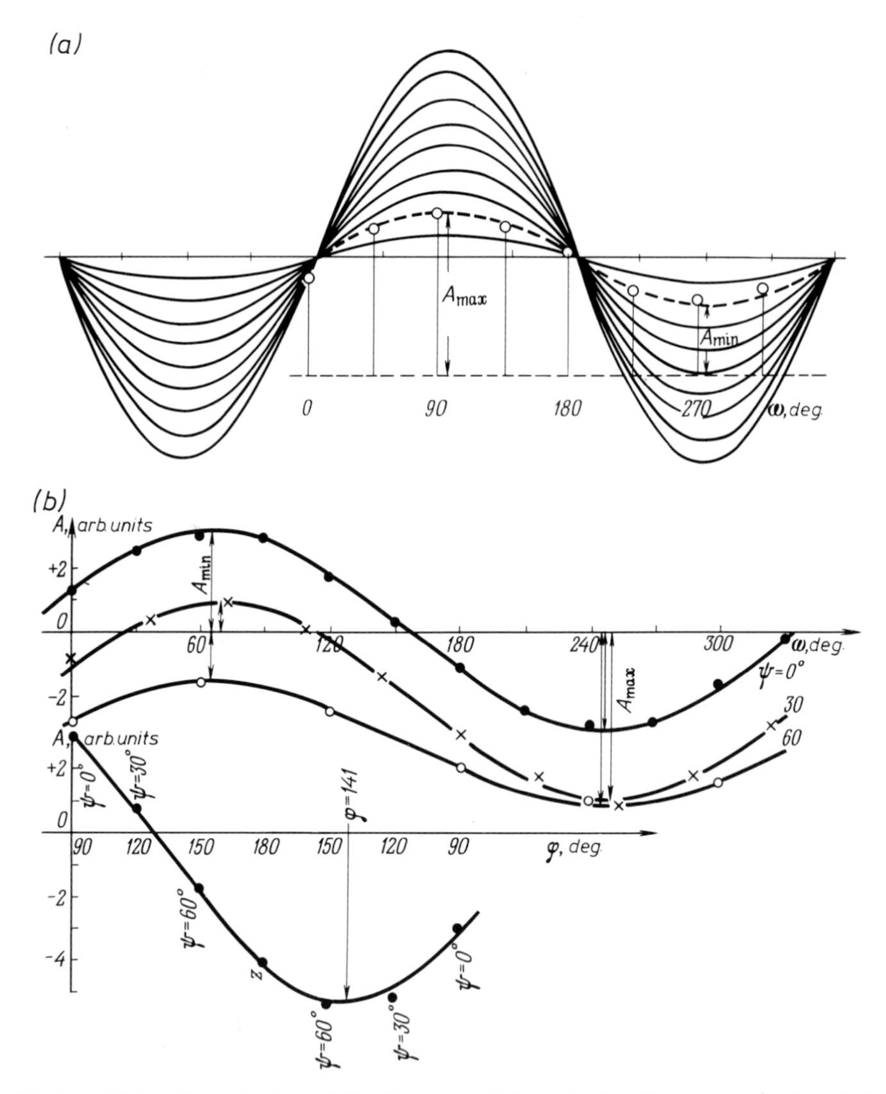

Fig. 35. Quantitative determination of direction of particle motion in a linearly-polarized oscillation. (a) Cluster of theoretical directional diagram; (b) determination of direction of particle motion with the aid of several conical sets.

The angle between the displacement vector and the vertical is determined from the ratio of the minimum and maximum amplitudes on the observed directional diagram. We may write the expression for the ratio of the sum and the difference of extremal values

$$\frac{A_{max} + A_{min}}{A_{max} - A_{min}} = \frac{A_0(p+q) + A_0(q-p)}{A_0(p+q) - A_0(q-p)}$$

$$= \frac{q}{p} = \frac{\sin\psi\cos\varphi}{\cos\psi\sin\varphi} = \tan\psi\cot\varphi,$$

or

$$\tan \varphi = \frac{A_{max} - A_{min}}{A_{max} + A_{min}} \tan \psi.$$

Denoting $A_{max}/A_{min} = C$, we may write this formula in the form $\tan \varphi = (C-1)/(C+1) \tan \psi$ and use this formula to find the angle φ for fixed values of ψ. The angle φ can also be determined graphically from nomograms calculated for fixed values of ψ [24].

In the particular case of $\psi = 45°$, the ordinate of the axis of the cosinusoid is proportional to the vertical oscillation component, and the amplitude of the cosinusoid (the distance from its axis to its peak) is proportional to the horizontal oscillation component, the proportionality factor being the same in both cases. Indeed, denoting the ordinate of the cosinusoid by Z and its amplitude by X, we may write $A_{max} = Z + X$, and $A_{min} = Z - X$. Substituting the values of A_{max} and A_{min} and putting $\psi = 45°$, we obtain $\tan \varphi = X/Z$.

The azimuth of the displacement vector ω may also be found from the extremal amplitudes on the observed directional diagram. Let us assume that the azimuth of the first seismometer of the set is the origin of the azimuth scale and graduate the abcissa of the observed diagram. The azimuth of the amplitude of the cosinusoid chosen to approximate the observed curve as shown in Figure 35a will correspond to the azimuth of displacement. Figure 35a depicts graticule of theoretical directional diagrams and the observed directional diagram of a wave recorded by a set with $\psi = 45°$ (indicated by the dashed line). The azimuth of the displacement vector will be seen from Figure 35a to be equal to 94°.

If the components of the polar seismogram form several conical surfaces, all the components may be used to determine the direction of motion. Figure 35b depicts observed diagrams for three conical surfaces ($\psi = 0°$, 30°, and 60°). All the extreme positive and negative amplitudes obtained lie in the wave's plane of incidence and are proportional to the cosines of the angles between the direction of particle motion and the components of the polar seismogram lying in the same plane. Accordingly the graphs of $\pm A_{max}$, $\pm A_{min}$ should be plotted as functions of φ ($\varphi = 90° - \psi$) and averaged to correspond to the theoretical directional diagram. The value of φ sought corresponds to the position of A_{max} on the φ scale (in Figure 35b, $\varphi = 141°$). The amplitude of the cosinusoid is equal to the modulus of the vector.

It should be kept in mind that the methods described may be used for determining the direction of the displacement vector from the three-component and multi-component observation data only, if such data have been obtained under conditions of identical sensitivity for all channels, or if the amplitudes on the records of all components have been reduced to a common sensitivity.

In the graphoanalytical method it is assumed that the axes of conical surfaces formed by the components of the polar sets are vertical. This condition is always fulfilled in surface observations. However, in drill-hole observations, the axis of the actual set will not necessarily coincide with the vertical. In this case the angles in the vertical plane should be measured not from the vertical, but from the set's axis, and corrections to compensate for the inclination of the set's axis, or what is tantamount to it, the drill-hole axis, should be introduced into the graphs. Suppose that the set's position in the borehole is known (Figure 36, I–III). The angle α_{I-R} between the direction of a seismometer,

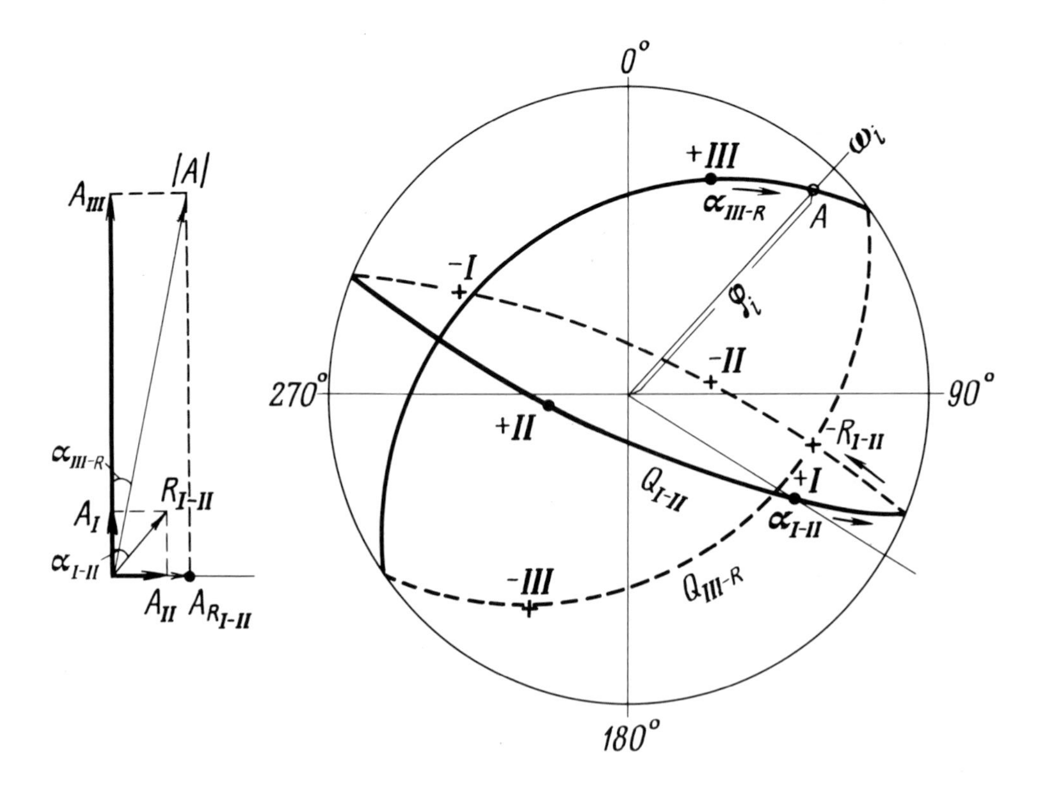

Fig. 36. Determination of direction of particle motion from records of a symmetrical three-component set.

for instance, I[20], and the direction of the projection of the vector of particle motion on the plane I–II (R_{I-II}) and the modulus of the projection $|R_{I-II}|$:

$$\alpha_{I-II} = \arctan \frac{A_{II}}{A_I}, \qquad |R_{I-II}| = (A_I^2 + A_{II}^2)^{1/2},$$

may be determined from the amplitudes recorded by any pair of seismometers.

The next step is the calculation of the direction and modulus of the vector

$$\alpha_{III-R} = \arctan \frac{A_{III}}{R_{I-II}}; \qquad |A| = (A_{III}^2 + R_{I-II}^2)^{1/2}$$

$$= (A_{III}^2 + A_I^2 + A_{II}^2)^{1/2}$$

where A_I, A_{II}, and A_{III} are amplitudes recorded by the three-component set.

These operations can conveniently be carried out on millimetre paper with rectangular coordinates (Figure 36). The signs of arrivals are not taken into account in the process. Subsequent plotting is performed on the stereo-net with account taken of the signs of arrivals (see Figure 36). First, the actual position of the set in the drill-hole, determined from the reading of position sensors, is plotted on the net. Any position is possible, depending on the drill-hole's inclination at the point of the set's location and on the set's orientation. Let the signs of arrivals on the seismometer records be $+A_{\mathrm{I}}$, $-A_{\mathrm{II}}$, $+A_{\mathrm{III}}$. Then the angle $\alpha_{\mathrm{I-II}}$ should be measured from the positive direction of the first seismometer (+I) in the plane $Q_{\mathrm{I-II}}$ towards the negative direction of the second seismometer's axis (–II) (indicated by arrows in Figure 36). Depending on the ratio of the amplitudes A_{I}, A_{II}, the point $R_{\mathrm{I-II}}$ corresponding to the direction of the total vector $R_{\mathrm{I-II}}$ may be located both in the upper ($+R_{\mathrm{I-II}}$) and in the lower parts ($-R_{\mathrm{I-II}}$) of the sphere. In our example the direction lies in the lower hemisphere $-R_{\mathrm{I-II}}$. Next, taking into account the position of point $R_{\mathrm{I-II}}$, we draw the plane $Q_{\mathrm{III}-R}$ in which the angle $\alpha_{\mathrm{III}-R}$ should be measured from the positive direction of the third instrument's axis (+III) in the direction of point $-R_{\mathrm{I-II}}$ (along the arrow in Figure 36). As the result, we obtain the direction of motion on the net (point A). With appropriate skill, such plotting can be done quite rapidly, which is quite important when a large number of seismograms have to be processed.

Analytical Method. The analytical method of determining the direction of particle motion is more unwieldy than the graphoanalytical. The analytical method developed by Bondarev [14, 17] for multi-component polar sets enables the accuracy of the results to be evaluated, the optimum components for determining directions of motion to be selected with the greatest accuracy, and other problems to be solved. As has been previouly demonstrated, the equation of the theoretical directional diagram has the form of a cosinusoid

$$Y_k = a \cos (fk + \beta) + c,$$

where Y_k are the observed amplitudes; a is the amplitude; f is the frequency; and β is the initial phase of the curve; c is a constant component; $k = 1, 2, \ldots, n$.

Because of random measurement errors, the observed diagram does not coincide with the theoretical. Least-square averaging of the points on the observed diagram is performed in accordance with the formula

$$\sum_{k=1}^{n} [Y_k - a \cos (fk + \beta) - c] = \min.$$

The equation of the observed diagram whose axis is displaced with respect to the abscissa is

$$Y_k = A \sin \frac{2\pi}{n} k + \beta \cos \frac{2\pi}{n} k + C, \qquad (k = 1, 2, \ldots, n),$$

where n is the number of the sets' components; $\beta = \arctan (A/B)$ is the initial phase.

Formulae for the determination of direction of motion may be obtained by averaging the observed directional diagram with the aid of the least-square method:

$$\omega = 57.3 \arctan\left(\frac{A}{B}\right) - \frac{360^\circ}{n},$$

$$\tan\varphi = \tan\psi\left(\frac{A^2+B^2}{C}\right)^{1/2}$$

where

$$A = \frac{2}{n}\sum_{k=1}^{n} Y_k \sin\frac{2\pi}{n}k; \qquad B = \frac{2}{n}\sum_{k=1}^{n} Y_k \cos\frac{2\pi}{n}k; \qquad C = \frac{1}{n}\sum_{k=1}^{n} Y_k.$$

The standard deviation of each value of the observed amplitude Y_k is found from the formula

$$m_Y = \pm\left(\frac{\Sigma_{k=1}^{n}\Delta_k^2}{n-3}\right)^{1/2},$$

where

$$\Delta_k^2 = \left(Y_k - A\sin\frac{2\pi}{n}k - B\cos\frac{2\pi}{n}k - C\right)^2.$$

The maximum absolute errors in the determination of coefficients A, B, and C will be equal to

$$m_A = \pm\sqrt{\frac{2}{n}}\,m_Y; \qquad m_B = \pm\sqrt{\frac{2}{n}}\,m_Y; \qquad m_C = \pm\frac{1}{\sqrt{n}}m_Y.$$

Using the maximum errors m_A, m_B, and m_C, we may obtain the maximum absolute errors in the determination of the direction of particle motion (of the azimuth and the angle with the vertical):

$$m_\omega = 57.2\sqrt{\frac{2}{n}}\,\frac{|A|+|B|}{A^2+B^2}\,m_Y,$$

$$m_\varphi = 57.3\sqrt{\frac{2}{n}}\,\frac{\tan\psi\cos^2\varphi}{C^2\sqrt{A^2+B^2}}\left[|CA| + |CB| + \frac{\sqrt{2}}{2}(A^2+B^2)\right]m_Y.$$

It will be seen from the above equations that the greatest accuracy is achieved when records of the horizontal seismometers are used to determine the azimuth.

The error in determining the azimuth increases rapidly as the inclination angle of the polar set's components increases. For each angle φ, an inclination angle ψ of the set's seismometers may be found for which the error in determination of angle φ will be minimum

$$\psi_0 = \arctan\left(\frac{\sqrt{2}}{2}\tan\varphi\right)^{1/2}.$$

Thus generally, the optimum conditions for determining the azimuth ω and the angle φ do not coinicde. If several sets are available, it is possible to select a separate set for each parameter. However, in the case of a single set, we may also choose the angle ψ so that the direction in space of the dispalcement will be determined with the utmost precision [13]. For inclination angles of the seismometers ψ, other than the optimum, the accuracy of simultaneous determination of ω and φ decreases 1.5–2 times as compared with separate determination under conditions of optimum angles ψ_0.

Determination of Displacement Directions in Surface Observations. As already mentioned, when determining directions we must keep in mind that the oscillation recorded on the surface is the result of superposition of the incident wave with waves reflected from a free surface. For this reason the direction of particle motion determined from three-component observations will not coincide with the direction of motion in the incident wave. To find the true angle of exit of seismic radiation with the apparent angle e_a, a correction should be made to the value of the apparent angle determined for an incident longitudinal wave from the formula in [114]

$$\cos e = \frac{v_P}{v_S} \cos \frac{(90 + e_a)}{2}.$$

In seismology, when long-period oscillations from distant earthquakes are being measured, the ratio v_P/v_S may be assumed to be equal to 1.71 with the angle e variable from 65° to 80°. In this case the apparent angle will be 5°–10° less than the true one. However, in the frequency range used in seismic prospecting the ratio v_P/v_S may vary from 1.71 to 5–6 with greater variations possible in individual cases, and the difference between the apparent and true angles will be substantially greater, so that in each case it should be decided separately whether this difference should be taken into account. Similarly, in drill-hole observations, especially in the vicinity of interfaces, the secondary reflected and refracted longitudinal and transverse waves appearing at such interfaces are superimposed on the longitudinal wave and change the direction of particle motion in it. The first extremum of a longitudinal wave is usually the least distorted, therefore it should be used for the qualitative determination of the displacement directions in the first arrivals. It should be noted that another case when the displacement directions do not correspond to the directions of the rays is that of waves being propagated in anisotropic media.

Determination of Parameters of Elliptically- and Multiply-Polarized Waves

The fundamental equations of polar correlation describe the phase and amplitude patterns of polar seismograms for elliptically-polarized waves. With the aid of the above formulae, we may calculate and plot a polar seismogram of an elliptically-polarized wave with arbitrary parameters of the polarization ellipse. This problem may be regarded as the direct one. Of much greater interest is the solution of the converse problem: the determination of the components of a complex oscillation from a polar seismogram (their number, displacement directions, phases, and intensity ratios). However, the converse problem is in essence an ambiguous one. The reason is that an elliptically-polarized oscillation may be the result of interference of any number of linearly-polarized oscillations with the same period and with arbitrary phases, amplitudes, and displacement directions. When studying the polarization of seismic waves and interpreting the trajectory of particle motion, we are able to determine the parameters of an elliptically-polarized oscillation uniquely from a polar seismogram. Let us agree to regard this as the converse problem.

The following parameters should be determined in the course of resolving the converse problem from a polar seismogram of an elliptically-polarized wave: the position in space of the polarization plane expressed in terms of the coordinates of the normal to it φ_n, ω_n (the polar angle and the azimuth); the shape of the ellipse m and its position in the polarization plane (the direction of the axes of the ellipse).

DETERMINATION OF THE POSITION IN SPACE OF THE POLARIZATION PLANE

Let us consider several methods, both analytical and graphical, of determining the position of the polarization plane from the shape of the line-up [15, 24, 41]. In many cases there is no need for a precise determination of the position in space of the polarization plane, a qualitative determination from visual analysis of the records being adequate. In such cases the position of the polarization plane is determined from the following factors.

THE PRESENCE OR THE ABSENCE OF AN EXTREMUM ON THE LINE-UP

As previously demonstrated, the presence of an extremum is determined solely by the mutual disposition of the polarization plane and the polar set's components. Extrema of line-up will be observed on the records, if the normal to the polarization plane lies inside the conical set, and this enables the position of the polarization plane to be qualitatively determined from a polar seismogram. For example, an extremum of the line-up for the S wave on the seismogram (Figure 34b) is observed only on the record of the set with $\psi = 60°$. This means that the normal to the polarization plane lies inside the cone with $\psi = 60°$, that is, the normal to the polarization plane of the S wave is oblique and close to the horizontal.

COMPONENT WITH A MINIMUM OR ZERO RECORDED AMPLITUDE

The direction of the component with minimum amplitude is close to the normal to the polarization plane. In the case of zero recorded amplitude the directions coincide. Thus, the minimum recorded amplitude of the P wave on the seismogram (Figure 37a) is observed on channel III of a three-component symmetrical drill-hole set. The plane perpendicular to the axis of seismometer III passes through the axes of the two remaining seismometers I and II and serves as the polarization plane of the P wave (see the stereogram).

Set of Radius-Vectors Plotted for Different Instants of Time. The plane of zero amplitudes may be drawn on the net through the components with zero recorded amplitudes for any instant of time. The direction of the radius-vector to be determined coincides with

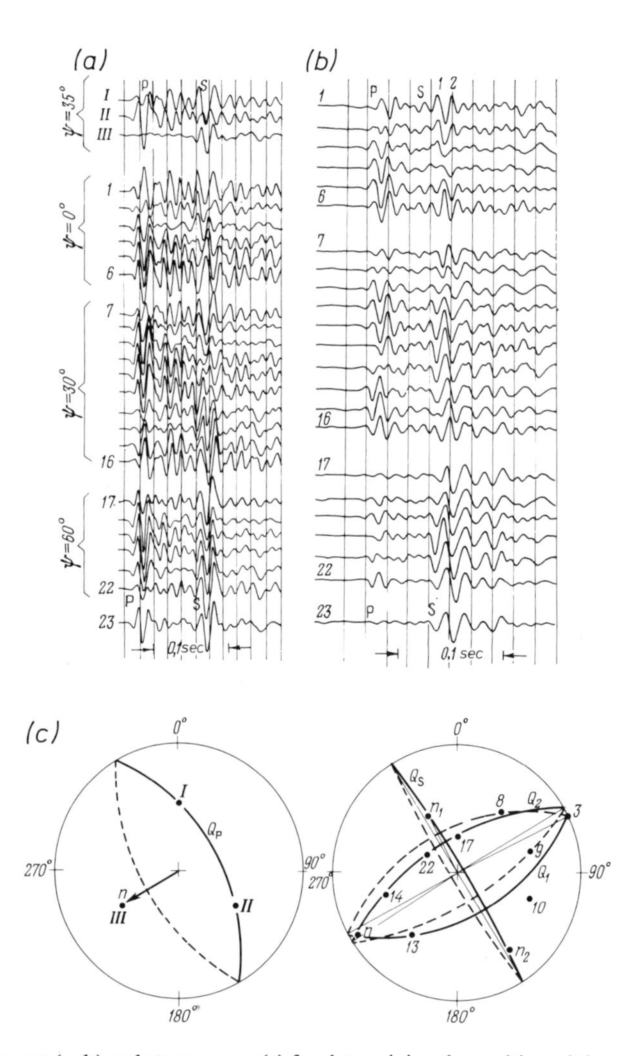

Fig. 37. Seismograms (a, b) and stereograms (c) for determining the position of the polarization plane (a) zero amplitude component; (b) set of radius-vectors

the direction normal to the plane of zero amplitudes. Such planes may be drawn for different instants of time, and the location of the corresponding radius-vectors can be determined. The polarization plane is drawn through the radius-vectors. Let us consider, for example, how to determine the position of the polarization plane for an *S* wave (Figure 37b). For the instant of time 1, zero recorded amplitudes are observed on traces 3 and between 9 and 10, and 13. The plane of zero amplitudes Q_1 (Figure 37b) is drawn through the directions of the axes of seismometers 3 and 13 on the seismogram. The normal n_1 to the plane Q_1 determines the location of the radius-vector at instant 1. In the same way, for instant 2, zero amplitudes are observed on channels 3, 8, 14, 17, and 22. The normal n_2 to plane Q_2 determines the direction of the radius-vector at instant 2. The oblique polarization plane Q_S of the *S* wave will pass through the directions

n_1 and n_2. The direction and modulus of the radius-vector can also be determined at any instant of time from the ratio of amplitudes on the polar seismogram with the aid of the same methods that were used to determine the direction of the displacement vector and the vector of a linearly-polarized oscillation.

Intersection of Equiphase Planes. Let us determine the orientation in space of the normal to the polarization plane as the line of intersection of equiphase planes.

The position in space of an equiphase plane can be defined by the directions in space of two pairs of components with equal phase shifts. Let the directions of each such pair be described by the following coordinates: $\psi_1\omega_1$; $\psi'_1\omega'_1$ and $\psi_2\omega_2$; $\psi'_2\omega'_2$. The phase shift for the first pair is β_1, and for the second β_2. Each pair of directions determines an equiphase plane. The equations of equiphase planes may be written in the form

$$\begin{aligned} 0 &= (\tan\psi'_1 \cos\omega_1 - \tan\psi_1 \cos\omega'_1)\sin\omega + \\ &+ (\tan\omega_1 \sin\omega'_1 - \tan\psi'_1 \sin\omega_1)\cos\omega - \\ &- \sin(\omega_1 - \omega'_1)\tan\psi, \\ 0 &= (\tan\psi'_2 \cos\omega_2 - \tan\psi_2 \cos\omega'_2)\sin\omega + \\ &+ (\tan\psi_2 \sin\omega'_2 - \tan\psi'_2 \sin\omega_2)\cos\omega - \\ &- \sin(\omega_2 - \omega'_2)\tan\psi. \end{aligned} \qquad (41)$$

We find the direction of the normal to the polarization plane (φ_n, ω_n) as the trace of intersection of equiphase planes. The azimuth of the normal is determined by the expression

$$\tan\omega_n = \frac{C_2B_1 - C_1B_2}{A_2C_1 - C_2A_1},$$

and the angle with the vertical by the expression

$$\tan\varphi_n = \frac{[(A_2C_1 - C_2A_1)^2 + (B_2C_1 - C_2B_1)^2]^{1/2}}{B_1A_2 - B_2A_1}.$$

Substituting the values of the coefficients A, B, C from Equations (41) in these equations, we obtain the following expressions for the direction of the normal to the polarization plane on a seismogram produced by a conical set (ψ = const.):

$$\tan\omega_n = \frac{\sin(\omega_2 - \omega'_2)\sin(\omega'_1 - \omega_1) - \sin(\omega_1 - \omega'_1)\sin(\omega'_2 - \omega_2)}{\sin(\omega_1 - \omega'_1)(\cos\omega_2 - \cos\omega'_2) - \sin(\omega_2 - \omega'_2)(\cos\omega_1 - \cos\omega'_1)}, \qquad (42)$$

$$\tan\varphi_n = \frac{\{[\sin(\omega_1 - \omega'_1)(\cos\omega_2 - \cos\omega'_2) - \sin(\omega_2 - \omega'_2)(\cos\omega_1 - \cos\omega'_1)]^2 + [\sin(\omega_1 - \omega'_1)(\sin\omega'_2 - \sin\omega_2) - \sin(\omega_2 - \omega'_2)(\sin\omega'_1 - \sin\omega_1)]^2\}^{1/2}}{(\sin\omega'_1 - \sin\omega_1)(\cos\omega_2 - \cos\omega'_2) - (\sin\omega'_2 - \sin\omega_2)(\cos\omega_1 - \cos\omega'_1)}. \qquad (43)$$

The inclination angle of the normal φ_n may also be expressed in terms of azimuth ω_n of the normal to the polarization plane

$$\tan \varphi_n = \frac{(\cos \omega_1 - \cos \omega_1') \sin (\omega_2 - \omega_2') - (\cos \omega_2 - \cos \omega_2') \sin (\omega_1 - \omega_1')}{[(\cos \omega_2 - \cos \omega_2') (\sin \omega_1' - \sin \omega_1) - (\cos \omega_1 - \cos \omega_1') (\sin \omega_2' - \sin \omega_2)] \tan \psi \cos \omega_n} . \tag{44}$$

Formulae (42) and (44) enable the orientation of the normal to the polarization plane to be calculated from the directions of two or more component pairs with equal phase shifts. In practice it is frequently convenient to employ graphical methods, because they are less cumbersome. Suppose we have several line-ups on the polar seismogram (Figure 38). We determine several pairs of points lying on them and having equal phase shifts,

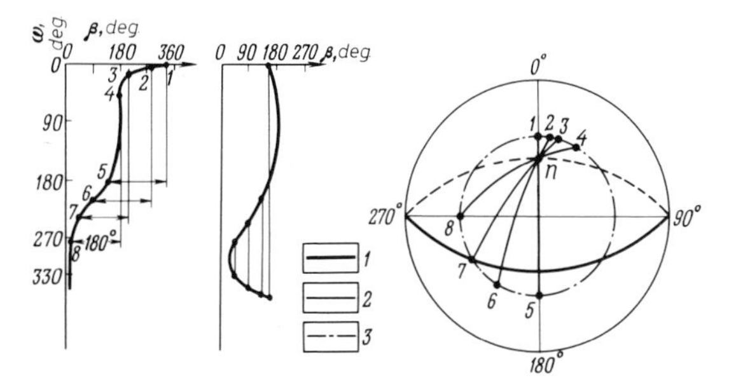

Fig. 38. Determination of position of polarization plane from the shape of a line-up of a conical set (ψ = 30°). (1) polarization plane, (2) equiphase planes, (3) conical set.

if the line-up has an extremum, or with phase shifts with a difference of 180°, if the line-up has no extrema. There are two directions in space to correspond to each pair of points on the line-up. If we draw a plane through each pair of such directions on the stereogram (Figure 38), we should obtain a family of equiphase planes whose line of intersection will determine the direction in space of the normal to the polarization plane, and the normals to the equiphase planes would lie in the polarization plane. A possible way to increase the accuracy of determination of the polarization plane's position is to use a greater number of equiphase planes than two. Methods of determining the position of the polarization plane may be of great interest to seismic prospecting practice, because it is based exclusively on phase criteria, which are the most stable, and completely ignores the amplitude characteristics of polar seismograms. This makes rigorous control of instrumentation sensitivity unnecessary.

Positions of Extremal Points on a Line-Up. Let us consider two methods of analytical determination of the polarization plane [15].

The first method is applicable to line-ups with an extremum. As has been demonstrated previously, the shape of the line-up, and accordingly the position of extremal points on it (the azimuths of components with maximum phase shifts), are determined solely by the position in space of the polarization plane. It may be demonstrated that, if the line-up has an extremum, the azimuth of the component responsible for the extremum is obtained from the formula

$$\omega_{ex} = \omega_n \pm \arccos(\cot\psi \cot\varphi_n)$$

or

$$\omega_n = \frac{1}{2}(\omega_{max} + \omega_{min}). \tag{45}$$

Hence it follows that the maximum phase shift between points of a line-up is independent of the azimuth of the normal to the polarization plane. The difference between the azimuths of the extrema is equal to:

$$\Delta\omega = [\omega_{max} - \omega_{min}] = 2 \arccos(\cot\psi \cot\varphi_n), \tag{46}$$

that is, for a definite conical polar set with angle ψ, the difference between the azimuths depends only on the inclination angle of the polarization plane φ_n. Expressions (45) and (46) enable φ_n and ω_n to be determined.

The second method is applicable to line-ups of arbitrary shape. The phase shifts are measured not between extrema, but between certain characteristic points on the line-ups, as in the graphical method described above [41]. If we choose two fixed points on the line-up (e.g. with the azimuths of the components equal to 0° and 90°), we will be able to determine the orientation of the normal from the following expressions:

$$\cot\varphi_n = \tan\psi \left(\frac{(X - XY - 1 - Y)^2}{4Y^2 + (X-1)^2 (Y+1)^2} \right)^{1/2},$$

$$\tan\omega_n = \frac{2Y}{(1-X)(1+Y)},$$

where

$$X = \tan\frac{1}{2}\omega_{0°}; \qquad Y = \tan\frac{1}{2}\omega_{90°}.$$

These formulae enable φ_n and ω_n to be determined simultaneously, but the calculations are rather lengthy.

DETERMINATION OF THE SHAPE OF THE POLARIZATION ELLIPSE AND ITS ORIENTATION IN THE POLARIZATION PLANE.

We shall consider methods of determining the shape of the polarization ellipse from line-ups of a polarization seismogram recorded from just two mutually-orthogonal seismometers.

Relative Phase Shifts of Points on a Line-Up. We have earlier divided all the line-ups into two families: with and without extrema. For the purpose of determining the shape of the ellipse, it will be more convenient to divide all the line-ups into two types according to their inclination angle at the point with azimuth $\omega = \omega_n$. All line-ups are at this point either ascending (the first type with a positive derivative) or descending (the second type with a negative derivative). Any line-up may be classified as belonging to one type only. For line-ups without an extremum, this is obvious (the line-ups either ascend or descend). For line-ups with an extremum, this follows from the fact that the point $\omega = \omega_n$, on account of the Equation (45), cannot be either maximum or minimum. In order to determine the shape of the polarization ellipse ($m = b/a$), we analyse the relative phase shifts for certain points on the line-ups for which the phase shifts can be determined with the greatest precision, that is, at the extrema and at point ω_n (it is assumed that ω_n has already been found). It should be kept in mind at this point that the maximum phase shift $\beta(\omega_{max}) - \beta(\omega_{min})$ is determined by the angles of inclination to the vertical of the normal to the polarization plane (φ_n) and the first component (φ_1), as well as by the shape of the polarization ellipse m, and is independent of the normal's azimuth ω_n.

Write the expressions for relative phase shifts between the points of a line-up with an extremum (15):

$$\Delta\beta_1 = \beta(\omega_{max}) - \beta(\omega_n) = \arctan \frac{m_1^{-1}\,[f(\omega_{max}) - f(\omega_n)]}{m_1^{-2} + f(\omega_{max})f(\omega_n)},$$

$$\Delta\beta_2 = \beta(\omega_n) - \beta(\omega_{min}) = \arctan \frac{m_2^{-1}\,[f(\omega_n) - f(\omega_{min})]}{m_2^{-2} + f(\omega_n)f(\omega_{min})},$$

$$\Delta\beta = \beta(\omega_{max}) - \beta(\omega_{min}) = \arctan \frac{m^{-1}\,[f(\omega_{max}) - f(\omega_{min})]}{m^{-2} + f(\omega_{max})f(\omega_{min})}.$$

Each equation may be used to find m:

$$m_1^{-1} = \frac{[f(\omega_{max}) - f(\omega_n)] \pm \{[f(\omega_{max}) - f(\omega_n)]^2 - 4f(\omega_{max})f(\omega_n)\tan^2\Delta\beta_1\}^{1/2}}{2\tan\Delta\beta_1},$$

$$m_2^{-1} = \frac{[f(\omega_n) - f(\omega_{min})] + \{[f(\omega_n) - f(\omega_{min})]^2 - 4f(\omega_n)f(\omega_{min})\tan^2\Delta\beta_2\}^{1/2}}{2\tan\Delta\beta_2},$$

$$m^{-1} = \frac{[f(\omega_{max}) - f(\omega_{min})] \pm \{[f(\omega_{max}) - f(\omega_{min})]^2 - 4f(\omega_{max})f(\omega_{min})\tan^2\Delta\beta\}^{1/2}}{2\tan\Delta\beta}.$$

Here m_1 has been found from the phase difference $\beta(\omega_{max}) - \beta(\omega_1)$, m_2 from the difference $\beta(\omega_n) - \beta(\omega_{min})$, and m from the difference $\beta(\omega_{max}) - \beta(\omega_{min})$. The most reliable determination of m may be found from the last equation containing the phase difference between ω_{max} and ω_{min}.

Consider two differences in phase shifts between three points of a line-up without extrema:

$$\Delta\beta_1 = \beta(\omega_n + 90°) - \beta(\omega_n),$$
$$\Delta\beta_2 = \beta(\omega_n + 270°) - \beta(\omega_n).$$

These phase differences depend not on the azimuth ω_n of the normal to the polarization plane, but on the shape of the ellipse. At the same time these phase shifts correspond to points distributed approximately evenly along the azimuth axis, so that their values are quite large. Using the expressions for the differences we obtain

$$m_1^{-1} = \frac{[f(\omega_n + 90^\circ) - f(\omega_n)] \pm \{[f(\omega_n + 90^\circ) - f(\omega_n)]^2 - 4f(\omega_n + 90^\circ)f(\omega_n) \tan^2 \Delta\beta_1\}^{1/2}}{2 \tan \Delta\beta_1},$$

$$m_2^{-1} = \frac{[f(\omega_n + 270^\circ) - f(\omega_n) \pm \{[f(\omega_n + 270^\circ) - f(\omega_n)]^2 - 4f(\omega_n + 270^\circ)f(\omega_n) \tan^2 \Delta\beta_2\}^{1/2}}{2 \tan \Delta\beta_2}$$

Because of cumbersome and lengthy calculations involved for line-ups of both types, it is advisable to use nomograms. Examples of such nomograms and calculations are contained in [15].

Records of Two Mutually-Orthogonal Seismometers. When processing observation data, one has frequently to calculate the polarization ellipse from the records of two mutually-orthogonal seismometers lying in the plane of the ellipse, that is, from the maximum amplitudes (C_x and C_y) and the phase difference $\Delta\beta$ (Figure 39). The angle between

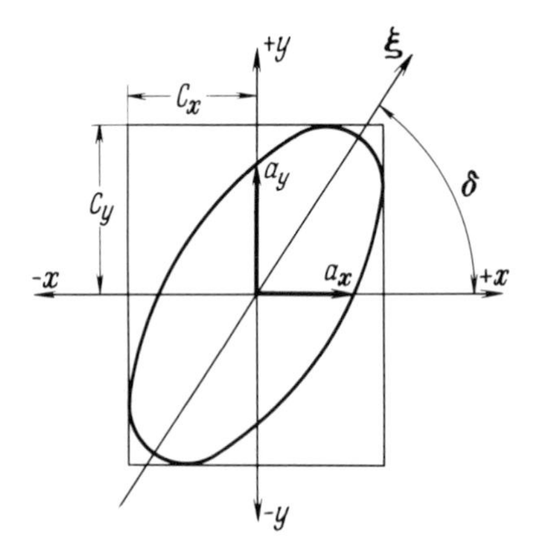

Fig. 39. Determination of polarization ellipse from records of two mutually-perpendicular seismometers.

the x axis and the principal axis of the polarization ellipse ξ is determined by the expression

$$\delta = \frac{1}{2} \arctan \frac{2C_x C_y \cos \Delta\beta}{C_x^2 - C_y^2},$$

and the ellipses' form-factor by the expression

$$m = \left(\frac{C_x^2 \sin^2 \delta - C_x C_y \sin 2\delta \cos \Delta\beta + C_y^2 \cos^2 \delta}{C_x^2 \cos^2 \delta + C_x C_y \sin 2\delta \cos \Delta\beta + C_y^2 \sin^2 \delta} \right)^{1/2}.$$

In radar the term for the problem of determining the parameters of the polarization ellipse from measured values of C_x, C_y, and $\Delta\beta$ is, 'polarization ellipse synthesis in rectangular coordinates'. In seismic prospecting, this problem may be of great interest in calculating the polarization ellipse from the records of two mutually-orthogonal seismometers. The polarization ellipse thus obtained may be used to calculate phase shifts between any two mutually-perpendicular directions (x, y) that lie in the polarization plane and do not coincide in direction with the principal axes of the polarization ellipse (Figure 39). The phase shift is determined from the formula:

$$\beta = \arcsin \frac{a_x}{C_x} = \arcsin \frac{a_y}{C_y},$$

where a_x, a_y are position vectors of the polarization ellipse in the chosen directions.

In order to find them, a rectangle with sides parallel to the chosen directions is described about the ellipse.

Values of the Motion Vector. The shape of the ellipse and its orientation in the polarization plane can be determined by plotting the trajectory of particle motion from the calculated values of the vector of motion. In order determine the modulus of the radius-vector and its direction, we may use methods devised for linearly-polarized oscillations. By plotting the moduli in the direction of the radius-vectors, we can construct the elliptical trajectory in the polarization plane. During bulk determinations, computerized methods of calculating the vector may conveniently be employed, for which purpose special programs are available [80].

By way of an example, let us consider the graphical method of finding the polarization plane and the trajectory of particle motion (the shape of the polarization ellipse) from the recorded amplitudes of three seismometers of a symmetrical three-component cluster A_{I}, A_{II}, and A_{III} (Figure 40). The amplitudes may be measured with any intervals of

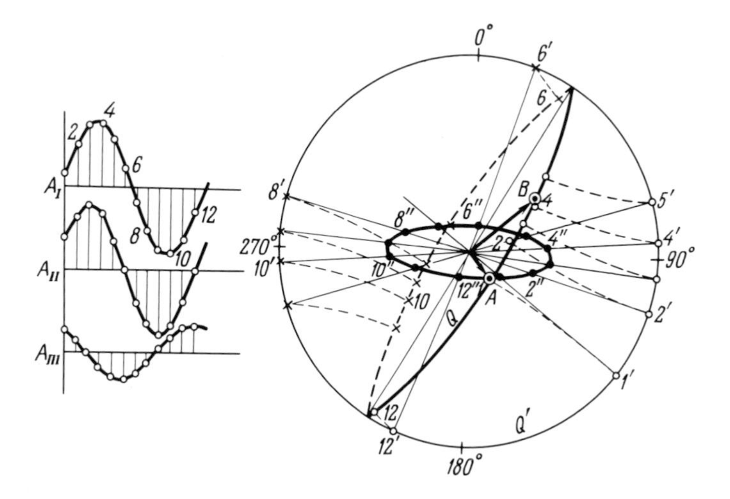

Fig. 40. Determination of orientation of polarization plane and construction of trajectory of particle motion.

time. In our example there are 12 points per period. Using any of the methods of determining the direction and modulus of the vector considered above, we may obtain the values of φ, ω, $|R|$ for each of the 12 points.

The directions are plotted on a stereo-net (Figure 40) and the polarization plane Q is drawn through points 1–12. To plot the trajectory of particle motion in the polarization plane, we need only plot the modulus of the motion vector in the direction of the radius-vector for each instant of time. As already mentioned, to this end we may conveniently use evolution on the horizontal plane. To do this, we should rotate any oblique vertical polarization plane to make it coincide with the horizontal and then plot the trajectory of motion in the horizontal plane. If the plane Q (Figure 40) is rotated until it coincides with the horizontal plane, points 1–12 of the oblique plane moving along the parallels of the stereo-net will occupy positions $1'$–$12'$ in the horizontal plane. We plot the moduli of the vector in the directions $1'$–$12'$, and by joining points $1''$–$12''$ thus obtained, draw the trajectory of motion. We find the orientation of the ellipse and the ratio of its axes from the trajectory.

The graphical method of plotting trajectories, in contrast to the preceding methods, employs only the amplitude characteristics of a polar seismogram. It may be preferable in the case of computer processing. However, this method requires rigorous sensitivity control of instrumentation. Its additional advantages are that it permits continuous plotting of the trajectory of particle motion across the entire polar seismogram, regardless of the type of waves and their polarization. The latter is especially important for determining the parameters of oscillations polarized in space.

CHAPTER 4

APPLICATIONS OF SEISMIC-WAVE POLAR CORRELATION

Polar correlation, now a well-established technique with a wealth of practical experience to its credit, has seen two periods in its development.

During the first period associated with the name of Gamburtsev, polar seismograms were produced by azimuthal multicomponent clusters. During the second, polar seismograms were derived from three-component observations through the use of polar analysers (see Chapter 2). In either case, polar correlation served the needs of both earthquake and exploration seismology.

The pioneer work of the first period was concerned with the detection of weak local earthquakes in various areas of Turkmenia and the Pamirs [24, 52–57]. Then followed a series of seismic surveys, but they were modest in scale and did not go beyond the experimental stage. A major limitation with them was basically one of technical nature, arising from the necessity to use multicomponent clusters at each observation point. In earthquake observations, the difficulties arose from the need to use multichannel recording under stationary conditions.

The second period began with the introduction of CDR–I and a technique for magnetic-tape recording and playback of multicomponent seismograms [27]. Now polar seismograms could readily be derived from three-component field observations at the time of seismic data-processing. Here, too, pioneer work was concerned with seismology, notably with distant earthquakes recorded by 'Zemlya'-type stations to detect transmitted converted waves [87]. It was followed by large-scale seismic surveys, first for ores, then for oil [37, 38, 40, 49].

The potentialities of polar correlation can best be illustrated by examples drawn from various seismic observations and related to various wave types. Of particular interest is the application of polar correlation to the analysis of shear (S) waves. In addition to the direct shear waves excited by explosions, we shall also discuss the shear waves induced by local earthquakes and transmitted converted waves produced by distant earthquakes. The emphasis on transmitted converted waves is due to the fact that they serve as the basis for regional studies widely employed at present. To a large extent, the efficiency of this method depends on the use of polarization in detecting and tracking converted waves.

Polarization of Explosion-Induced Direct P and S Waves

In seismic exploration, direct waves figure prominently, because they give birth to the entire wave process and because they control to a marked degree the parameters of each wave, the number of secondary waves, and their relative position on a seismogram, that is, the structure of the entire record [36]. Variations in the shape and intensity of a direct wave as it is propagated in a medium give a fairly accurate clue to the properties of the propagating medium. Also, as VSP work has shown, observation of direct waves in the interior parts of the medium enables one more accurately to analyse and monitor the conditions in which the waves are most intensely excited – a factor practically ignored until quite recently, althought its effect on the wave field and the effectiveness of seismic surveys is beyond any doubt.

As a matter of record, it is the VSP method that has led to a more rigorous definition of the direct wave and offered a means for a quantitative study of this wave under conditions where the distortion caused by the upper part of the section is held to a minimum. This alone might be rated as a major achievement of the VSP method. A still better opportunity to analyse direct *P* and *S* waves excited by explosion sources normally used in seismic prospecting is now opened up by polar correlation. This development seems to be especially encouraging now that ever more ground is being covered by the polarization method capable of using not only *P*, but also *S* and *PS* waves.

It appears that the application of VSP to both *P* and *S* waves may provide a sound physical basis for the direct search for oil and gas.

DIRECT *P* WAVES

A peculiarity of direct *P* waves basic to polarization studies is that they are often the first arrivals on seismograms. As such, they can give a more accurate idea about wave polarization and lead to a more accurate quantitative interpretation of records. As experience with polar correlation shows, the first *P* wave is often polarized in a way other than linearly, even though on the vertical component record the *P* wave may appear as a relatively simple and short pulse. For example, the *P* wave appearing as the first arrival on all vertical profiles recorded in the Nura-Taldy ore field (Central Kazakhstan) is, strictly speaking, non-linearly polarized, which points to superposition of the waves. Among the factors that may disturb the polarization of the *P* wave is its interference with other waves.

Superposition of Direct P and S Waves. When observed near the source and when the direction of approach of the first arrival is close to the vertical, the *P* wave appears as a relatively simple and short pulse on the vertical component record (Figure 41a), because the *S* wave is not recorded on the *Z*-component seismogram in such a case. As the polarization changes, it can be seen on the polar seismograms with $\psi = 60°$ and $\psi = 30°$. When $\psi = 0°$, the initial-phase amplitude of the *P*-wave pulse is a minimum, and the arrival of a horizontally-polarized *S* wave can clearly be seen against such a background. As the distance between the shot point and a shallow bore-hole increases, the direction from which the first *P* wave comes deviates from the vertical, and the move-out time between the *P* and *S* waves increases (Figure 41b). Now the first arrival on the vertical-component record has a multiplicity of phases, because the record contains the vertical components of both *P* and *S* waves.

Superposition of Other Waves. Sometimes, the polarization of a *P* wave can be disturbed by factors other than the superposition of direct *S* waves. Over large distances where the *P* wave is registered outside the area of interference with an *S* wave, polar correlation

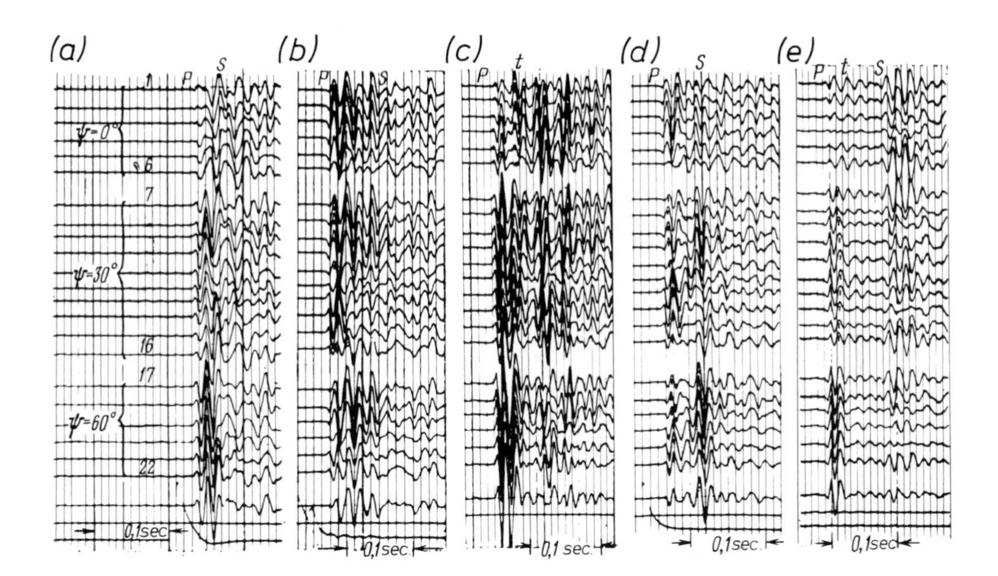

Fig. 41. Polarization of a P wave distrubed by superposition of other waves (DH 188, Nura-Taldy area, Central Kazakhstan). (a) l = 45 m, H = 180 m; (b) l = 125 m, H = 60 m; (c) l = 125 m, H = 520 m; (d) l = 125 m, H = 120 m; (e) DH 213, l = 260 m, H = 560 m.

can in some cases detect yet another horizontally-polarized wave, the t-wave. As a rule, the t-wave is registered in the tail of a P-wave pulse, but its time position may vary somewhat for different shot-points. On the seismograms in Figures 41c–e, the tail of the direct-wave record displays a deviation from linear polarization, which is manifested as changes in the relative intensity of adjacent wave phases (Figure 41c) and a curved line-up at ψ = 30° (Figures 41d, e). The superposition of a horizontally-polarized t-wave can be clearly seen on trace 16 where the P-wave has smaller amplitudes.

On the seismogram in Figure 41e, the pulse on the Z-component record is simple and persists on all components of the polar seismogram. Although the line-ups of the first arrivals are practically straight lines, the first wave is non-linearly polarized, because on the clusters with ψ = 0°, 30°, and 60° there are marked shifts between adjacent phases. Whereas the first peaks line up on all the clusters, the second peak on the cluster with ψ = 0° lags behind by about 0.005 s. Outwardly, it appears as if the various clusters have recorded the first arrival at different frequencies.

Thus, although the Z-component record of the first P wave may be relatively simple in shape, polar correlation shows that the P wave is complexly-polarized; this is revealed by the changed shape of the P wave on the off-normal and horizontal component records and by the curved line-ups. In all these cases, multiple polarization of the P wave has been caused by the superposition of the t-wave. As will be shown later, polarization-position correlation enables the t-wave to be positively tracked along the entire traverse.

Initial Portion of the P-Wave Pulse. From a study of a large number of records, it is seen that the first P-wave arrival may also be non-linearly polarized. Let us consider polar seismograms for the direct wave, which appears as a simple pulse on the Z-component record. Multiple polarization is manifested by distortion in the various component records (Figure 42a) and by the curved line-up of the first arrivals (Figure 42b).

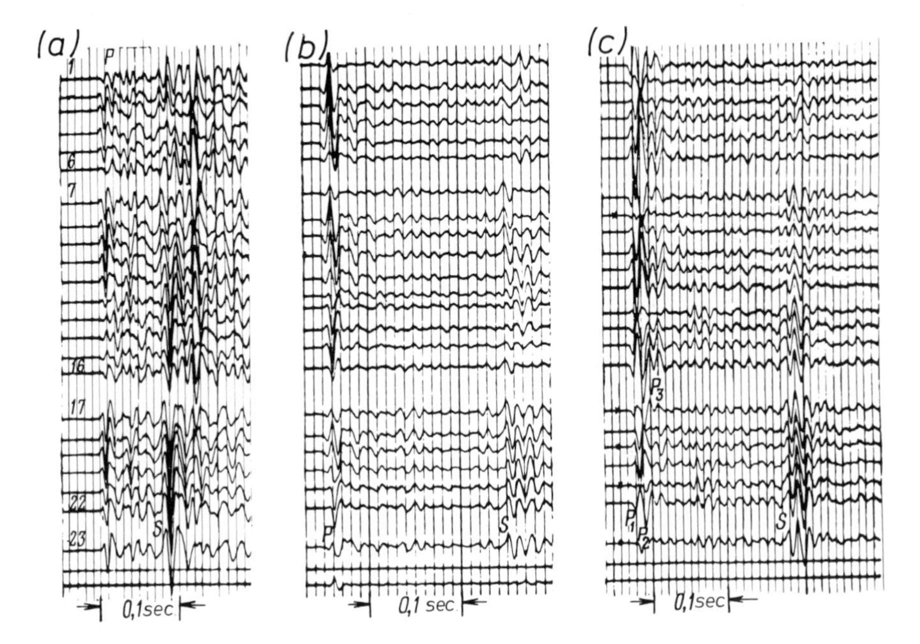

Fig. 42. Polarization of the pulse head. (a) DH 188, l = 440 m, H = 30 m; (b) DH 198, l = 160 m, H = 430 m; (c) DH 198, H = 390 m.

Sometimes, polar correlation brings out two or more pulses in the direct wave. In the polar seismogram in Figure 42c, traces 18 and 23 on which the P_1 wave has a minimum amplitude display a second pulse, P_2 shifted somewhat in time. Traces 8, 17 and 21 where the second pulse has a minimum amplitude bring out one more pulse, P_3.

Complex polarization of the first arrival has also been noted in other localities. For example, observations at Zhairem (Central Kazakhstan) where waves were excited and recorded under conditions different from those obtaining at Nura-Taldy a direct non-linearly polarized wave was registered in many cases.

Thus, polar correlation of the first arrival over a wide range of depths and distances to the source seems to have shown that as often as not the P wave is a product of wave interference. We still lack knowledge about the nature of the interfering waves and the mechanism(s) by which they are generated. In order to fill the gap, it will be necessary to conduct experiments, which mainly throw light on wave formation near the source. So far as deviation from linear polarization in the head of the direct-wave pulse is concerned, and even during the first phase itself, it may arise from the superposition of secondary waves forming on the discontinuities of the geologic section.

In the light of the above analysis of direct-wave polarization, it may be argued that polar correlation offers a versatile tool for studies into excitation conditions. With it, we may monitor not only the wave form, but also the directional effects of the source, that is, the distribution of energy along various directions in space.

DIRECT SHEAR WAVES

The polarization of direct shear waves is of special interest, because it is very sensitive to discontinuities in the medium. Also, it may help greatly with the study of reflected *S* waves, transmitted *PS* waves, and the anisotropy of media. Extensive work on the polarization of shear waves has been conducted under the guidance of Puzyrev and Brodov [6, 102, 130]. The polarization of transmitted *PS* waves has received much attention from Pomerantzeva [100, 101], Egorkina [68–71], and Obolentzeva [65, 78, 92–94]. Among other things, Obolentzeva has investigated the polarization of plane *P* and *S* waves and has traced the relationship between deviation in the direction of particle displacement from the ray path and the properties of anisotropic media. She has shown that for *S* waves, this deviation may be so large that they may be recorded by vertical seismometers immediately following the *P* wave.

Consider the polarization of direct *S* waves recorded at the Sayak orefield (Northern Balkhash area). The direct *S* wave recorded everywhere was usually multiphase and non-linearly polarized. The polar seismograms display both curved and oblique line-ups (Figure 43a). The *S* wave owes its non-linear polarization to the superposition of several waves. In some cases, the interfering pulses may be readily discerned on the polar seismogram for the various components and their move-out times can be accurately measured. On the seismograms produced by a symmetrical three-component cluster (Figure 43b, 1), the first pulse of the *S* wave reaches seismometer II with a lead of about $3T/4$. The other two seismometers scarcely respond to the wave, so that the next pulse can clearly been seen against the background of minimal amplitudes. A similar separation of the component pulses can be seen on the records produced by the top and bottom channels (H = 940 m). In most cases, however, the records of the *S* wave are extremely complex. A minimum of interference is only observed at the start of the *S* wave record (as a rule, this is the first peak), because the interfering waves are usually displaced in time from one another. Because of this, data-processing mainly yields polarization parameters for the early phases of the *S* wave. For the succeeding phases of the *S* wave where the polarization pattern becomes very complicated, the direction of particle displacement can be determined mainly for some dominant phases. The manner in which the waves superimpose on one another has a direct bearing on the character of *S* wave polarization revealed on polar seismograms by the shape of the line-up pattern. In 70 to 80% of the recorded cases, the tilt of the line-up pattern remains the same at the various observation points from all shot-points. This gives grounds for assuming that basically the polarization of the direct *S* wave is related to the conditions of excitation. Consider an example of polar correlation as applied to an *S* wave.

The direction of particle displacement for a *P* wave is found by reference to stereogram components 6 and 11 (see Figure 43c) as the normal to the zero-amplitude plane, and is seen to be close to the direction of stereogram component 3. This is so, because on the polar seismogram it is seen that the *P* wave has a maximum amplitude on trace 3. The *P* wave arrives in a nearly horizontal direction, so it is scarcely seen on the *Z*-component record. Note the polarity of the *P* wave on trace 3. The downward propagation of the *P* wave (the angle between the particle displacement in the *P* wave and stereogram component 3 is less than 90°) corresponds to the downward deviation of trace 3 from the equilibrium position. The zero amplitude plane Q_{0P} is tangent to the *P*-wave front.

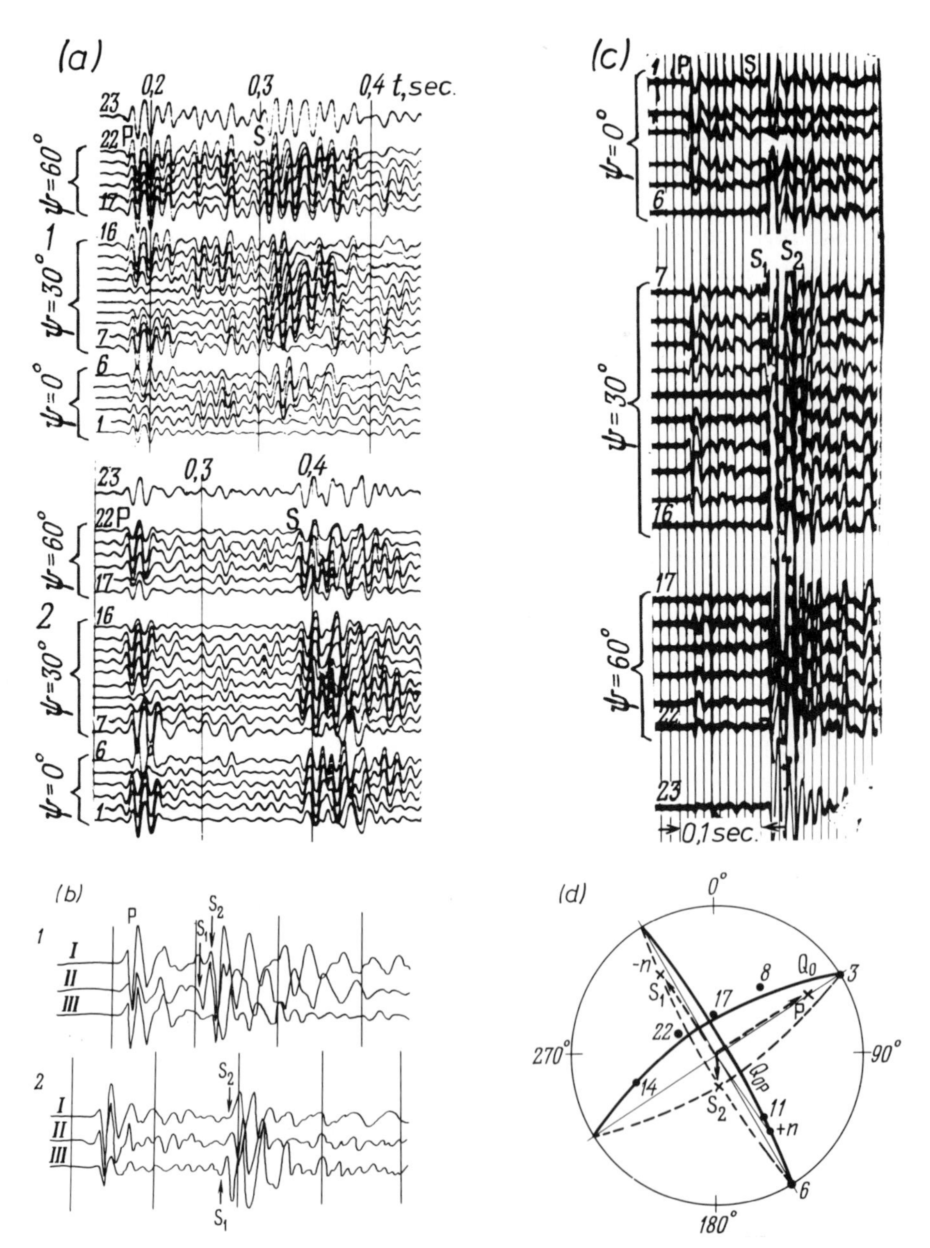

Fig. 43. Polar seismograms of P and S waves (DH 243). (a) SP3, l = 575 m; (1) H = 600 m; (2) H = 980 m; (b) three-component seismograms, SP2, H = 680 m (1); SP3, H = 940 m (2); (c) SP4, H = 120 m.

The S wave presents a complex waveform recorded as a multiphase wave by the Z cluster arm and as a short pulse by the cluster arms with ψ = 0° (see Figure 43c). The inclined component records (with ψ equal to 30° and 60°) show curved line-ups. The plane of polarization for the S wave may be positively located from the component records with $\psi = 0°$. For one thing, the S wave is scarcely registered on trace 3; for another, its line-ups are vertical. Therefore, the plane of polarization of the S wave is perpendicular to stereogram component 3 and is very close to the vertical. On the stereogram it is seen that the above arguments apply to the orientation of the Q_{0P} plane, so that the S wave may well be polarized in that plane.

From a careful comparison of S-wave records on the $\psi = 0$ and Z-component seismograms (traces 1–6 and 23), it is seen that, although the records are identical in shape, they show a phase shift of $\Delta\beta \approx \pi/2$ between the wave peaks. This is an indication that in the beginning of the S-wave record wave interference causes the particles to vibrate along an elliptical path or, more precisely, along a vertical ellipse. Because, as already noted, the component pulses are registered with a time shift, it is an easy matter to determine the direction of particle displacement for the first of the interfering waves. The first S-wave peak with a minimum amplitude is recorded on traces 3, 8, 14, 17, and 22. If we pass the zero-amplitude (Q_0) plane through the directions of the respective stereogram components, then the direction of particle displacement for the first S-wave arrival S_1 will be along the normal to the positive ($+n$) or negative ($-n$) half-plane Q_0. The actual situation may be ascertained from the signs of the arrivals. The traces deflect downward from the equilibrium position for stereogram components 1, 2, 7, 15, and 16 whose directions make an angle of less than 90° with the direction of particle displacement in the S_1 wave. Consequently, the direction of particle displacement in S_1 is in the $+n$ half-plane. Referring to the stereogram, it is seen that the direction of particle displacement in S_1 lies in the zero-displacement plane of the P wave.

The components on which S_1 has a minimum amplitude (traces 8, 14, 17, and 22) clearly show S_2 arrivals. The direction of particle displacement in the S_2 pulse is difficult to determine because of the interference that takes place at the same instant as the pulse arrives. The vertical ellipse of polarization may, however, be produced by the superposition of waves arriving from mutually-perpendicular directions close to the horizontal and the vertical. The direction of particle displacement in the S_1 arrival does lie close to the horizontal. So we may only assume that the direction of particle displacement in the S_2 arrival is close to the vertical and perpendicular to that of the S_1 arrival in the Q_{0P} plane. The signs that the S_2 arrival makes on traces 8, 14, 17, and 22 do not contradict this assumption.

Thus, the polar correlation of the S wave establishes that it is polarized in a plane tangent to the P-wave front and also brings out, in the beginning of the wave record, two component pulses S_1 and S_2, which are identified as SH and SV waves. We have discussed the above example in detail in order to demonstrate both the potentialities and the techniques of polar correlation in analysing a wave at an observation point.

In many cases, the superposition of several waves produces a complex interference pattern, so that the polarization parameters cannot be determined from the record shape in a straight-forward manner.

Applying polar correlation to the S wave on the VSP seismograms recorded at Sayak, we have determined the direction of particle displacement for the first phase of the S wave, as shown on the seismograms in Figure 44. For better visualization, the P waves have been aligned in the same direction for all depths. As is seen, for shot-point 4, most of the directions of displacement (70%) lie within $\pm 15°$ of the Q_{0P} plane of the P-wave

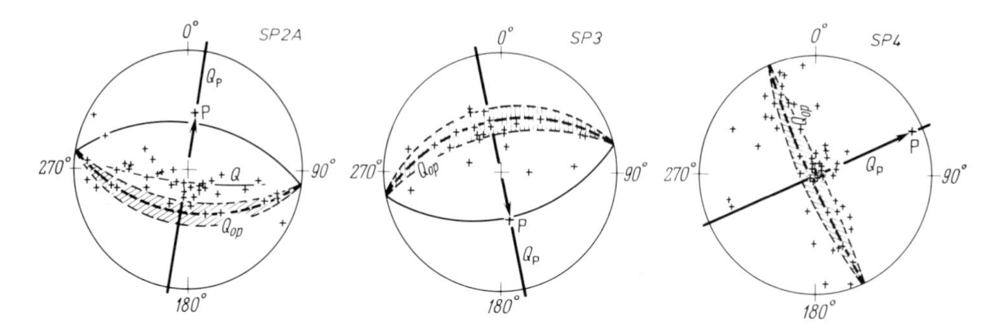

Fig. 44. Combined stereograms for the direction of particle displacement in *S* waves (DH 243, Sayak deposit). IIB2a–SP2a; IIB3–SP3; IIB4–SP4.

front, being grouped predominantly on one side of the vertical ray plane Q_P of the *P* wave.

For shot-point 2a, the directions of displacement are localized around the *Q* plane which deviates from the Q_{0P} plane, with an equal spread on either side of the ray plane Q_P. Although less pronounced, a similar picture emerges on the stereogram for shot-point 3. The pattern in which the directions of displacement are oriented is traceable to the geologic structure and suggests some asymmetry of the medium. A detailed analysis may well yield further data about the structure of the medium. This, however, calls for more investigations.

Polarization of Direct Shear and Other Waves Induced by Weak Local Earthquakes

The polarization of *S* waves induced by earthquakes evokes considerable interest for several reasons. For one thing, the polarization of *S* waves can help with precise timing of their arrivals. The point is that *S* waves are often non-linearly polarized, so the time at which they arrive depends on the spatial orientation of their components. Use of components not related to the plane of incidence (such as those oriented along the cardinal points of the compass) may introduce an appreciable error in determining the arrival time for *S* waves. On the other hand, precise knowledge of the arrival time for *S* waves is of paramount importance in many applications. This, above all, applies to temporal variations in the velocity ratio v_P/v_S as used in earthquake forecasting. For another, the polarization of *S* waves strongly depends on the anisotropy of the medium, so that two shear waves, *SV* and *SH*, appear, and the time shift between them may be indicative of velocity anisotropy. According to recent data [145], the time shift between *SV* and *SH* waves is indicative of the stressed state of the material and may well serve as a forewarning of an incipient earthquake. We leave out the traditional aspects such as the determination of the move-out time between *S* and *P* waves necessary for the calculation of the epicentral distance, studies into the focal mechanism, etc.

DIRECT SHEAR WAVES

In the study of *S* waves induced by weak local earthquakes, we have used the records produced by conical polar arrays consisting of eight seismometers making an angle of 45° with the horizontal and spaced 45° apart in azimuth [55]. Records were made of the high-frequency (15–30 Hz) aftershocks of the Khait earthquake, which occurred in the Hissar Range in the Pamirs in 1949 [53, 56, 57]. The records were produced with a magnification of 10^7.

The manner in which the S waves were polarized was analysed from a total of 50 records of weak local earthquakes. On 40 records, the move-out time for P and S waves was $\Delta t_{S-P} < 3$ s, and on 10 records it was $\Delta t_{S-P} > 3$ s. The number and arrangement of waves on the seismograms vary from earthquake to earthquake and with the hypocentral distance. As a rule, the seismograms of the nearby earthquakes are simpler (Figure 45b, $\Delta t_{S-P} \approx 0$–97 s). However, even a slight increase in Δt_{S-P} complicates the seismogram. This also applies to the pulse waveform of the first P wave. The shortest P-wave pulses are seen on the records of the nearest earthquakes with $\Delta t_{S-P} \approx 1$ s. An increase in Δt_{S-P} makes the pulse waveform of the P wave more complicated in most cases.

The polarization parameters (azimuth and angle with the vertical) were found from

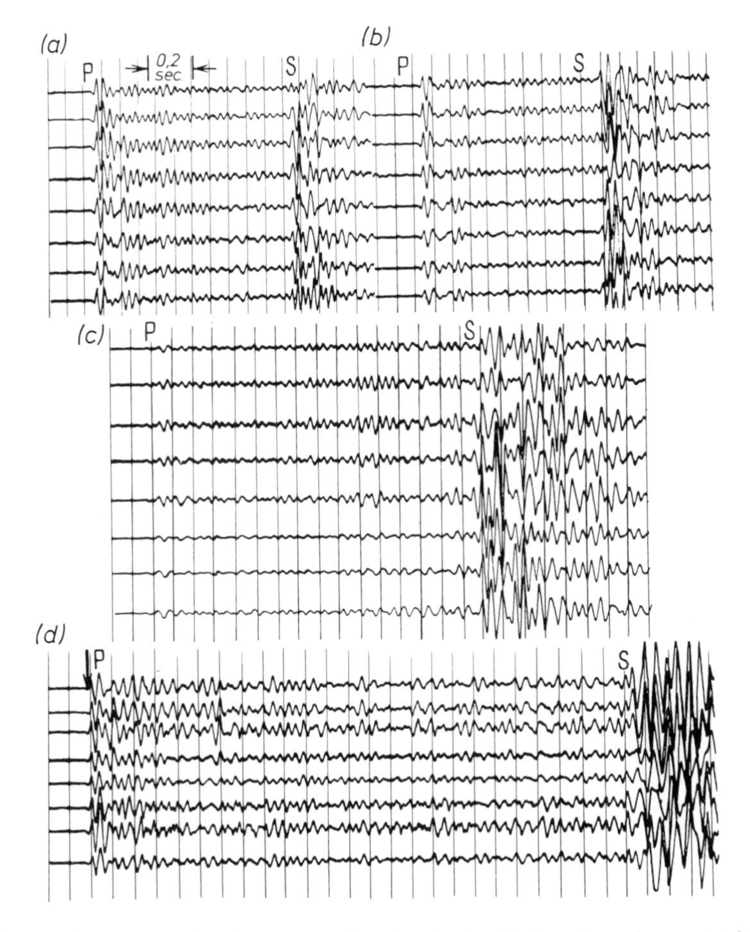

Fig. 45. Polar seismograms of weak local earthquakes in the Khait epicentral area, produced by an eight-component polar array with $\psi = 45°$.

the relative amplitudes for the dominant phases of the S waves and for the first peak of the P wave on the various component records. Attempts were also made to determine the direction of particle displacement for the first phase of the S wave, if it could be detected against the background of the preceding record in a sufficiently pure and readable form. The directions of particle displacement were plotted on a stereo-net for each earthquake and were then combined into a single stereogram by type of wave and type of polarization of S waves. The polarization of direct shear, and also PS and SP waves, is shown relative to that of the P wave. As a rule, the P wave is polarized linearly only at the head of the pulse, that is, over the first two or three phases even in records of the simplest form. The succeeding phases of the P wave show a more complex polarization, and the polar seismograms often display curved line-ups.

Direct shear waves differ widely in both shape on the record and polarization. In all the records, the frequency of the shear waves is below that of the compressional waves. The predominant frequencies for S waves extend between 18 and 24 Hz, and those for the P waves, from 26 to 30 Hz. In terms of shape on the record, S waves induced by an earthquake may arbitrarily be classed into three groups as follows.

Earthquakes of the First Group. The earthquakes in this group are characterized by non-linear polarization of the S waves, as revealed by inclined line-ups and the absence of peaks. The angle of inclination for the line-ups is the same for all the earthquakes in the first group (Figure 45d). In most cases, the inclined line-up is preceded by an almost linearly-polarized phase. The directions of particle displacement are determined for that preceding phase, but the scatter is appreciable because the recorded S wave is a product of wave interference. The directions of particle displacement as found for S waves on the basis of 18 earthquakes in the first group are shown on the combined stereogram in Figure 46a. As is seen, 16 directions of displacement are scattered on either side and within $\pm 20°$ of the zero-displacement plane Q_{0P}; 11 points are uniformly scattered on either side and within $\pm 30°$ of the vertical ray plane Q_P of the P wave. In some cases where the first phases of the S wave are linearly polarized, the respective displacements are arranged precisely in the vertical ray plane of the P wave. The respective scatter ranges are shaded on the stereograms.

Thus, in terms of polarization, the first more linearly-polarized phase of the S wave is analogous to an SV wave. With some time delay, the records show another pulse within the S wave. On interfering with the first, it produces a stable form of polarization with an inclined line-up, so that the succeeding phases of the S wave are, as a rule, non-linearly polarized. In some cases, however, the waves are almost linearly polarized. Their directions of displacement are shown in Figure 46a II. As is seen, the predominant direction of displacement is that of an SH wave.

On the combined stereogram, SH waves take up 50% of the shaded sector that they could have occupied. Obviously, the range of scatter for the interference-affected phases in the later part of the record may well be greater than for the early phases.

Earthquakes of the Second Group. The group covers the earthquakes for which the records are less consistent in regard to the shape of S waves and show a large number of linearly polarized dominant phases. Stereograms for the early phases of the S wave and also for the dominant phases in the later part are shown in Figure 46b. As with the first

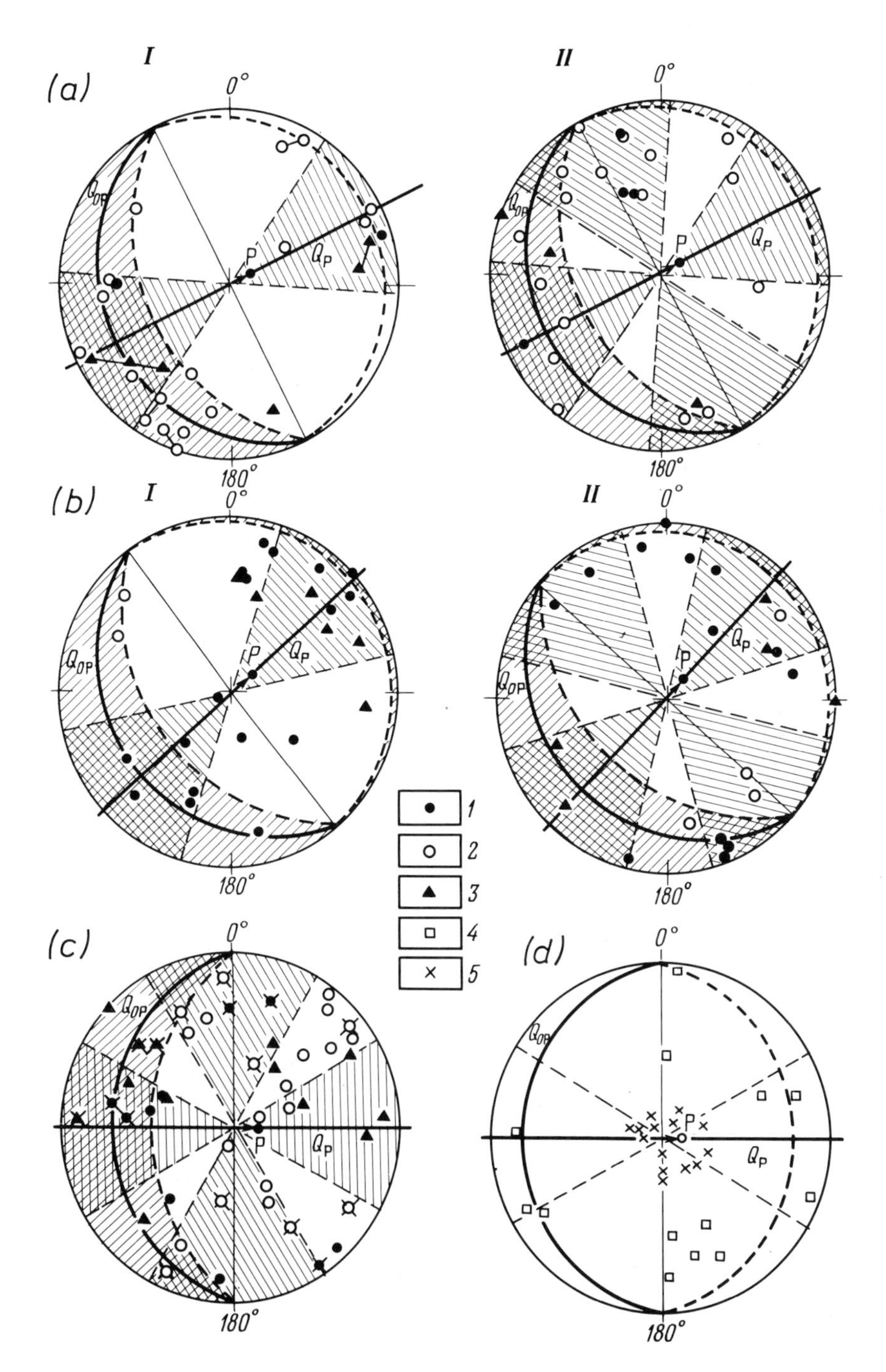

Fig. 46. Combined stereograms for the direction of particle displacement in *S* waves. (a) Earthquakes in the first group; I – first phase, II – late phases; (b) earthquakes in the second group; I – first phase, II – late phases; (c) earthquakes in the thrid group; (d) converted *PS* and *SP* waves; (1) $t_{S-P} < 1.8$ s; (2) $t_{S-P} = 1.8$ to 3.1 s; (3) $t_{S-P} > 3.1$ s; (4) *PS* waves; (5) *SP* waves.

group of earthquakes, most direction of displacement are close to the horizontal. For the early phases (Figure 46b I), in 50% of the cases, particle displacement is in the ray plane of the P wave, that is, within what we have arbitrarily labelled the SV sector. For the later phases, no clear delineation between SV and SH waves in terms of particle displacement can be noted.

Earthquakes of the Third Group. For the earthquakes of the third group, the records show a more complex interference pattern for the S wave. Because of this, there is an appreciable scatter between the directions of particle displacement in both vertical angles and azimuths (Figure 46c). For the most linearly-polarized first phase of the S wave, the points are concentrated (with some scatter) in the wavefront plane, but such records are few in the third group.

In classifying the earthquakes into groups, we have used the shape of S-wave records as a basis. Obviously, this is no rigorous criterion, because we have had to deal with records of complex multiphase waves rather than a single pulse. Yet, it is of interest to look at the breakdown of each group in terms of the move-out time Δt_{S-P} (Table II).

TABLE II

Δt_{S-P}, s	Number of earthquakes per group		
	1st group	2nd group	3rd group
1.8	2	7	5
1.8–3.1	13	3	9
over 3.1	3	3	4

As is seen, the predominant earthquakes in the first group are those with Δt_{S-P} = 1.8 to 3.1 s, and in the second with Δt_{S-P} = 1.8 s (see Figure 45c, d).

TRANSMITTED *PS* WAVES

On most seismograms, PS waves were observed about 0.2 s behind the first P-wave arrivals. On some records, they are practically linearly polarized with directions of particle displacement markedly different from those in the P wave. On other records, the PS waves show a more complex polarization with curved line-ups on seismograms. A stereogram of particle displacements for linearly-polarized PS waves is shown in Figure 46d. As is seen, in 50% of the cases particle displacements are in the plane of the P-wave front, and only in one case does it lie precisely on the intersection between the ray plane and the wavefront plane, that is, as in an SV wave.

INTERMEDIATE WAVES

In addition to PS waves, the seismogram shows other converted waves in the interval between the P and S waves, for which the direction of particle displacement is different from that of the *P* wave. Such waves are best seen on the seismogram in Figure 45d. The arrival times for the converted waves lie in the interval 0.55–0.6 s and 1.1–1.2 s following the first arrivals. Most seismograms, however, show no intermediate converted waves. The recorded intermediate waves are linearly polarized, but they do not retain the same shape on the various component records.

SP CONVERTED WAVES

On many seismograms, the *S*-wave group is preceded by *SP* waves. As a rule, the *SP* waves dominate what we have called the intermediate waves. The directions of particle displacement for the *SP* waves are shown in the stereogram in Figure 46d; they are seen to be clearly grouped around the directions of particle displacement for the *P* wave.

PS and *SP* waves may best be singled out on the records of earthquakes of the first group. In the second group, they are seen on two seismograms only.

Polar Correlation of Transmitted Converted Waves Induced by Distant Earthquakes

The past 10 to 15 years have seen an appreciable expansion in the use of distant sources (explosions and earthquakes) for the structural studies of large regions (regional surveys). Such studies have been based on a combined analysis of both *P* and transmitted *PS* waves induced at the boundaries located within the area of reception [5, 17, 18].

A marked stimulus in the use of transmitted converted waves induced by distant earthquakes for regional surveys was given by the advent of the specialized 'Zemlya' equipment. A good deal of work has been done by Pomerantzeva, Moszhenko, Egorkina and others at VNIIGeofizika [68–71, 100–101, 123]. Further progress in regional studies has taken place in the more recent years with the development of improved equipment to record earthquakes (such as 'Cherepakha' stations) and distant explosions ('Taiga' stations). For all the advantages offered by this method, its application still meets with many difficulties both fundamental and technical in nature [2, 50, 95, 127].

Difficulties of a fundamental nature arise, because the wave fields of interest are complex, and also because there are no physically-substantiated criteria by which to identify converted waves and to correlate the conversion boundaries stratigraphically.

Difficulties of a technical nature arise, because there are no standardized and approved procedures that produce field data enabling a concurrent analysis to be made of waves polarized in various directions and recorded by the various seismometers in a three-component array. More specifically, there is no technique that would show whether the channels are performing identically (in terms of both response and sensitivity) in three-component observations. Also, in the three-component *XYZ* arrays used for traditional observations, the horizontal seismometers are usually oriented along the cardinal points of the compass. Because earthquake foci may be located in various directions, the horizontal components thus recorded are in no way related to the direction of wave arrival and are, in a sense, accidental. Furthermore, the polarization of the converted waves under conditions of interference may be other than linear, so that the time of wave arrival, as already noted, varies with the spatial orientation of a given component. Errors in determining the arrival time of a converted wave on the basis of an accidental component (transverse or longitudinal) may lead to an error in determining the depth of the conversion boundary. All of these technical limitations make the waves detected on the vertical and horizontal components of a record far more difficult to identify.

Taken together, the above limitations lead in the final analysis to discrepancies in data interpretation. In many cases, the wave fields observed on seismograms do not fit into generally-accepted theoretical concepts, notably in regard to the intensity of converted waves. This above all applies to studies concerned with the principal interfaces in the Earth's crust. This may well be the reason why there is little agreement (in fact, a good deal of controversy) on the interpretation of waves in the initial part of seismograms despite the effort put in. This is true not only of wave nature, but also of wave characteristics (e.g. the polarization of converted waves).

As an example of the urgency of the matter, it may be mentioned that two diametrically opposite opinions exist concerning the polarization of transmitted *PS* waves. According to one [69], *PS* waves are non-linearly polarized, mainly owing to the anisotropy of the medium. According to the other [101], *PS* waves are essentially linearly-polarized and would become non-linearly polarized only when the transmitted *PS* waves polarized in different planes on different boundaries are superimposed on one another.

In the present author's opinion, this disagreement in views arises mainly from imperfections in observation techniques and procedures. On the other hand, there are publications [2, 95] that state that the directions of particle displacement in *PS* waves may make various angles with the direction to the source and that in the general case the two are in no way related. It is also argued that the horizontal waves in the initial part of seismograms are not, in most cases, transmitted *PS* waves and that the general aspect of seismograms is to be a considerable extent decided by transverse waves associated with the block structure of the upper part of the Earth's crust and also by secondary surface waves induced by interaction of the incident wave with the surface.

Naturally, there is also no consensus concerning the explorational capabilities of the method. In the circumstances, the best approach appears to perfect observation techniques and procedures and to apply polar correlation to the analysis of the waves in the initial part of seismograms. The new techniques and equipment for three-component measurements (see Chapter 2) and the polar correlation method developed at IEP AS USSR make it possible to treat the problem of converted waves from a new angle. The early experiments using 'Zemlya' stations were conducted by the Ili Geophysical Expedition, of the Kazakh Geophysics Trust [87]. Later, similar surveys were conducted, using 'Cherepakha' stations. Let us take a closer look at some examples and results of using polar correlation in the analysis of transmitted *PS* waves.

POLAR CORRELATION OF *PS* WAVES

A major obstacle to detecting transmitted *PS* waves is the superposition of the strong compressional waves at the beginning of a seismogram. On seismograms from distant earthquakes, the first arrival usually is a linearly-polarized *P* wave. Immediately upon the arrival of the converted (*PS*) waves, the linear path of the *P* waves changes to an elliptical one, owing to the superposition of the *PS* waves on the *P* wave. The efficiency of CDR–I as a technique for identifying converted waves lies in the very fact that these waves may now be observed in planes of zero displacement for *P* waves where *P* waves are not registered and cannot therefore affect converted-wave records. *P* and *PS* waves may be polarized in both the vertical and an inclined plane for which the orientation in space is mainly decided by the polarization parameters of the interfering waves. Although the interfering *P* and *PS* waves may be in any phase relationship, the elliptical form of polarization is not markedly affected, but there may be changes in the shape and spatial orientation of the ellipse and the direction of particle motion (with $\Delta\beta < 180^\circ$, the particles move in one direction, and with $\Delta\beta > 180^\circ$, in the opposite direction). The principal axis of the ellipse will be along the particle-displacement vector for the *P* wave only when the particles in the *P* wave move vertically, and $\Delta\beta = 90^\circ$, so that the interference produces a vertical elliptical path (which is usually the case with ground observations in the low-velocity upper part of the section).

As the succeeding phases of the *PS* wave gain in intensity, the ellipse changes in shape, and its principal axis rotates towards the direction of particle displacement in this wave. As will be recalled, one and the same ellipse of polarization will result from various phase and amplitude relationships between the interfering waves. Because of this, the inverse problem (determination of original wave components from the path of particle motion in the interference-produced wave) has no solution. If, however, the parameters of one of the interfering waves (say, the *P* wave) prior to its interference with *PS* waves are known, the problem is amenable to solution. To this end, it is necessary to obtain a record of one of the interfering waves, not distorted by a projection of the other wave, that is, in a plane of zero displacements for that wave. Since the displacement parameters of a *P* wave prior to its interference with *PS* waves can readily be obtained, the zero displacement plane Q_{0P} for the *P* wave can be as easily located. It will be recalled that the components lying in a plane of zero displacements for the *P* wave will be projections, and not moduli, of the displacement vector for *PS* waves least distorted by the superposition of the *P* wave.

Obviously, the extremal values of projections of a *PS* wave in the Q_{0P} plane will be at zero displacements ($A_P = 0$) in the *P* wave. At those instants, the arms of a polar array will register projections of a *PS* wave, and the direction of particle displacement for this wave may be ascertained from the relationships between these projections. This form of analysis calls for a very careful alignment of the directional pattern, which can only be ensured by the use of CDR–I analysers. Where polar correlation utilizes outline polar seismograms with a fixed set of components, rigorous analysis of waves in a zero-displacement plane will, as a rule, be unfeasible, so one has to be content with a qualitative analysis of the components close to the direction of particle displacement in the *P* wave, not necessarily lying exactly in its zero-displacement plane. Records of the *P* component and of the components in the zero-displacement plane for the *P* wave may show various phase shifts, depending on how much the principal axis of the polarization ellipse deviates from the direction of particle displacement in the *P* wave. Where interference involves *P* and *PS* waves with the same phase relationships, the resultant path is a straight line, but again the component records in the Q_{0P} plane will only show projections of the *PS* wave.

The above reasoning holds for both *SV* and *SH* waves. If the particles in the *P* wave move vertically, the *P* and *SV* waves will interfere in the same way as the *P* and *SH* waves. If the particles in the *P* wave move in an inclined plane, the plane of polarization for the *P* and *SV* waves will usually be vertical, whereas the *P* and *SH* waves will ordinarily be polarized in an inclined plane. This is displayed in various ways on polar seismograms.

In the former case, the component records with $\psi = 0$ show vertical line-ups through points of minimum amplitude on any component record; in the latter case, the line-ups with $\psi = 0$ will be inclined. The zero-amplitude component will always be directed along the normal to the plane of polarization. These differences between component records with $\psi = 0$ may be used to determine the orientation of the plane of polarization for *P* and *PS* waves and also to detect *SH* waves. Where *P, SV*, and *SH* waves are recorded simultaneously, the component records with $\psi = 0$ show inclined line-ups.

The interference of *SV* and *SH* waves produces an elliptical path in the plane of polarization. Irrespective of the orientation and the angle of inclination of the plane of polarization, the component records with $\psi = 0$ will display inclined line-ups; with ψ equal to 30° and 60°, the pattern of line-ups is decided by the relative position of the components and the planes of polarization [24, 41].

In polar correlation, all constructions are carried out in a local coordinate system. To this end, one locates the direction of the wave vector in space for the first compressional wave by one of the methods described earlier and according to the desired degree of accuracy and plots it on a stereogram (Figure 47a–c). The vertical plane Q_P passing through the direction of particle displacement for the *P* wave locates the x_P axis of the local coordinate system (the $+x_P$ direction being along the azimuth of the *P* wave). Then a plane perpendicular to the direction of the *P*-wave vector and tangential to the *P*-wave front is constructed – this is the zero-displacement plane Q_{0P} for the *P* wave.

In homogeneous and isotropic media, *SV*-converted waves are polarized so that the direction of particle motion coincides with the line of intersection between the Q_P and Q_{0P} planes (let it be called the radial direction *r*). The position of the Q_{0P} plane decides which components are optimal from the viewpoint of detecting and establishing continuity for converted waves. The directions of particle displacement in *P* and *SV* waves are projected on to the same direction of the x_P axis. The $+y_P$ axis lies along a normal to the Q_P plane. It is in this direction that *SH* waves will be polarized in inhomogeneous and isotropic media.

Converted waves can often be polar-correlated by visual methods, namely, by determining the direction of particle displacement on the basis of maximum amplitudes, by constructing zero-displacement planes with reference to traces displaying minimum amplitudes or phase inversions, etc. In all of these cases, the constructions are carried out in stereographic projection, using the procedures explained in Chapter 3.

In all of the polar seismograms shown, the upper three traces are records produced by a traditional *XYZ* cluster, with the *X* member of the cluster being aligned in the northerly direction (the 0° direction on the stereo-net). The next 23 traces make up a multicomponent polar seismogram in which the traces are oriented as follows: traces 1–6 are horizontal with $\psi = 0°$, and the difference in azimuth between adjacent components is $\Delta\omega = 30°$; traces 7–16 are inclined with $\psi = 30°$ and $\Delta\omega = 36°$; traces 17–22 are inclined with $\psi = 60°$ and $\Delta\omega = 60°$; and trace 23 is vertical (*Z* component) with $\psi = 90°$.

To illustrate the application of polar correlation to converted waves, we shall use records obtained in Central Kazakhstan and on the northern slopes of the Zailiisk Alatau.

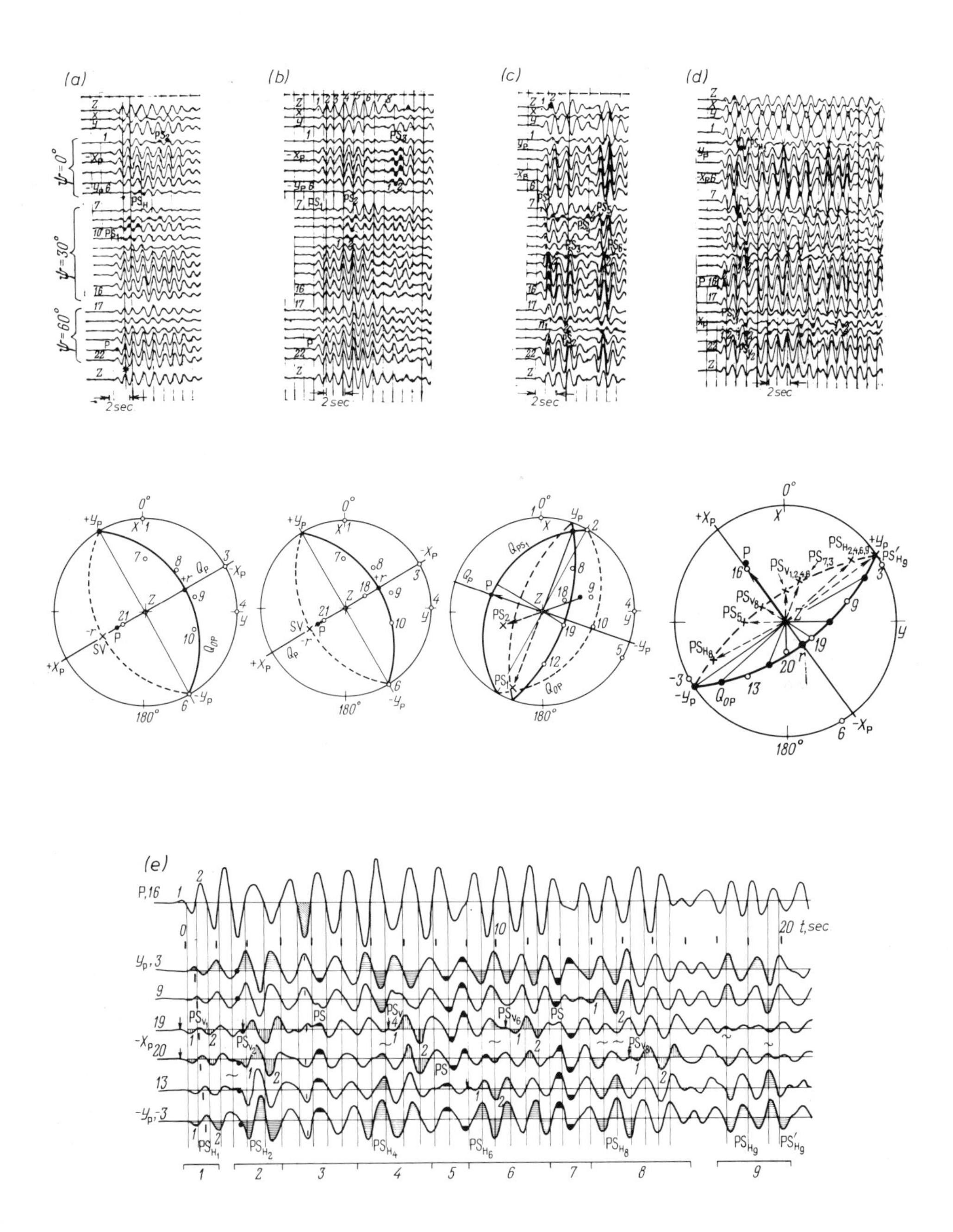

Fig. 47. Polar correlation of converted waves excited by distant earthquakes and recorded on 27 September 1974 by a 'Cherepakha' station (Central Kazakhstan). (a) T = 0556 hrs., point 124; (b) T = 0556 hrs., point 136; (c) T = 0531 hrs.; (d) T = 0531 hrs., point 111; (e) seismogram in the zero displacement plane Q_{0P}.

OBSERVATIONS IN CENTRAL KAZAKHSTAN

As applied to these observations, polar correlation was carried out as follows.[1]

Example 1. The direction of particle displacement in the P wave (Figure 47a) is such that $\psi = 36°$ and $\omega = 240°$, so that trace 21 is optimal with respect to its detection (see the stereogram). The line of intersection between the Q_P and Q_{0P} planes determines the radial direction r in the local coordinate system associated with the direction of particle displacement in the first P wave. Let X_P designate the horizontal component in the Q_P plane, and Y_P designate the component perpendicular to the Q_P plane. In this example, X_P coincides with trace 3 of the polar seismogram and Y_P with trace 6, that is, with components 3 and 6 taken in reverse polarity (to be defined later). The X_P and Y_P components of the local coordinate systems are marked on the respective traces of the polar seismogram.

Traces 8, 9, and 10, lying practically in the Q_{0P} plane, do not display the P wave. This can readily be ascertained by comparing traces 8, 9, and 10, on the one hand, and trace 21, on the other, at the instant of time when the P wave has a maximum amplitude. As is shown by the mark applied to the respective traces on the seismogram, the amplitudes of components 8, 9, and 10 are zero at that instant. As a consequence, even the extremal amplitude of the P wave does not distort traces 8, 9, and 10, and one can readily identify on them, in a most pure form, the PS_1 wave, which has about the same form as the P wave on trace 21 in reverse polarity.

Absence of marked phase shifts on the traces of components 8, 9, and 10 is a qualitative indication that the PS_1 wave is linearly polarized. The point is that the angle between components 8 and 10 is in excess of 60° and, if the PS_1 wave were non-linearly polarized, the traces of these components would show far greater phase shifts.

Let us compare projections of the PS_1 wave on to various components of the polar seismogram at times when the particle displacement in the P wave is zero. Projections comparable in terms of intensity are to be noted on components 3 and 9 close in direction to the Q_P plane. Thus, the PS_1 wave may be identified as an SV wave, although the direction of particle displacement in the PS_1 wave does not lie precisely in the Q_{0P} plane, but deflects towards the horizontal.

Depending on the problem in hand and the desired accuracy, the direction of particle displacement in PS waves may be determined qualitatively or quantitatively. Quantitative determinations are of particular interest, because they throw additional light on velocity relationships above the conversion boundaries and, therefore, enhance the accuracy with which their depth may be determined. In many cases, even a visual comparison of the PS wave in respect of amplitude on the component records produced by a polar array with $A_P = 0$, may bring out the deviation in azimuth of the direction of particle displacement in PS waves from the vertical ray plane (see Figure 47c).

In addition to components 8, 9, and 10, the Q_{0P} plane also contains component 6 (Y_P). As is seen from the seismogram, the Y_P component does not display the PS_1 wave, which again confirms that the PS_1 wave is an SV, that is, a vertically-polarized wave. Also, this helps to identify a PS_H wave on the Y_P trace against a virtually quiet

[1] After Frolova.

background. As already noted, the interference of *SH* and *SV* waves results in elliptical polarization with phase shifts of $\Delta\beta = 90°$ along the principal axes of the ellipse. The same phase shifts are to be seen on the seismogram in question: with $A_P = 0$, maximum projections of the PS_1 wave on to the X_P component coincide with zero projections of the PS_H wave on to the Y_P component, and vice versa. This also bears out the fact that the wave registered on the Y_P component is an *SH* wave. The elliptical polarization of the PS_1 and PS_H waves is corroborated by the inclined line-ups on the component records with $\psi = 0$ (see Figure 47a). Here, the PS_H wave experiences interference with both the PS_1 wave and the PS_2 wave arriving next. As a result, the line-ups show a changed inclination. The detection of converted waves in the late arrivals is based on a similar analysis of the amplitude and phase relationships between the components recorded on a polar seismogram.

Example 2 (Figure 47b). This is a polar seismogram of the same earthquake as in Example 1 recorded at point 136, at about 30 km from point 124. Here, the zero-displacement plane for the *P* wave Q_{0P} can readily be located from traces 6 and 10 showing a minimum amplitude. Such an approach is convenient in cases where the line-up pattern formed by the first phases of the *P* wave on a polar seismogram is curved, and the direction of particle displacement found from such a curve would be in error. The direction of particle displacement in the *P* wave ($\varphi = 40°$, $\alpha = 240°$) coincides with the direction of component 21, as at point 124 (see Figure 47a). Component 6 lies in, and components 7, 8, and 9 are near, the Q_{0P} plane. Without going into details of polar coorelation at this observation point, its basic results are as follows.

The record of components in the zero-displacement plane shows both the PS_1 and the PS_2 waves. The PS_1 wave is very weak and will therefore be omitted in the subsequent discussion. In the closing phases (components 6–8), the displacement of particles in the *P* wave is closer to vertical than in the leading phases, so the components that are optimal from the viewpoint of detecting the PS_3 wave are those with $\psi = 0$, and not components 10, 9, and 8, which were optimal for the detection of the PS_1 and PS_2 waves. The PS_2 and PS_3 waves are shorter than the *P* wave registered on trace 21 and are linearly polarized. Their polarization is such that projections of the *P* and *PS* waves on to any horizontal component take the same sign. There is no azimuthal deviation in the direction of particle displacement in the PS_2 and PS_3 waves, because their maximum projections at $A_P = 0$ are noted on components 3 and 9 lying close to the Q_P plane. Thus, the PS_2 and PS_3 waves are polarized as *SV* waves, but not in the Q_{0P} plane; their particles are displaced closer to the horizontal (projections of the *PS* waves on to components 3 and 9 are about the same in magnitude). The wave registered by the Y_P component is not an *SH* wave, because on the traces with $\psi = 0$ the line-up is vertical. If it were an *SH* wave, its interference with the *P* wave would produce in their plane of polarization an elliptical path, which would be projected on to the horizontal components and give an inclined line-up. In all probability, this is a projection of a direct wave for which the direction of particle displacement varies from phase to phase.

It is interesting to note that the PS_2 wave recorded on traces 8, 9, and 10 differs in terms of frequency from traces produced by the *XYZ* array and from any of the 20 other traces on the polar seismogram. This difference provides an important clue to the identification of the PS_2 wave among the strong low-frequency phases 4–6 of the

P wave registered simultaneously. The higher frequency of the PS_2 wave is an indication that, first, this wave is not related to the conversion of strong P-wave phases and, second, it is not a simple projection of phases 4–6 of the P wave, which dominate all the other traces. Obviously, the PS_2 wave could not have been identified and analysed on the basis of records produced by the three-component XYZ array alone.

Example 3. Let us consider the case (see Figure 47c) in which the direction of particle displacement in the PS waves deviates in azimuth, so that the interference of P and PS waves results in a non-vertical plane of polarization.

The direction of particle displacement in the P wave is such that $\varphi = 64°$ and $\omega = 268°$. The zero-displacement plane for the P wave Q_{0P} contains components 8, 12, and 19 on which the P wave has a minimum amplitude. The first two extrema of the P wave are registered as a linearly-polarized wave with phase inversions on traces 1, 2, 8, 12, and 19, because the early phases of the PS_1 wave are very weak and scarcely affect the linear path of the P wave.

As the later phases of the PS_1 wave gain in intensity, the vertical line-ups on the polar seismogram become curved, and the only indication that the plane of polarization for the P and PS_1 waves is close to the horizontal is the presence of an extremum line-up on the component records with $\psi = 60°$.

The interfering P and PS_1 waves show a minimum amplitude on component 19, which lies along the normal to the plane of polarization Q_{P+PS_1}. The PS_1 wave is registered optimally and with the same amplitude on traces 8 and 12, but in antiphase. On trace 19, the PS_1 wave is scarcely seen, which makes this component ideally suited for the detection of the PS_2 wave. The fact that the direction of particle displacement in the PS_1 wave lies in the Q_{P+PS} plane is an indication of an appreciable azimuthal deviation of the direction of particle displacement in the PS_1 wave from the Q_P plane.

Maximum projections of the PS_2 wave may be seen on components 8, 9, 10, and 18. The directions of particle displacement are shown in the stereogram also with azimuthal deviations from the Q_P plane.

The PS_3 wave is registered practically simultaneously with the PS_2 wave, but on traces 12 and 19 they appear well separated in time. The PS_4 wave in which the displacement of particles is perpendicular to the direction of component 19 is optimally registered on trace 8. Like the PS_2 and PS_3 waves, the PS_5 and PS_6 waves are separated in time on traces 19 and 12. On trace 8, all the waves appear least resolved. This example illustrates the capabilities of polar correlation and the advantages it offers over the traditional observations based on XYZ clusters.

Example 4 (Figure 47d). In the previous examples, the zero-displacement plane contained as few as two or three traces of a polar seismogram. A greater number of such traces would materially improve the capabilities and detail of analysis, and the resultant combined seismogram in the zero-displacement plane would enable a more rigorous determination of polarization for the interfering waves. Such a combined seismogram is shown in Figure 47e along with the usual polar seismogram and a stereogram (Figure 47d).

Component 16 on the combined seismogram (see Figure 47e) is in a direction very close to that of the particle displacement in the first P wave. Vertical lines have been drawn for instants of times $A_P = 0$ when the amplitudes on the traces of the combined seismogram cannot be projections of P waves. Within 2 s of the first arrivals, the combined seismogram shows only one interval 0.5 s long where the traces of all components are in-phase with that of the P wave (the shaded portion on the top trace) and may be the projection of a compressional wave. In all the other cases, the times, $A_P = 0$ correspond to extremal amplitudes on the traces of the X_P and Y_P components, which are not projections of a P wave.

According to the form of polarization, the following intervals may be singled out on the combined seismogram (marked from 1–9 in Figure 47e).

Interval 1. The arrival of a weak PS_1 wave may be seen on traces 19 and 20; its polarity is well defined (it is in antiphase with the P wave), as is its move-out time Δt relative to the P wave. There is also a PS_{H_1} wave, registered with a delay of less than $T/4$ (optimally on the Y_P component). On interfering, they form an elliptically-polarized resultant wave with an inclined line-up.

Interval 2. The strong PS_{H_2} wave leads the PS_{V_2} wave, so that the first phase of the PS_{H_2} wave is linearly polarized. Linear polarization gives way to elliptical polarization just as the PS_{V_2} wave arrives.

Interval 3. Only an interference-produced record can be seen, with no distinct individual waves.

Intervals 4, 6, and 8. The PS_{V_i} and PS_{H_i} waves arrive with a time shift and interfere with one another. The PS_{V_i} wave lags increasingly behind with increase in recording time. All the detected waves are not precisely polarized as SV or SH waves. There are azimuthal deviations of particle displacements from the Q_P plane for the dominant phases. The directions of particle displacement are shown on the stereogram (see Figure 47d).

Intervals 5 and 7. Here, the PS waves are linearly polarized. So far as the directions of particle displacement are concerned, they cannot be identified with either SV or SH waves.

Interval 9. The record shows linearly-polarized PS_{H_g} waves for which the azimuthal deviations of particle displacement are the same as for $PS_{H_{2,4,6}}$ waves, and the PS_{H_9} wave is precisely polarized in the Y_P direction.

Referring to the combined seismogram, the shape of the recorded PS_{V_i} waves deserves special mention. As is seen from Figure 47e, they are short, mainly three-phase waves with a stable dynamic relationship from phase to phase. The PS_H waves are not so well resolved on the Y_P component, probably because the Y_P component (3) does not lie precisely in the Q_{0P} plane.

The components that are optimal with respect to the detection of PS waves do not usually lie along the directions of particle displacement in the respective waves. For example, in the $PS_{V_{2,4,6}}$ waves the directions of particle displacement lie along component 20, whereas the optimal component so far as their detection is concerned is component 19, owing to the fact that it is at right angles to the particle displacement in the $PS_{H_{2,4,6}}$ waves, which precede the $PS_{V_{2,4,6}}$ waves whose projections are registered on component 20.

The principle by which a combined seismogram is obtained in the zero-displacement plane for a P wave underlies the design and operation of the high-selectivity CDR–I analyser.

OBSERVATIONS IN THE ZAILIISK ALATAU

Consider the polar seismograms of earthquakes registered by a 'Zemlya' station. Figure 48 shows a polar seismogram for a local earthquake with $\Delta t_{S-P} = 1.65$ s. The first P wave is almost linearly polarized. The slight departure from linearity appears as the curved line-up on the traces with $\psi = 0°$ and $\psi = 30°$. The P wave is seen to have a maximum amplitude on traces 19 and 20, and a zero amplitude on trace 3. From comparison of the amplitudes on the various traces it may be concluded that the direction of particle displacement in the first P wave is such that $\varphi = 30°$ and $\omega = 150°$. The Q_P plane of the P wave is shown in Figure 48.

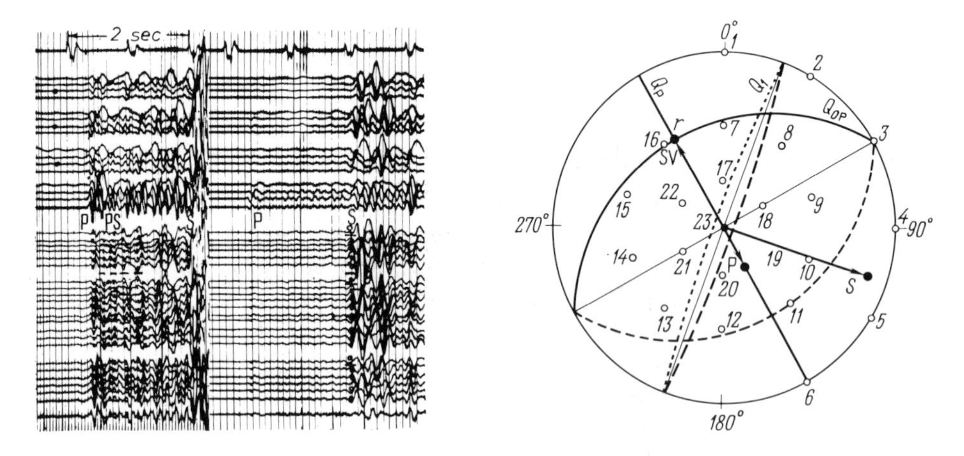

Fig. 48. Polar correlation of converted waves excited by a local earthquake recorded by a 'Zemlya' station in the Zailiisk Alatau at 0506 hrs.

The Q_{0P} plane of the compressional wave, which is at right angles to the direction of particle displacement in that wave, contains a radial component r, which lies along the intersection of the Q_P and Q_{0P} planes. The direction of r in space is such that $\varphi = 60°$ and $\omega = 330°$ ($\pm 180°$) and is very close to that of trace 16 on the polar seismogram (see Figure 48). Thus, the latter component is optimal with respect to the detection of PS waves. Several PS waves can be seen the first of which is registered with a move-out time of 0.15 s relative to the P wave. In all the converted waves, the particles are dispalceed in nearly the same direction, namely, along r, as is borne out by the fact that the zero-displacement planes coincide and these planes are at right angles to the direction of the P wave.

Because the P wave is obliquely incident on the observation plane, the compressional and converted waves are registered on both the vertical and the horizontal components. As a consequence, it is difficult to detect the converted waves on the record produced by an XYZ cluster (traces 1, 4, and 23). This can, however, be done with certainty, using traces 19 and 20 for the *P* wave and trace 16 for the *SV* wave. Yet, the interpretation of waves will be ambiguous, unless all of the components of the polar seismogram are analysed.

The direct *S* wave stands out clearly owing to its high intensity. For comparison, we may refer to the record made at a lower gain on the right of Figure 48. The first phases of the *S* wave are linearly polarized whereas the later more intensive phases are non-linearly polarized. In the first phases, the displacement of particles is closer to the horizontal. Because the amplitude is maximum on traces 4 and 5, the direction of particle displacement appears normal to the zero-displacement plane, Q_1. The latter is identified from the phase reversals between components 1 and 2, 7 and 8, 12 and 13, 17 and 18, and 20 and 21. The direction of particle displacement in the *S* wave is such that $\omega = 180°$ ($\pm 180°$) and $\varphi = 84°$, so that it makes an angle of 42° with the ray plane of the first compressional wave.

USE OF THE SELECTIVE CDR–I ANALYSER

In the above examples, the polar correlation of transmitted converted waves was done on the basis of outline polar seismograms. On the other hand, converted waves can be far more effectively detected and analysed in the zero-displacement plane for the *P* wave, because of the presence of strong dominating compressional waves, and this requires that the directional pattern of the CDR–I analyser be carefully oriented, since otherwise the projections of the strong *P* waves would corrupt records of *PS* waves. This analyser is described in Chapter 2, and here we shall only illustrate the application of polar correlation to converted waves in the zero-displacement plane of *P* waves.

Figure 49a shows a combined seismogram, which incorporates the components lying in the zero-displacement plane of the P_2 wave. The orientation of the components is shown in the stereogram. Traces 2–4 display a t_1 wave with a vertical line-up and a progressively diminishing amplitude. Traces 5–7 show a curved line-up. Traces 8–11 scarcely show the t_1 wave, and this wave re-appears only on traces 15 and 16, but with a phase reversal. On the other hand, traces 10–12 perpendicular to the t_1 wave and close in direction to component 9 on the combined seismogram display a t_2 wave different from the t_1 wave in shape and phase shift. A similar picture is to be observed in the zero-displacement plane for the P_3 wave in which the directions of particle displacement differ from those in the P_2 wave, which may clearly be seen on the stereogram (Figure 49b). The many components lying in the zero-displacement plane serve to detect two linearly-polarized waves in two mutually-perpendicular directions, namely, 11, 12, and 13, and 1, 2, 22, and 23 shifted from each other in time.

Experience with the selective CDR–I analyser has proved its effectiveness and has shown that transmitted converted waves may best be analysed in the zero-displacement planes of *P* waves.

SOME CHARACTERISTICS OF *PS* WAVES

Polar correlation has brought out more precisely some of the characteristics of *PS* waves, such as record form, arrival polarity of converted waves, polarization, and azimuthal deviations. Consider these matters in more detail.

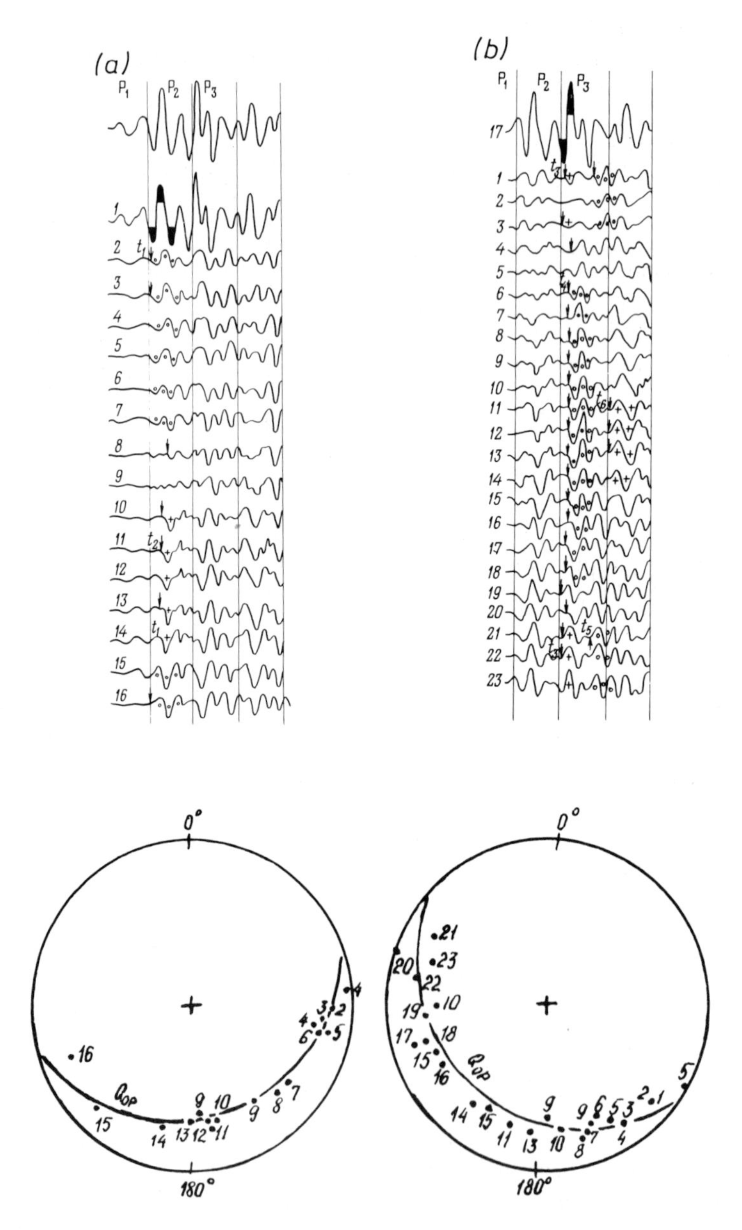

Fig. 49. Use of a CDR–I selective polar analyser. Polar correlation of *PS* waves in the zero displacement plane for (a) P_2 waves and (b) for P_3 waves.

The record form of *PS* waves is one of the basic factors used in their identification. Because the initial part of a seismogram will usually show a complex extended train of *P* waves, converted waves can be reliably detected only if their recorded form is precisely known. Many authors [18, 19, 100, 101] maintain that *PS* waves are analogous in form to *P* waves. Yet, analysis of converted-wave records in the zero-displacement plane of the P wave where the P wave does not corrupt the PS wave-record has shown that in many cases PS waves are simple in shape. This is true of both the first converted wave PS_1 and the later converted waves. Referring to the seismogram of Figure 50, the first

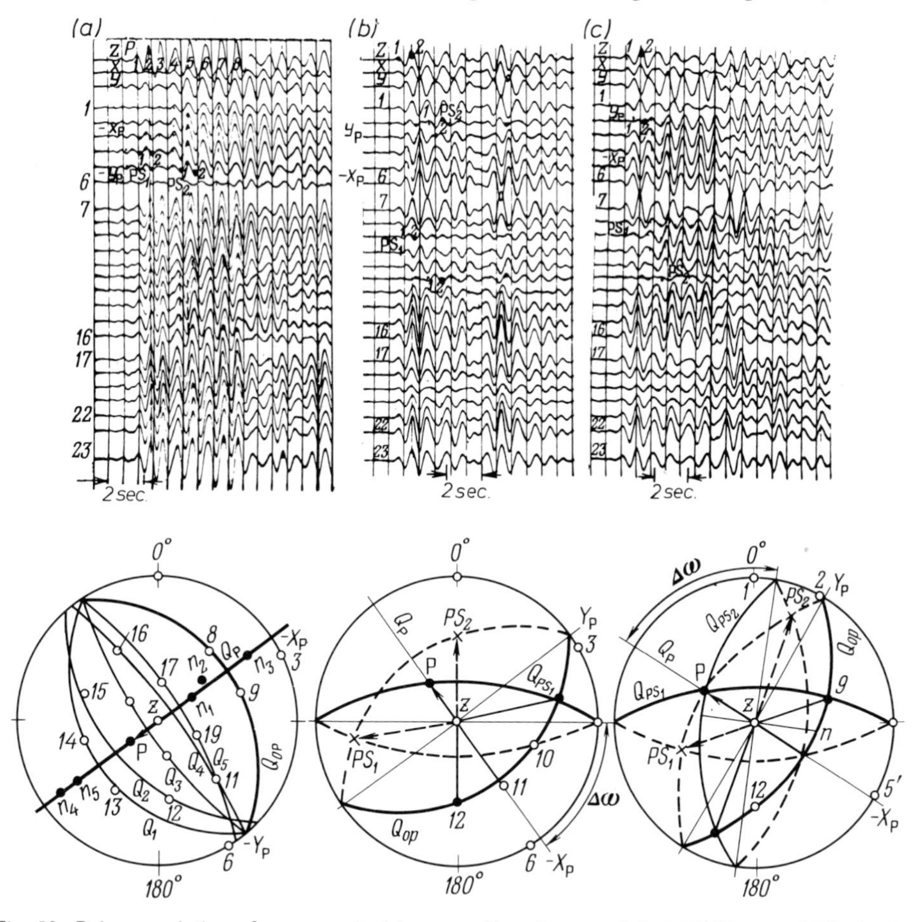

Fig. 50. Polar correlation of waves excited by an earthquake recorded at 0556 hrs. on 27 September 1974: (a) point 114 (vertical plane of polarization); (b, c) points 122 and 125 (inclined plane of polarization).

converted wave PS_1 of the SV type can clearly be detected on traces with $\psi = 0$ in reverse polarity with respect to the P wave recorded on the Z component. However, only three phases of the P and PS_1 waves can be identified, because already the fourth phase of the P wave is practically nonexistent in the PS_1 wave. On the other hand, none of the six traces with $\psi = 0$ show strong phases of the P wave. These traces would optimally register the converted wave, if it were more extended. Thus, it may be concluded that the PS_1 wave is half as long as the P wave.

On the seismogram in Figure 47a, the markedly shorter PS waves are detected on traces 1 and 6, although the later record has a minimal amplitude, and nothing would prevent recording an extended converted wave similar to the P wave, if the entire group of compressional waves had taken part in the conversion.

On the seismogram of Figure 47b, the traces with $\psi = 0$ show the PS_2 wave interfering with the P wave. In a purer form it may be seen on trace 10. Within the interval between phases 7 and 8, the intensity of the recorded wave on traces with $\psi = 0$ rapidly decreases almost to zero, which would not occur if the PS_2 wave replicated the entire train of P waves. The PS_3 wave is likewise much shorter, although its final phases fall within the time interval where the P wave is no longer registered. It is to be noted that the PS_2 and PS_3 waves are identical in form.

To sum up, the records examined above and many other similar data seem to suggest that only the first wave out of a group of P waves takes part in the conversion, and that the PS waves detected in the succeeding part of the seismogram may have been formed on different boundaries.

The arrival polarity of PS waves is essential in determining their move-out times relative to P waves. Any error in move-out time determination may give rise to serious errors in subsequent constructions. One of the criteria for the identification of PS waves on earthquake seismograms is that the polarity of P and PS waves on any horizontal component (X or Y) must be the same. Because polar correlation involves comparison of 23 components, including the vertical and horizontal ones, we shall examine the polarity of PS waves on the various traces of a polar seismogram. To this end, we determine the directions of particle displacement for the P wave induced in the earthquakes registered by 26 stations disposed along the profile. In Figure 51a these directions are plotted on rectangular coordinates. Then we plot the directions of displacement for the P wave on a stereogram (Figure 51a) and construct the vertical ray planes Q_P and the

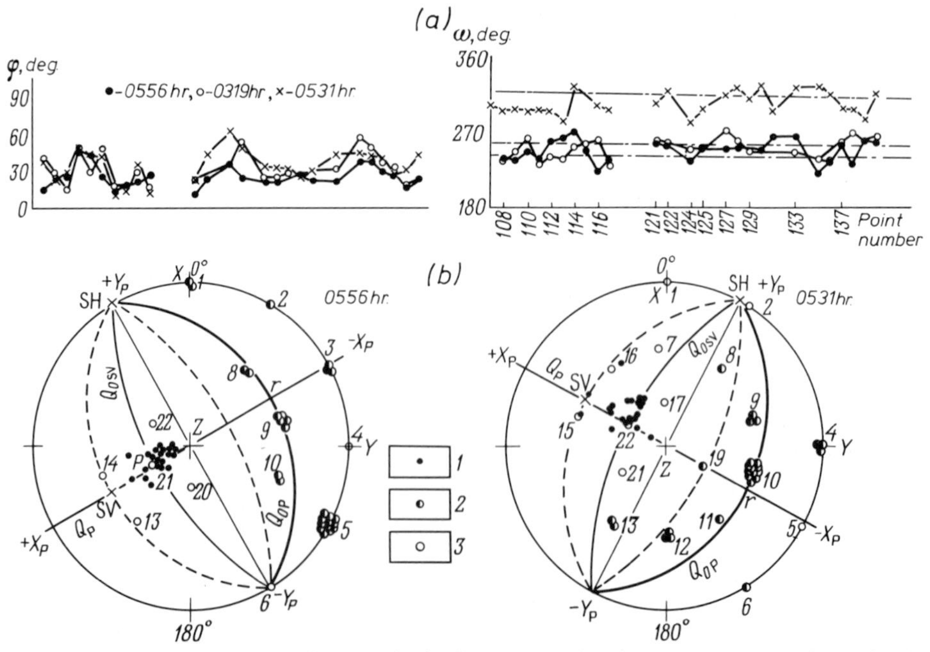

Fig. 51. Direction of particle displacement in the first compressional wave in (a) rectangular projection and (b) stereographic projection. (1) – direction of particle displacement in a P wave; (2) – direction of particle displacement in the tracking components; (3) – direction of particle displacement in the fixed components.

zero-displacement planes Q_{0P} for the P wave. The directions of displacement for the SV wave lie in the lower half-space and run along the intersection between the Q_{0P} and Q_P planes. For the SH wave, they lie in a plane tangent to the Q_{0P} front and perpendicular to the Q_P plane. This construction carried out for each earthquake and for each reception point helps in comparing the polarity of the P and PS waves. The polarity of the P and SV waves on any horizontal component will always be the same, whereas the P and SH waves may be in phase and in antiphase. It is to be recalled that compressional waves are best registered on P components, and PS waves – on the components in the zero-displacement plane for the P wave. Thus, for the earthquake registered at 0556 hrs. (Figure 51b), these are components 21 and 8, 9. Therefore when identifying P and SV waves on records, it should be borne in mind that they are in antiphase. Similarly, SH waves on component 6 and P waves on component 21 (or 23) will be in reverse polarity with respect to each other, whereas on component 21 (Z) and component 1, the P and SH waves will be in the same polarity. Change of sign occurs in the Q_{0P} plane in the case of P waves, in the Q_{0SV} plane – in the case of SV waves, and in the Q_P plane – in the case of SH waves. In the subsequent discussion of PS waves, the polarity answering the adopted criteria will be referred to as normal. All converted waves whose records permit reliable determination of arrival polarity show it to be normal. Therefore, for the PS waves whose polarities are difficult to ascertain and appear ambiguous, the displacement azimuth ($\pm 180°$) is taken on the assumption of normal polarity.

The polarization of PS waves is of interest, because it has a direct bearing on such important characteristics as the velocity anisotropy and the stressed state of the material. At present, there is no consensus on the matter of polarization, nor is there a complete clarity about the factors determining the type of polarization. Difficulties in investigating the polarization of converted waves mainly arise from the fact that PS waves are, as a rule, registered under conditions of continuous interference with compressional waves. This is the reason why the polarization of PS waves proper can be judged solely from records of components lying in the zero-displacement plane for the P wave. In most cases no more than three components have been found to occur in the Q_{0P} plane, which is not sufficient for the form of polarization to be ascertained rigorously. At some points where the Q_{0P} plane is close to the horizontal, it is possible to utilize as many as six horizontal components for which $\psi = 0$ in determining the form of polarization of the recorded PS waves. In such cases, the records usually show vertical line-up patterns displaying good reproducibility for PS waves. The amplitudes decay cosinusoidally, which is an indication that the PS waves are linearly polarized. It is to be stressed that the absence of phase shifts on the records produced by the horizontal members of a three-component cluster is not sufficient evidence of linear polarization.

PS_H waves registered outside the range of interference with P and PS_V waves are usually linearly polarized. Within $\Delta t \approx 10$ s of the first arrivals, pure PS_H waves can be noted only at some points. More frequently, PS_H waves interfere with PS_V waves to form elliptically-polarized resultant waves. In such cases, the Q_{0P} plane displays well-defined inclined line-ups (see Figure 47a).

Except for the first extremum, all other phases of the P wave are seen to interfere continuously with other waves. In the general case the waves forming within the interference range are both plane- and spatially-polarized. It is not unlikely that the interference of regular stable waves produces waves with stable polarization characteristics.

For example, on most seismograms of the earthquakes registered at 0556 hrs. and 0536 hrs., the stability of plane polarization for the interfering waves manifests itself in the minimum amplitudes on one of the horizontal components of the polar seismogram. In the former case, this is an indication of polarization in a vertical plane, which on some seismograms scarcely alters its orientation over considerable time intervals.

Figure 50a shows a polar seismogram on which the lowest-amplitude component stands midway between traces 1 and 6. The later arrivals of strong compressional waves change the orientation or tilt of the plane of polarization very little, if at all, because their amplitudes on trace 6 are not more than 10% of those recorded for the strongest phases of the P waves. Thus, the P waves remain polarized in a vertical or nearly vertical plane over the entire range of interference with PS waves.

The horizontal component with a minimum-amplitude record usually corresponds to the Y_P component of the seismogram in a local coordinate system referenced to the direction of arrival for the first P wave. As a rule, a minimum amplitude is observed on the Y_P component only within certain time intervals, because the Y_P component registers SH waves, laterial waves, and SV waves with azimuthal deviations. For example, on the seismogram of Figure 47a, the Y_P component has zero amplitude (trace 6) before and after a PS_H wave is registered.

The plane of polarization is inclined, if any inclined component (see Figure 50b, c) shows a minimum amplitude over a more or less extended time interval. In any case, a component showing a minimum amplitude is normal to the plane of polarization.

Azimuthal Deviations. In homogeneous or axial-symmetrical media, the direction of particle displacement in PS waves lie in a vertical plane aligned with the source, which is another way of saying that there are no azimuthal deviations in such cases. Here, azimuthal deviations are defined as the angle between the direction of particle displacement in PS waves and the vertical plane.

Polar correlation shows that the converted waves registered at one point may show azimuthal deviations of different signs. This may be traceable to the difference in the orientation of the conversion boundaries. Detection of such deviations calls for a specially careful quantiative analysis for which a well-developed technique is still lacking. As an example, we shall examine the results of a quantiative evaluation of azimuthal deviations for two recorded converted waves PS_1 and PS_2 (Figure 51a).

On the seismograms of the earthquake registered at 0556 hrs., the PS_1 and PS_2 waves are polarized in a vertical or a nearly vertical plane along the entire profile, and azimuthal deviations in the displacement of particles are practically nonexistent, except for point 127. The orientation of the planes of polarization changes but little as manifested by the directional stability of the Y_P component; on the polar seismogram of Figure 50a, these are mainly components 1 and 6. As is seen from the plot of Figure 51a, the P wave also shows no azimuthal deviations.

In the case of the earthquake registered at 0531 hrs., the PS_1 and PS_2 waves show azimuthal deviations over the entire profile except at points 130, 131, and 138. Examples of seismograms for PS_1 and PS_2 waves and stereograms are shown in Figure 52. The azimuthal deviations of converted waves are assessed in terms of the angle $\Delta\omega$ (the difference in azimuth between the normals to the vertical plane aligned with the source and the plane of polarization for the $P + PS$ wave). The maximum values of $\Delta\omega$ are

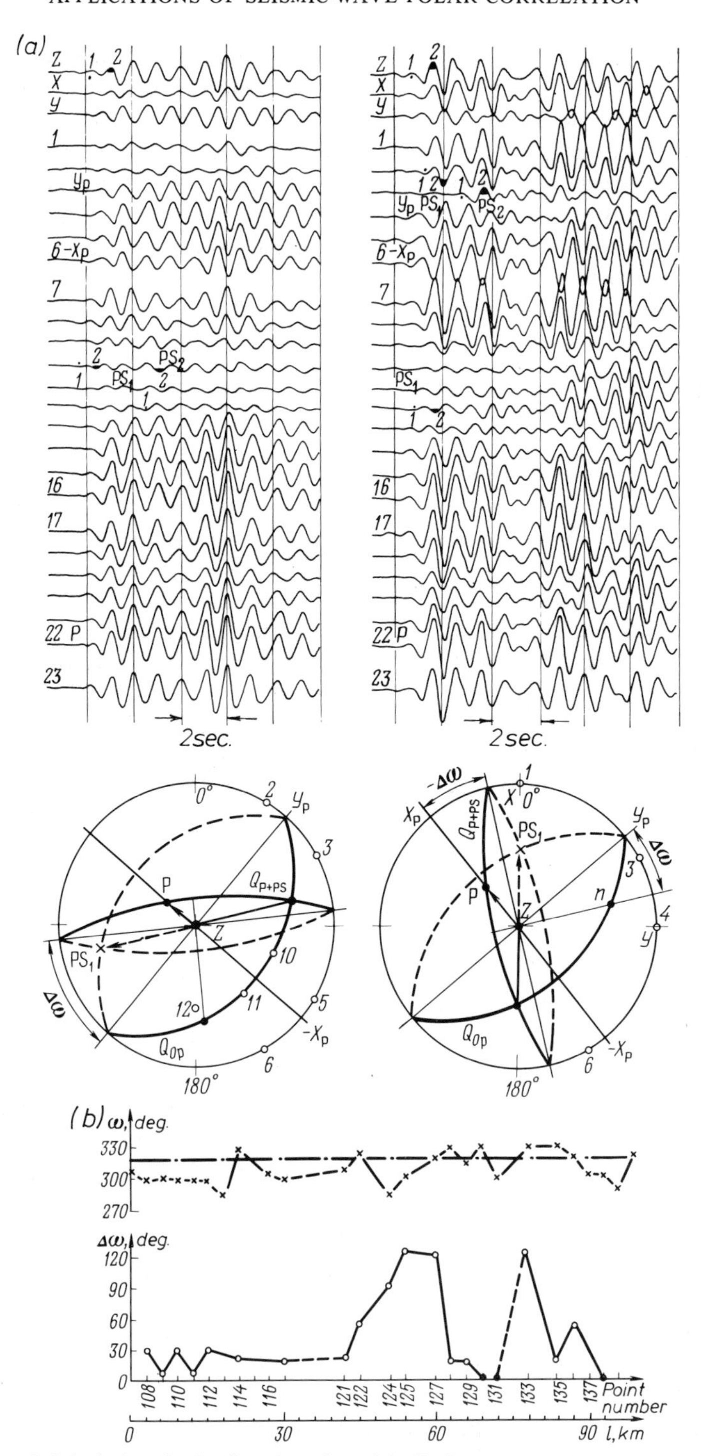

Fig. 52. Azimuthal deviations in the direction of particle displacement in the converted waves excited by the earthquake recorded at 0531 hrs on 27 September 1974.

observed in the vicinity of points 127, 124, and 133 where the waves are registered optimally on the Y_P components (Figure 52a). Near point 121, the angle $\Delta\omega$ descreases and attains a constant value.

In the case of the earthquake registered at 0531 hrs. (see Figure 51a), the same intervals show azimuthal deviations in particle displacement for the first *P* wave. Obviously, the azimuthal deviations in displacements for the *P* and *PS* waves may well be traced to the existence of inclined boundaries in the section. The high values of $\Delta\omega$ (90–120°) on the right of the plot in the absence of azimuthal deviations for compressional waves still await a special explanation.

Use of Polar Correlation in Studies Concerned with the Effect of Observation Conditions on the Shape of Wave Records

The differences in shape between converted and compressional waves disclosed by polar correlation and, notably, the fact that a *PS* wave can well be simple and short, although the associated *P* wave is complex and multiphase, have spurred the interest in the factors that control the wave shape and the structure of the initial portion of records for distant earthquakes. Knowledge of such factors is important because the difficulties in interpreting the observations obtained by the transmitted converted-wave method arise at present from the complex nature of the wave fields forming the initial portion of seismograms and lack of understanding of how these fields are formed. A wealth of VSP data seem to indicate that the wave field is strongly affected by sharply-defined boundaries located in the top part of the section. The ground surface is defined best of all. Observations at considerable depths have made it possible to investigate the ground surface as a reflecting boundary and to determine its effect on the wavefiled. It has been found that at practically any depth within the investigator's reach the bulk of the energy is returned to the medium upon reflection from the ground surface. In the case of a horizontal terrain, this effect shows in the excitation of a large number of incident waves being propagated downward into the medium at about the same velocity. Should, however, the terrain have topographical discontinuities, the wave field would be deformed, the rays would be focused and defocused, etc. It is only natural to assume that the terrain effect would be as significant in surface observations. To verify the idea, the techniques and capabilities of polar correlation have been used. Experiments have been carried out on the northern slopes of the Zailiisk Alatau. Owing to the sudden change from mountains to plains, the stations can be set up short distances apart, and records for the same earthquakes can conveniently be compared (Figure 53).

As a rule, the records produced by the stations positioned on the plain (Figure 53a) are relatively simple. The strongest events in the initial part of the seismograms are compressional waves in which the particle displacements are close to vertical. The components lying in the zero-displacement plane of the compressional wave display converted waves. For example, each strong *P* wave in Figure 53a is followed at a move-out time of $\Delta t = 1$ s by *PS* waves correlated with the top of the Palaeozoic basement.

The data collected over an extended period of time at a station located about 3 km above sea level (Figure 53b) show that the initial part of the seismogram is a complex wave field formed by a large number of waves. As a rule, the first relatively weak *P* wave in which the particle displacement is in a nearly vertical plane is followed by a train of

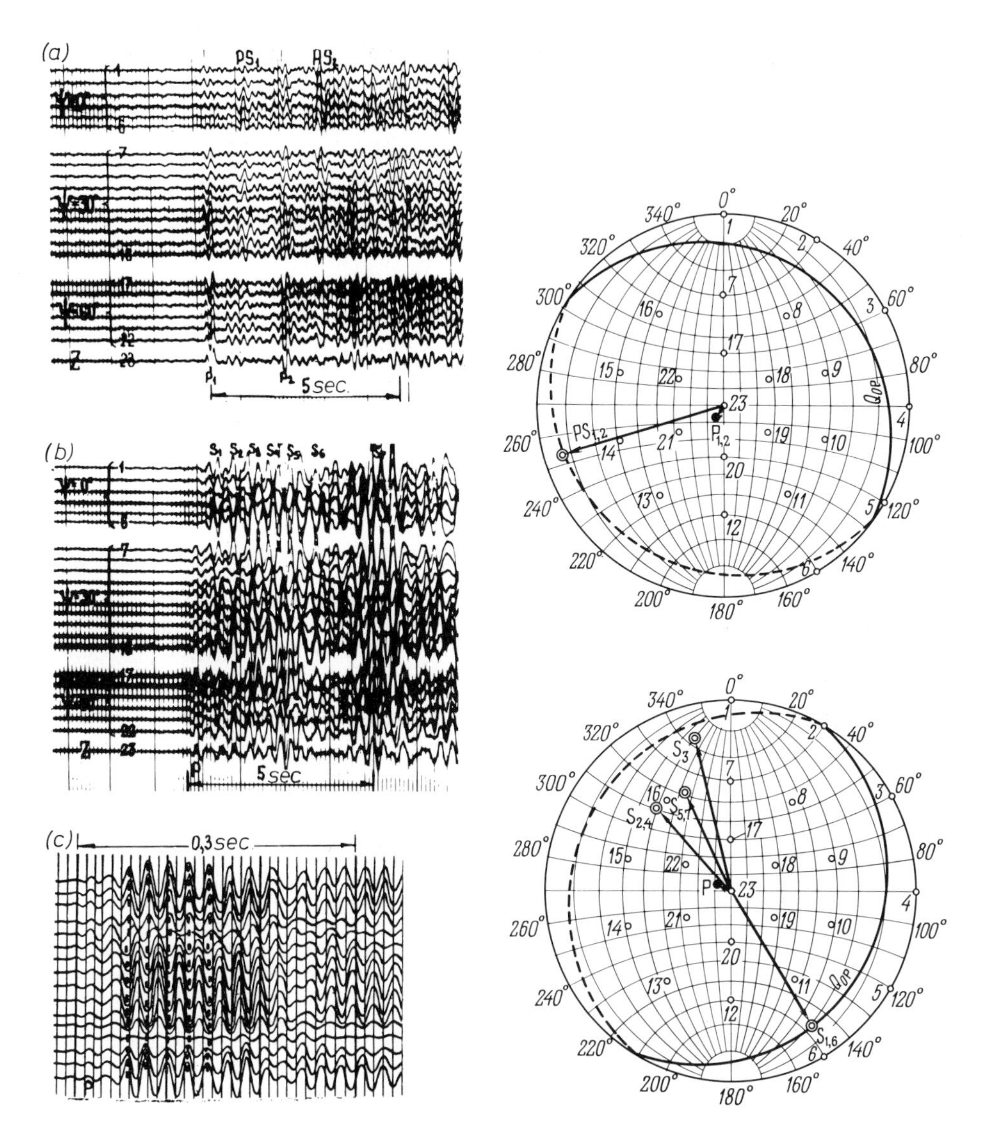

Fig. 53. Effect of observation conditions on the structure of the initial part of the record of an earthquake registered at 1139 hrs. on 1 September 1972 by (a) a station in the plains; (b) a station in the mountains; and (c) the shot-point.

substantially stronger waves polarized in various directions and not only in the plane of the compressional wave front (S_1 and S_2), but also away from it ($S_{2,4}$ and $S_{5,7}$). As special experiments have shown, these waves are not related to any particular recording technique and are characteristic of records produced by a station located in the mountains.

In all probability, the complex wave field in the initial part of the siesmogram recorded by stations in the mountains is related to the surface irregularities rather than any subsurface structure. In addition to reflection and refraction, this effect may, as investigations have shown [56], be due to the interaction between the incident wave and the terrain irregularities, causing some of the energy carried by the compressional body wave to be converted into that of surface waves. In the circumstances, converted waves cannot, as a rule, be detected.

To get better insight into the effect of surface terrain on the shape of the *P*-wave record and the structure of the initial part of seismograms, a series of observations were carried out under the conditions of a sudden transition from mountain to plain. Observations were made at 10 points of a meridional profile with a total length of about 25 km, with its southern end terminating in a typically mountainous locality on the northern slope of the Zailiisk Alatau mountains and with its northern end lying in the Ili depression, which is a typical plain.

Let us consider the polarization of waves in the initial part of a seismogram, taking as an example the earthquake registered at 2348 hrs. on 13 February 1974 (Figure 54b). In this seismogram traces 1 and 2 apply to the mountains, trace 3 applies to the foothill area featuring the outcrops of bedrock; and trace 4 applies to the plain with a

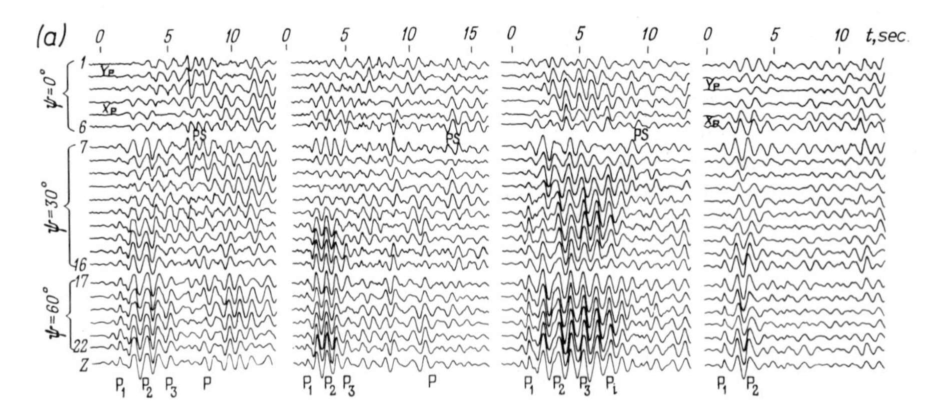

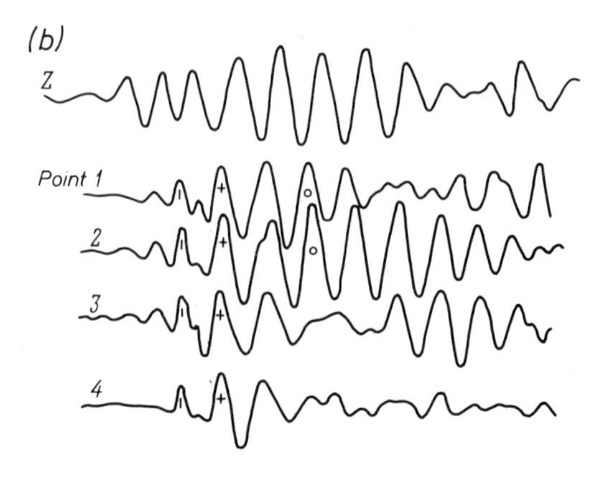

Fig. 54. Effect of observation conditions on the shape of the record for the first arrivals. (a) Polar seismograms of the earthquake recorded at 2348 hrs. on 13 February 1974; (b) combined vertical component seismogram.

thick sequence of sediments. The first two compressional waves P_1 and P_2 are seen to produce records of about the same shape at all the stations, except that the P_1 wave shows a substantially lower intensity and a larger proportion of high-frequency components. The records of the later compressional waves, however, drastically differ from station to station. The P_3 wave prominent on the records produced by the moutain stations (traces 1 and 2) is conspicuously absent in the record produced by the station on the plain (trace 4). Because of this, the records for the first train of compressional waves obtained by the mountain and foothill stations strongly differ from those produced by the stations on the plain. In the latter case, the record is relatively short and simple; in the former case it is complicated by additional phases of the P_3 wave.

A similar difference may be noted between the waves in the remaining part of the record. Whereas the mountain stations recorded several more compressional waves comparable in strength with the first wave and polarized in a plane other than the vertical ray plane passing through the direction of particle displacement for the first P wave, the station on the plain scarcely recorded such waves. For ease of comparison, Figure 54b shows seismograms for the vertical components.

The difference is more conspicuous in the polarization of the later waves recorded by the horizontal members of the array ($\psi = 0$). In the mountains a complex polarization is predominant, in various inclined planes not related to the direction of wave motion. On the plain the waves are mainly polarized in the vertical ray plane.

It is to be noted that the records produced at the bedrock outcrops in Central Kazakhstan likewise differ from those with about the same degree of complication in the seismogram obtained for sedimentary rocks. For comparison, Figure 54b shows a Z-component record of an earthquake registered in Central Kazakhstan. It is still unclear how and why the records grow in complexity. One possibility is the interaction between the incident waves and the ground surface, or it may be due to the conversion of body waves into secondary surface waves.

It is not unlikely that, apart from surface terrain, discontinuities in the top part of the section may also be responsible for the increased complexity of seismograms, especially where sruface terrain is gentle. No observations under such conditions have yet been made in the earthquake frequency range. In default of other data, we offer a polar seismogram recorded in the Volga valley near Saratov (Figure 53c) in the seismic-prospecting frequency range. In structure, the seismogram scarcely differs from those recorded in the mountains. The relatively weak first P wave is followed by an extended and poorly resolved train of waves polarized in a horizontal plane.

We have dealt with the matter at length because if it is true that the initial part of seismograms, in addition to conversion, is affected by the nature of the ground surface and the top part of the section, this must be taken into consideration when interpreting data produced by the transmitted converted wave method. The differences in form between records of compressional and converted waves described above may well be related to these factors.

CHAPTER 5

POLARIZATION-POSITION CORRELATION (PPC) OF SEISMIC WAVES

Polarization-position correlation (PPC) is based on the simultaneous use of parameters defining the spatial behaviour of the wave field; namely, the direction of particle displacement and the direction of wave travel. Thus, it involves the discrimination of waves at a point in terms of wave polarization and in a volume or plane in terms of the apparent velocity, v_a. The capabilities of either technique are utilized jointly and completely. The first step in this method is to detect what we call the tracking component for each wave; the second step is to follow (establish continuity for) this component along the line or over the plane of traverse. Thus, like polar correlation, PPC is an extension of the polarization principles widely used in all forms of seismic prospecting. The discrimination of waves at a point is opposed to the traditional forms of discrimination. Rather, it is intended to be used together with them in order to enhance the efficiency of wave-field analysis. Already, early experience with the method has shown this combination to be promising and fruitful.

An Outline of PPC

The peculiarities of PPC arise from the fact that correlation is executed both at a point and in space.

POLAR AND POSITION CORRELATION COMBINED

The manner in which the two forms of correlation are combined is varied to suit a particular wave field and the objects of study. Where the wave field is relatively simple and the waves are easy to correlate by a technique based on only one type of waves (say, reflected *P* waves), position correlation is predominant, and polar correlation only supplements it, mainly by giving a general idea about the wave field and in analysing interference zones. This form of combination was used in seismic prospecting during the first period [24]. Basically, polar correlation has been used solely at certain points in the profile spaced a sufficiently large distance apart. In this way it has yielded additional data about the variously-polarized dominant waves not recorded on the position seismograms of the traditional fixed component, and has in some cases pin-pointed the cause of failure in position correlation. However, progress in the use of polar-position techniques has been hampered by the fact that during the first stage polar seismograms were obtained by multicomponent azimuthal clusters, and a large number of channels had to be used in observations at a point.

In contrast, polar correlation is predominant in cases where observations are carried out at isolated points not tied into a continuous profile. This is true, for example, of earthquake observations, notably weak local earthquakes. When used during such observations, polar correlation helped to locate seismic activity areas in Southwestern Turkmenia [57] and study the epicentral area of the Khait earthquake in the Pamirs [56].

Having been proposed and applied to earthquake studies for the first time by Gamburtsev, the combination of polar and position correlation has led to the development of the combined earthquake study method (CESM) [53, 54]. It is to this method that we owe a good deal of progress in seismology in recent years.

In recent years, combined polar and position correlation has undergone further extension and development, in a large part spurred on by such practical tasks as the study of complex media and the simultaneous use of various wave types in order to identify the composition of geologic sections. As a result, a unified method of polarization-position correlation has come into being. It has already been used in surface and drill-hole studies in the especially complex ore fields of Central Kazakhstan and in oil fields with intensely deformed strata (the southern side of the West Kuban's trough) and extensive diapiric folds (the Taman area). These studies have borne out the efficiency of PPC as applied to complex wave fields and have served to accumulate practical experience in the use of the method.

PRINCIPAL FEATURES

Polarization-position correlation owes its effectiveness to the fact that instead of the usual vertical component, we keep watch on what we call the tracking (or optimal) component. As already defined, the tracking component is that spatial component that shows a maximum signal-to-noise ratio for a given wave. As with other forms of discrimination, the tracking component of interest may be selected by a polarization-filtering system varying in selectivity according to the task on hand and the complexity of the wave field to be analysed.

For relatively simple wave fields, it may in many cases be sufficient to follow components having predetermined polarization characteristics. In more complex cases, PPC utilizes optimal polarization characteristics for which the signal-to-noise ratio is a maximum over their range of values and which are determined directly from the wave field. In such a case recourse is made to self-turning polarization filters. A special case of this form of filtering in which the parameter sought is the direction of particle displacement is CDR–I.

In the above two forms of filtering, we detect and discriminate between signals. As an alternative, filtering with a view to determining the polarization characteristic of the valid signal may be based on interference-produced waves, that is, waves formed by the superposition of interfering waves on the valid signal. In this case, the valid signal is detected against the background of noise.

In each form of filtering, we may use processing levels differing in detail. For example, if the tracking components are detected by CDR–I, the first processing level may utilize the components fixed in space over a certain correlation interval, with the length of the correlation interval decreasing as the wave field increases in complexity. In the case of very complex wave fields, it may be necessary to select tracking components in the vicinity of each observation point. In PPC, as in other form of discrimination, resort to the more selective systems will be warranted only where less selective systems have failed.

PPC, involving analog data, is carried out mainly by the CDR–I method which can readily be implemented in analog form and is more comprehensible to the observer.

Traking Component. In the general case, the spatial orientation of tracking components of a wave depends on two factors, namely, the manner in which the direction of the wave vector varies along the line of traverse and the superposition of spurious waves at each observation point. The part played by these factors and their influence on the effectiveness of PPC vary from observation to observation. The direction of the wave vector at each point depends on the direction of wave travel, which is in turn dependent on the relative position of the source and reception point and the structure of the medium. In moving from one point of observation to another, the direction of the wave vector varies in a regular fashion and, as a rule, gradually. In contrast, the conditions under which waves interfere at each point are in the general case random and may even differ from point to point. So, although the direction of arrival or, rather, the direction of the wave vector, varies little and gradually along the line of traverse, the directions of the tracking components may vary to a more considerable degree. Indeed, as the angle that the direction of particle displacement in a given valid wave makes with that of the spurious wave decreases or, in other words, as the interference ellipsoid becomes more extended in space, the tracking component will in the general case, deviate more and more from the direction of the wave vector – the spurious wave will 'drag' the tracking component increasingly more away from the wave vector. For simple linearly-polarized waves registered against a quiet background, the tracking component is directed along the wave vector at a given instant.

As already noted, both factors affect the effectiveness of PPC. However, their effect may be different in different seismic observations. For example, in VSP, even in the case of a homogeneous or horizontally-layered medium the direction of the wave vector varies from one observation point to another for the same shot-point. Yet, the wave is registered against a quiet background. In the circumstances, the orientation of the tracking component at each point is solely determined by the direction of the wave vector, and the power of the method arises from correlation of the components directed along the wave vector, rather than some fixed (say, Z) component. In the later part of the record where the waves are recorded in the presence of interference, the efficiency of PPC is controlled by both factors, that is, by variations in the direction of the wave vector and by the superposition of spurious waves.

In surface observations in relatively simple media the direction in which a given wave travels along the line of traverse changes but little. In surface observations in complex structures (such as in ore prospecting) the direction of wave arrival may change appreciably, but not so much as in VSP. Therefore, in surface observations, the efficiency of PPC depends in most cases on whether the spurious waves can be suppressed at the observation point. In observations under low-velocity layers where the direction of particle displacement agrees with the direction of wave arrival, we must reckon with both factors again.

An important distinction of PPC is also the fact that it permits any waves, irrespective of their kind and velocity, to be detected and tracked along the line or plane of traverse simultaneously. Thus, with PPC all types of waves (compressional, shear and converted) may be used in the study of the medium, and this promises a further enhancement in the efficiency of the polarization method. According to VSP data, shots in shallow holes usually employed in seismic prospecting excite both compressional and shear waves, the latter often being comparable with, or even exceeding, compressional waves in intensity. A joint analysis of the basic wave types will undoubtedly serve to raise the effectiveness of seismic exploration and yield further data essential for predicting the composition of geologic sections, apart from handling structural tasks. From this point of view, PPC may be recommended for use not only in the study of complex media where the conventional techniques are of little value, but also in the seismic exploration of relatively simple media where the traditional methods based on a single wave type, say compressional waves, can readily cope with structural problems.

With respect to some waves (say, shear waves), PPC compares favourably with other techniques also, because the orientation of the wave vector is strongly dependent on the structure of the medium and the amplitude of fixed-component (XYZ) records might vary appreciably. This impairs the selectivity of the directional pattern in the case of interference reception, notably when use is made of the common-depth point (CDP) method. PPC detects the tracking component of the shear wave for subsequent summation and thus augments the strength of the CDP method in the detection of shear waves.

Another advantage of PPC is the increase in the signal-to-noise ratio without loss of resolution. In many cases, spurious waves can successfully be suppressed and the regular waves separated by the traditional methods used in seismic exploration, notably CDR–II. This, however, entails a loss of resolution, impairs the quality of exploration data and necessitates further processing. In contrast, CDR–I does not affect the wave form, and the pulse is not attenuated. This is why it will be good practice to begin an analysis of waves on seismograms with PPC and to proceed to any other techniques for signal separation, and then only to the extent absolutely necessary in each particular case.

Tracking components for a given wave are identified on the basis of v_a and polarization as follows. First, position correlation is used to detect the regular waves passing at a given instant and within the profile part adjacent to a given point, then polar correlation is used to determine the path of particle motion at the chosen point for the given wave, and the components are selected for which the signal-to-noise ratio is a maximum. The decisive factor in choosing the tracking components is the proper orientation of the CDR–I directional pattern with respect to spurious waves. In the presence of a strong spurious wave, the desired tracking component is that which is normal to the direction of polarization of the spurious wave, if it is linearly polarized or to its strongest component if it is nonlinearly polarized. This is true, for example, of cases where transmitted converted waves are detected in the presence of strong compressional waves.

How much a tracking component deviates from the wave vector is immaterial (see Chapter 1), because the selectivity of the CDR–I analyser along the directions close to the axis of the directional pattern is low. The gain is mainly due to the suppression of the unwanted wave. The unwanted waves can best be traced down from an analysis of modulograms on which the unwanted waves appear as dominant waves. The principal unwanted waves thus detected can be subtracted in advance, thereby enhancing the value of PPC.

At some points along the profile where several regular waves differing in v_a interfere there may be several tracking components – one for each of the interfering waves. This will inevitably require more effort in identifying the tracking components, and this strongly limits the use of polarization-position correlation in analog form.

We have examined the use of CDR–I in detecting the tracking components in such detail because it can readily be implemented in analog form. It must be added, however, that in the polarization method wave components may well be separated by other polarization filtering techniques widely differing in selectivity, including pre-tuned and self-tuning filters, which adapt themselves to the wave field on hand. Polarization filtering offers a wide choice of capabilities and, as with other forms of discrimination, the final decision must be taken to suit a specific situation.

In contrast to CDR–I, polarization filtering can mainly be implemented with a digital computer. We have used polarization filtering to analyse the initial part of distant-explosion seismograms recorded in deep drill-holes with a view to detecting transmitted converted waves [1]. The filter was of the Flinn type [139, 140]. On the basis of records of three mutually perpendicular components of complexly-polarized waves (with known direction of the ray plane and known angle of incidence for compressional and shear waves), this filter made it possible to detect *P, SV*, and *SH* waves on the components directed along the polarization axes at a given instant, if the directions of these axes fluctuated with time. It should be noted that filterning might cause non-linear distortion of the signals, but the distortion would be negligible at peak amplitudes.

The data were processed by an M–20 digital computer using the programs compiled by the computer centre of SB AS USSR. Figure 55 shows seismograms for the *X* and *Y* components and *SV* and *SH* waves in the depth range 1300–1400 m. As is seen, the *SV* and *SH* records contain markedly fewer waves, and the detected waves are relatively simple in shape. Analysis of earthquake data has confirmed the validity of space-time filtering.

In the examples that follow polarization-position correlation is carried out by comparing a set of combined fixed-component seismograms produced by a computer. On each seismogram, continuity is established for the waves, and their line-ups are then utilized as combined travel-time curves for the fixed components. In more detail the actual techniques will be taken up in the examples illustrating wave-field analysis.

It is interesting to compare PPC and polarization in optics. In seismics an unpolarized wave field is formed by a multiplicity of waves with all likely directions of particle displacement, existing all at the same time or following one another rapidly and in a random manner. To some degree, such an unpolarized field is depicted by a modulogram.

In optics light is termed 'polarized' if the vibrations of the light vector have been confined to a few or even one plane by some technique. In seismics, a wave field is referred to as 'polarized' for a given wave, if the directions of particle displacements in that wave have been suitably confined in some way. To extend the analogy, polarization correlation brings the individual seismic waves out of the complex wave field in much the same manner as polarized light brings out a particular structure in a rock sample when viewed under a polarization microscope.

Common-Depth-Point Polarization Method. In an analysis of complex wave fields, the need arises for more powerful discrimination techniques. At present, by far the best choice in this respect is the CDP multiple-coverage technique. This has been borne out by its use in the CDP polarization method, although it poses some specific problems.

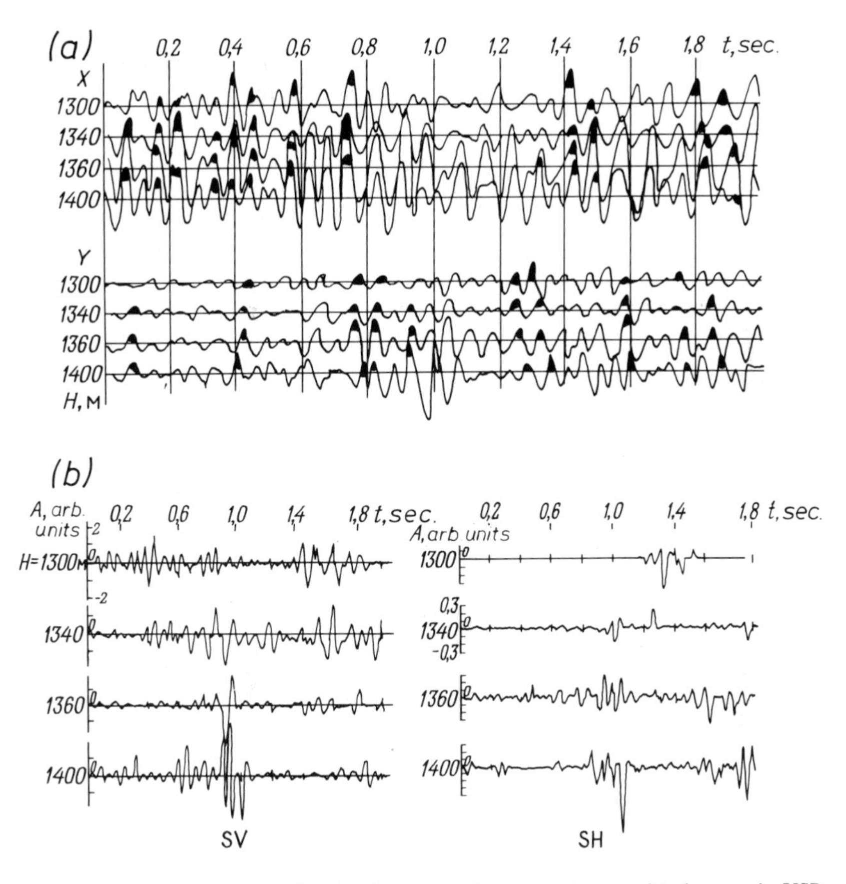

Fig. 55. Use of polarization filtering for the detection of converted transmitted waves in VSP work (Volgograd Region). (a) *X*- and *Y*-component seismograms; (b) *SV*- and *SH*-component seismograms.

Wave discrimination by polarization may be utilized to full advantage in areas where the upper part of geologic sections is made up of high-velocity rocks. This is especially true of ore deposits where the upper part of the geologic section consists of igneous or intensely deformed high-velocity rocks, and localities buried under permafrost and traps. In the USSR, there are vast oil- and gas-prospecting areas falling in this class. For the most part, however, the presence of low-velocity layers (LVL) is typical.

As will be shown later, in passing through a low-velocity layer the particles in a compressional wave tend to move in a direction close to the vertical, which strongly limits the reliability of compressional-wave discrimination by polarization in surface observations. On the other hand, immediately beneath a low-velocity layer, particle displacement is mainly in the direction of wave travel. Experience has shown that in situations where the conventional techniques show poor performance, wave discrimination in terms of all the three characteristics (frequency, polarization, and v_a) may be performed even in the presence of a low-velocity layer by the CDP polarization method. In addition to complex structures, the CDP polarization method may be of value under relatively simple platform conditions in the presence of strong multiple waves. The

CDP polarization method can readily be applied, if each shot hole is first utilized as an observation hole [63]. If necessary, detector grouping may be replaced by shot grouping.

To carry out sub-LVL observations, Obrezha and Mirzoyan at the 'Krasnodarneftegeofizika' Trust have developed three-component seismometers, which can readily be lowered into and aligned in drill-holes and lifted upon shot recording.

Level Observations. In cases where sub-LVL observations prove insufficient, local structural studies can be carried out in drill-holes beneath the main marker boundaries. This is a modification of the VSP method inaccurately called the inverted time-depth curve method [119]. As has been found, polarization-position wave correlation can appreciably improve analysis of a complex wave field. Figure 56 shows tracking-component seismograms for nine waves detected in the initial part of the record made during

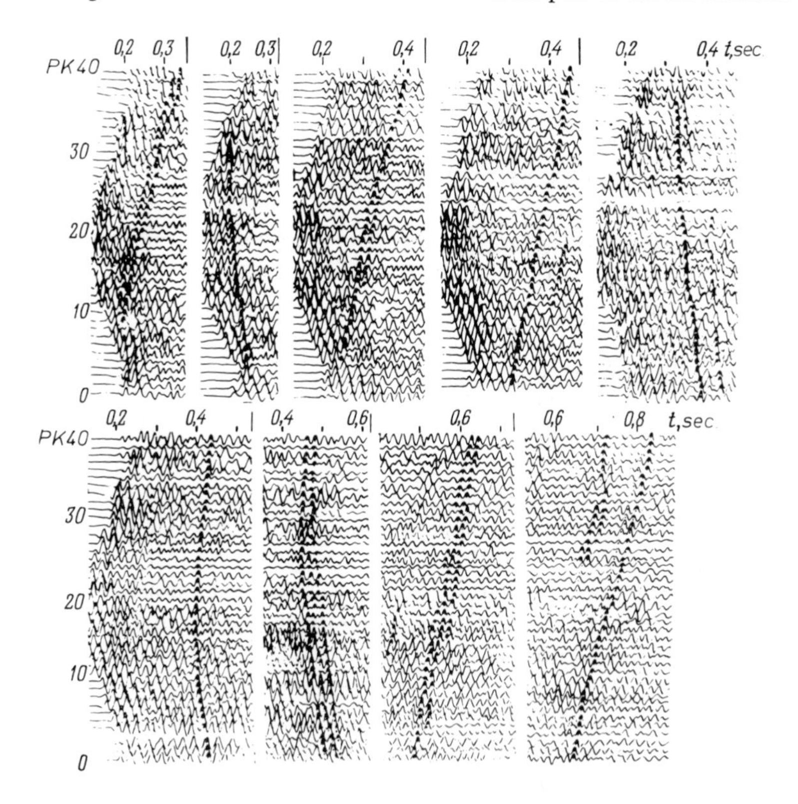

Fig. 56. Tracking component seismograms in VSP work by the inverse travel-time curve method (DH 128, Akshagat deposit, after R. N. Khairutdinov).

observations at a depth of 440 m and with shot-points spaced 50 m apart. The waves detected by PPC are seen to move at different apparent velocities and in different directions and are associated mainly with steeply-dipping boundaries.

Because of the complex structure at the top of the geologic section, the initial part of the surface seismogram shows a multiplicity of strong unwanted waves against which the reflected waves cannot be detected in practice.

It is to be stressed, however, that deep drill-hole observations cannot be of high exploration value; they should preferably be used in tackling local problems. It is not unlikely that with PPC they may, in many cases, be replaced by surface or sub-LVL observations. The explorational value of such observations is naturally higher, because they are not tied to any points (drill-holes) and can be carried out over large areas. Also, they will make it possible to set up widespread observation systems and to utilize the capabilities of polarization-position correlation in full.

It may so happen that in some regions it will be advisable and, indeed, feasible to use only one type of wave. This, for example, is true of the large 'adyr' areas in Middle Asia, known for their very complex wave field. VSP work has shown that the primary sources of noise are direct and multiply-refracted shear waves. The vertical components of such waves are very strong and upset the correlation of the reflected compressional waves. For a long time, all seismic surveys have been based on the suppression of such waves. To this end, use has been made of multi-element groups of sources and detectors. Although in some cases such schemes may serve their purpose, a good many regions have remained inaccessible for seismic exploration. On the other hand, as VSP observations have shown, the shear waves generated by conventional explosions are very strong, reach appreciable depths, and may therefore be used for seismic surveys. It is only necessary to record reflected shear waves over substantially longer time intervals and at a materially lower level of spurious waves. Already early reconnaissance surveys have confirmed the feasibility of such a technique [36]. The application of polarization-position correlation in the circumstances will go a long way towards making seismic exploration in Middle Asia more fruitful. It will be a good plan first to try out the usefulness of shear waves excited by conventional sources (explosions), reserving directional sources for cases where shear waves generated by conventional explosions fail.

The polarization method does not in any way preclude the use of shear waves generated by directional sources. Indeed, it envisages a wider use for them on the basis of various combinations of sources. Undoubtedly, directional sources may well be indispensable in many cases. Yet, given certain structures in the top part of a section, some tasks can successfully be handled by using shear waves from ordinary sources. The type of source to be used for preference and how it is to be used must be decided in each specific case. Whatever the choice, however, polarization-position correlation may markedly facilitate the detection and analysis of shear waves.

EARLY EXPERIENCE WITH POLARIZATION-POSITION CORRELATION

The advent of vertical seismic profiling brought with it the need for analysis of the complex wave fields observed in the interior points of the medium. At relatively shallow depths, the existing methods of position correlation were still effective. As the depth range of seismic surveys increased, however, the applicability of VSP became increasingly more limited by difficulties in analysing the complex wave field. As will be recalled [36], wave correlation in VSP runs into difficulties because of two basic factors. First, there are a large number of waves propagating in two mutually opposite directions, from and towards the ground surface. On interfering with one another, these waves produce a complex interferential field against which the individual waves may in most cases be followed with difficulty, if at all. Second, the direction in which waves arrive changes according to the relative positions of the observation point, vertical profile line and source, and with the velocity characteristics of the medium. Because of this, the intensity of a particular wave may vary strongly along the profile if we observe only one fixed (say, Z) component. Since such variations are different for different waves, they may lead to an erroneous estimate of wave intensity.

The situation is still more complicated in cases where the waves are polarized in different directions (e.g. compressional and shear waves) and registered by detectors differently oriented in space. Under such conditions, the conventional methods of wave correlation often prove insufficient. Such a situation exists in the Volga valley near the city of Volgograd. To evaluate the exploration value of the transmitted converted-wave method widely used there, a series of three-component VSP observations were carried out. In the processing stage it was found that the transmitted converted waves associated with boundaries in a Devonian terrigenious sequence and with the top of the crystalline basement could be detected and analysed with difficulty, because of their interference with compressional waves. This difficulty could, it was hoped, be avoided by using interference-type spreads and arrays, which have proved their worth in surface observation. Because of the complex wave pattern produced by the multiplicity of sharply-defined interfaces in the alternating carbonate-terrigenous sequence, this proved insufficient and led to a considerable loss of resolution. To achieve better results, we used a correlation technique based on tracking a variable component along the line of traverse. Initially, this component was chosen to be the wave vector determined by analysing the path of particle motion at each point [1]. From the wealth of data processed on a digital computer we were able to: 1) develop a technique for the computerized directional reception utilizing the polarization properties of waves (CDR–I) and based on the wave records produced by an unoriented three-component drill-hole cluster; 2) use this technique to analyse wave polarization and to construct wave-field components optimal for the detection and establishment of continuity for various waves (compressional, shear, and converted) along a vertical profile; and 3) evaluate theoretically the intensity and the conditions for the detection of various waves (notably transmitted converted waves) at the interior points of the medium and on the ground surface, using the data about the seismic structure of the medium yielded by VSP.

It was during these studies carried out in 1966–67 that controlled particle-displacement reception was implemented and polarization-position wave correlation was applied for the first time [1]. The path of particle motion was established by applying polar wave correlation to computer-compiled multicomponent seismograms. The application of CDR–I to VSP involved the following steps: a) the orientation of the drill-hole cluster was determined at each point (by measuring the angles between the sensitivity axes of the seismometers and the vertical ray plane); b) polar seismograms referenced to arbitrarily assumed directions in an oriented coordinate system related to the ray plane were constructed; and c) the optimal components of compressional and shear (converted) waves were detected.

The algorithm for CDR–I uses a digital representation for three-component seismograms in the form of tables of $A_x(t_k)$, $A_y(t_k)$, and $A_z(t_k)$ values at discrete instants $t_k = t_0 + k\,\Delta t$ ($k = 0, 1, 2, \ldots, n$). Initially, the direction of particle displacement is determined in a coordinate system referenced to the recording cluster.

The orientation of the drill-hole cluster is determined relative to the direction of particle displacement for the first compressional wave. In doing so, it is assumed that the medium has symmetry and that the particles in the first compressional wave vibrate in a vertical plane passing through the points of excitation and reception.

In order to determine the orientation of the cluster, a time slot $t_0 \leqslant t_k < t_n$ ($k = 0, 1, 2, \ldots, n$) of width n sufficient to accommodate several compressional and converted waves is chosen on the horizontal-component records. According to the assumption made, the orientation of the plane in which the first waves are polarized is determined as follows. This plane is taken to pass through the z-axis and the vector $\mathbf{l}(\alpha_n) = \cos\alpha_n\,\mathbf{i} + \sin\alpha_n\,\mathbf{j}$ ($n = 1, 2, \ldots$) lying in the xy-plane (here, $\mathbf{i}$ and $\mathbf{j}$ are the unit vectors of the x- and y-axes), such that the angles α_1 and α_2 between the x-axis and vector $\mathbf{l}(\alpha_1)$, as reckoned clockwise from the x-axis, correspond to the maximum of the function

$$F^2(n, \alpha) = \sum_{k=0}^{n} [f_k(t_k)\cos\alpha + f_y(t_k)\sin\alpha]^2$$

over the time slot n.

The function $F^2(n, \alpha)$ is a periodic function of the angle α with a period of 180°, so the difference between the angle α_1 and α_2 at which this function is a maximum is 180° (Figure 57a).

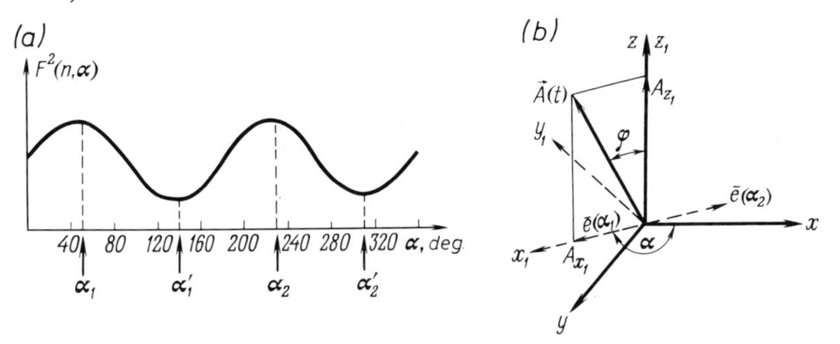

Fig. 57. Orientation of borehole cluster with reference to a direct wave. (a) Plot of the function $F^2(n, \alpha)$; (b) xyz, original coordinate system, $x_1y_1z_1$, coordinate system referenced to the vertical plane of polarization.

In order to convert the seismograms from the original xyz coordinate system to the $x_1y_1z_1$ system referenced to the vertical ray plane, the new x_1 axis must be rotated through an appropriate angle so that it is oriented towards or away from the source (Figure 57b). If x_1 be oriented away from the source, then of the two angles, α_1 and α_2, we shall select that for which the arrivals of the compressional wave on the z_1 and x_1 components take the same sign (if the z-axis points upwards). For consistency in determining the signs of arrivals, it will be good practice to consider the signs of the sums $\Sigma_{k=0}^{m} A_{x_1}(t_k)$ and $\Sigma_{k=0}^{m} A_{z_1}(t_k)$ within the time slot accommodating the first half-cycle of the compressional wave within the first-arrival interval.

The components of the wave vector in the plane of incidence are given by

$$A_{x_1}(t_k) = f_x(t_k)\cos\alpha + f_y(t_k)\sin\alpha\ ,$$
$$A_{y_1}(t_k) = f_x(t_k)\sin\alpha + f_y(t_k)\cos\alpha\ ,$$
$$A_{z_1}(t_k) = f_z(t_k)\,.$$

The wave component arbitrarily oriented in space is given by

$$A(\varphi, \alpha, t_k) = A_{z_1}(t_k) \cos \varphi + A_{x_1}(t_k) \sin \varphi ,$$

where $A(\varphi, \alpha, t_k)$ is the desired component.

By varying the angle φ from 0 to 180°, with an arbitrary α, we obtain a polar seismogram lying in a plane with α = const. If the angle α locates the vertical ray plane, the set of seismograms for different ψ will locate the polar seismogram in that plane.

The basic operations comprising the above program are: seismograms are presented in the form of tables of amplitudes for the three components (sampling interval Δt in seconds, channel gain K, and depth of observation points); the cluster is located in the coordinate system referenced to the vertical ray plane (orientation of the cluster); the components of polar seismograms arbitrarily oriented in space and also in any plane, including the ray plane, are computed.

Polar correlation at each observation point yields components optimal for the detection of transmitted converted waves. The criterion of optimality is the zero amplitude of the first strong phase of the compressional wave. It is also assumed that the particles vibrate in about the same direction in the first and the succeeding phases of the compressional wave. In the circumstances, the compressional waves show a minimum amplitude on the optimal component record. It should be noted that the compressional waves are comparable or superior in strength to the converted waves even under such optimal conditions. This is an indication that the compressional waves are substantially stronger than the converted waves. Therefore, even a minute deviation in the direction of particle displacement from the vertical in the *P* wave (and this deviation may be as great as 20° to 30° in high-velocity layers) may give rise to an appreciable horizontal component in the compressional waves.

As the next step, combined seismograms for the tracking components along the vertical profile are constructed. To establish continuity on these seismograms, the waves are discriminated in terms of the direction of wave travel (CDR–II), which is done on a computer using the program compiled by Gel'chinsky and Krauklis [81].

The above scheme of polarization-position correlation substantially extends the line-ups for the converted *PS* waves and has often enabled us to detect and follow such waves under conditions where the correlation of fixed components would be of little use [1, 50]. Apart from confirming the value of the new method, more reliable data are obtained about the characteristics and nature of the recorded waves. Among other things, it has been found that the complex wave fields on the horizontal component records result from the superimposition of the transmitted *PS* waves associated with the boundaries in the top part of the section. This explains why the converted waves associated with deep boundaries are so difficult to detect. In fact, such detection is practically impossible without a rigorous analysis of the complex wave field. The results thus obtained cast doubt on the exploration value of the transmitted converted wave method as applied to the study of the sedimentary formation in the Volgograd area where carbonate and terrigenous sediments typically alternate [1, 36, 50].

Practical Procedure for PPC

In principle, polarization-position correlation can be accomplished on analog and digital computers. Preference for digital computers is motivated by the tediousness of correlation in analog form. Recently, improved programs have been compiled for the generation of source data for PPC to be performed by a digital computer. Yet, it appears worthwhile to discuss the PPC procedure in analog form first. In many cases, especially under field conditions and during preliminary data-processing, PPC in analog form may prove more feasible. Also, this discussion will enable the reader to obtain a clearer idea about polarization-position correlation.

ANALOG PROCESSING

Wave correlation may be visualized as consisting of three steps. The first step concerns the formation of a general idea about the wave field and the detection of the strongest and dominant waves. This step is carried out by reference to the original fixed-component seismograms recorded by a three-component cluster. In complex media, the detection of regular waves on such seismograms runs into appreciable difficulties, so instead of the original seismograms the second step detects and establishes continuity for regular waves by reference to seismograms of fixed components suitably oriented and uniformly distributed in space. Experience has shown that the use of records of different components in the different parts of a profile for the detection of a particular wave can appreciably help with detecting and establishing continuity for the basic wave types (compressional, shear, and converted) polarized in different directions. This is the basic step in analog correlation. The profile intervals over which continuity may be established for a particular wave on a fixed-component seismogram depend on the degree of complexity of the wave field and may vary from wave to wave.

The third step is concerned with establishing continuity for the tracking component of each wave. Because this step is labour-consuming in analog form, it is applied only to the most complicated parts of records, usually where the waves detected during the second step interfere.

Thus, the various steps use different kinds of seismograms.

Original Fixed-Component Seismograms. According to the type of three-component cluster employed, this step uses seismograms of *XYZ* components or of symmetrical-cluster components I, II, and III. In surface observations, the clusters are oriented along the cardinal points of the compass or along the line of traverse, and the fixed components are referenced to space coordinates. VSP observations call for further orientation of the records (see Chapter 2).

Oriented Records. For polarization-position correlation, it is convenient and for quantitative interpretation it is essential, that the paths of particle motion at all observation points be referenced to a single system of space coordinates. In borehole observations, the alignment of the clusters may be achieved either by suitable cluster-alignment sensors, which determine the actual position of a cluster in its drill-hole, or approximately, on the basis of the direct wave (see Chapter 2).

The direction of particle motion in a *P* wave may be determined at each point of a vertical profile by reference to oscilloscope records of signals from the three-component cluster or to a polar seismogram. Using the angles through which the coordinate system must be rotated in order to align the cluster, the coefficients are computed for the orientation circuit, and from its output the signals oriented relatively to the P wave are recorded on magnetic tape or applied to an analyser in order to generate an oriented polar seismogram.

Fixed-Component Seismograms. When field data are given a preliminary processing, it is usual to analyse them for the following fixed components: components lying in the vertical plane of polarization of the first compressional wave (traces 1, 7, 12, 17, 20, and 23); the four inclined components not lying in that plane (traces 8, 11, 13, and 16), and also the horizontal component 4. As will be recalled, components 1, 4, and 23 correspond to the respective members of the XYZ cluster. Depending on the objective sought, the remaining components may be analysed as well.

Seismograms in a Local Coordinate System. On combined component seismograms in a local coordinate system, it is possible to isolate waves polarized in the direction of the first compressional wave or in a plane tangent to its front. In this respect, the local coordinate system is equivalent to the XYZ coordinate system in surface observations when the direction of arrival for the first wave is the same and close to the vertical at all points.

Within the profile intervals where the compressional wave is registered free from interference with other waves, the P component in the local coordinate system coincides with the vibration vector and is the tracking component. In VSP the R component may display different waves, depending on the distance from the source. At short distances when the direction of arrival for the compressional wave is close to the vertical, the R components may be used as the tracking components for shear waves. In the case of oblique wave incidence the R components may serve as the tracking components for the reflected waves been propagated towards the ground surface, because they do not display incident compressional waves, which usually tend to upset the correlation of reflected waves.

A combined seismogram for the Q component characterizes the field of tangential waves. It must be admitted, however, that for the late arrivals in complex structures with steeply-dipping interfaces the P, R and Q components may be random, and not tracking ones, but this may, incidentally, also happen with the XYZ components. Yet, in order to form a tentative idea about a wave field, it is convenient to begin component-wave analysis in the local coordinate system.

To detect waves whose tracking components do not run along the axes of the local coordinate system, it is necessary to analyse the seismograms of individual fixed components both lying in the vertical plane, which contains the compressional wave, and not lying in that plane, but distributed more or less uniformly in space.

Combined Seismograms for Various Components. The first step of polarization-position correlation involves the construction of a wave-field diagram. On this diagram, the line-ups are plotted by transfering the extrema of all the waves detected on the combined seismograms for various fixed components [40]. The seismograms thus produced are analogous to combined travel-time curves for the waves detected on the various traces, but they are more attractive, because they require less time for construction and are more graphic in depicting the complexity of the wave field. In addition to these diagram-type combined seismograms produced for the wave of particular interest, it is customary to derive combined tracking-component seismograms from the fixed-component seismograms.

Thus, the general sequence in carrying out PPC in analog form is as follows. Regular waves are first detected, and their continuity established on those parts of the profile where possible, using fixed-component seismograms and appropriate criteria. Then, a combined travel-time curve is constructed for the waves detected on all the traces. On the basis of the travel-time curve, the basic regular waves are identified, their apparent velocity v_a is measured, and the direction of travel is determined. For each wave appearing simultaneously on several adjacent traces, a seismogram is chosen on which the wave is optimal at a given point, and a combined tracking-component seismogram is compiled. It is interesting to note that with short correlation intervals, the tracking component of a wave will retain its orientation over the profile in most cases, and that this orientation changes but seldom, and then only within localized parts of the profile.

As with wave discrimination in terms of other wave-field charactersitcs (say, frequency), polarization-position correlation may employ systems varying in the degree of selectivity. The desired selectivity is specified according to the complexity of the task in hand and must never exceed a reasonable limit. The selectivity of PPC can be controlled by varying several parameters. One such variable is the degree of detail in the choice of tracking components. For example, it will suffice to analyse as few as three original fixed components in a record during the first step of processing. Should subsequent examination show that this has been insufficient for signal detection, a more selective system may be used for the profile parts involved, and the correlation of the tracking components can be carried out. Generally, the tracking components may be observed over a fairly broad interval (part of a profile), and only when this is impossible, should a tracking component be detected and followed over shorter intervals.

The selectivity of a system can also be controlled by the choice of a method for the identification of tracking components. In some cases, it may prove sufficient to use linear transformations related to a change in the spatial orientation of the wave components. In the more complex cases, recourse may be had to trajectory filtering (various polarization filters) or still more rigorous discrimination systems. Also, the effectiveness of PPC may be improved by the choice of an observation system ensuring the requisite selectivity in terms of apparent velocity. This, however, involves the use of specialized systems, which adds to the complexity and cost of work. Therefore, it appears more attractive to utilize the advantages offered by wave discrimination at a point.

Present-day computers can effect any of the above approaches, so the task reduces mainly to the choice of an economical and streamlined processing sequence. The underlying logic appears to be this: the selectivity of a system must be increased only where necessary and to the extent actually required.

AUTOMATED DIGITAL PROCESSING

Work on automated digital processing systems has been going on concurrently with the use of analog computers. During the first period, a program was compiled for the BESM–4 digital computer [80]. This program can determine the following quantities:

the trajectory of particle motion in spherical coordinates in the form of modulograms, angular records, and azimuthograms; radial and tangential components of the vibration vector; and a stereographic projection of the vibration vector. During the second period, an automated system built around the BESM–4M digital computer was developed, in association with the 'Krasnodarneftegeofizika' Trust (V. S. Starodvorsky), for processing three-component observation data (known in the USSR as the 'Polyarizatsiya' system). This system, specifically designed to process seismic observations by the polarization method, has already been in operation for several years. It has processed a very large amount of surface and drill-hole observation data gathered in areas with complex diapiric folds. We shall only give a brief outline of the system.

The 'Polyarizatsiya' system consists of two subsystems, namely, a polar correlation subsystem and a polarization-filtering subsystem. The first is operating on a routine basis, and the second is being tried out.

The polar correlation subsystem consists of ten special-purpose programs and several service routines. The subsystem accepts 110 records produced by three-component arrays (a total to 330 traces). At the time of input, it discards any invalid traces present on the film and applies static and dynamic corrections in the usual manner, as in the case of the CDP method. Then the subsystem determines the time of the first extremum and the direction of particle motion in the first wave and converts records from a symmetrical array into a rectangular coordinate system formed by two horizontal (X and Y) detectors and one vertical (Z) detector; the azimuth of the x-axis is the same as that of component I of the symmetrical array.

Then the subsystem orients the records produced by the cluster. This can be done in a precise form in space coordinates on the basis of signals from a cluster-alignment sensor, or approximately, relatively to the compressional wave. In the latter case, the direction of particle motion is found for the first compressional wave. The subsystem computes the vibration vector, that is, its modulus and its direction as specified by the angle φ in a vertical plane and by the angle α in a horizontal plane. The presentation of the trajectory of particle motion (seismograms) in spherical coordinates is convenient for the subsequent polarization analysis. The subsystem aligns the axes of the local coordinate system in space, converts the trajectory of particle motion from the rectangular to the local coordinate (prq) system referenced to the direction of arrival for the first wave, and generates combined seismograms for the components in the local system along the line of traverse.

Should the polar correlation subsystem prove insufficient, the polarization filtering system goes into action. It incorporates an assortment of polarization filters, which can raise the selectivity of the entire system owing to wave discrimination based on the trajectory of particle motion. Wave discrimination based on polarization at a point ends with the generation of combined tracking-component seismograms. These seismograms provide a basis for wave discrimination according to the direction of wave travel, to which end use is made of various space-time filtering devices not considered in this book.

Also, the system can isolate any wave component at each point and generate combined fixed-component seismograms. The generated data can be put out in the form of seismograms at any stage of processing. This system is the basis for a complex of PPC programs, which permit the isolation of tracking components for each wave.

Use of PPC in Seismic-Prospecting for Ores

High-velocity media, steep boundaries with complex relief and weak differentiation in terms of velocity combine to produce extremely complex wave fields in localities with ore deposits. For this reason the capabilities of PPC were first tried out in seismic prospecting for ores in Central Kazakhstan in 1969–75. Carried out by IEP AS USSR in cooperation with KazVIRG, the work was concentrated on the analysis of the wave fields observed in VSP.

However, analysis of wave fields in the inner parts a medium poses serious problems because, as will be recalled [36], such fields are characterized by a large number of waves. With accumulation of experience and with the growing confidence in the applicability of PPC to complex wave fields, it was realized that PPC could be used in surface observations as well. Recent years have seen a rapid expansion in the scope of such observations. What follows is a brief review of the use of PPC in VSP and surface observation.

VSP OBSERVATIONS

Apart from the appreciably larger number of waves involved as compared with surface observations, the wave fields analysed in the interior points of a medium owe their distinction to the fact that the waves come from various directions, and that these directions may vary from point to point along the profile differently for different waves. It may be added that for offset profiles these variations may be considerable even in a horizontally-layered medium. Records of vertical components display not only variations in strength from zero to a maximum, but also polarity reversals for the arrivals. Also, sudden changes in the direction of wave arrival can be caused by the complex geometry of boundaries. Changes in the direction of arrival bring about changes in strength on the records of fixed components, including the vertical component, which may lead to the loss of, or erroneous, correlation. It is under such conditions that PPC may prove of value in the correlation of the first and late arrivals.

DIRECT WAVES

In the polar correlation of a direct wave (see Chapter 3), it is shown that the first wave may be a complex interference-produced wave. Polarization-position correlation offers a means for a better insight into the first wave and for establishing continuity along the line of traverse.

Let us consider a combined Z-component seismogram along a vertical profile, recorded in borehole 188 from shot-point 4 for which l = 340 m (at I in Figure 58a). It is seen that in the interval 480–540 m the P wave is a pulse consisting of several peaks whose records show good reproducibility. However, in the interval 0–200 m the first wave differs in both shape of record and strength from those obtained at great depths; also, there are variations from point to point in both. Obviously, it is practically impossible to track the first wave on such a record. In contrast, the correlation of tracking components produces identical records (Figure 58a, II) over the entire profile. For ease of comparison, the arrival times for the first wave are shown aligned on the seismogram. The tracking components used to identify the P wave at the top-most part of the profile

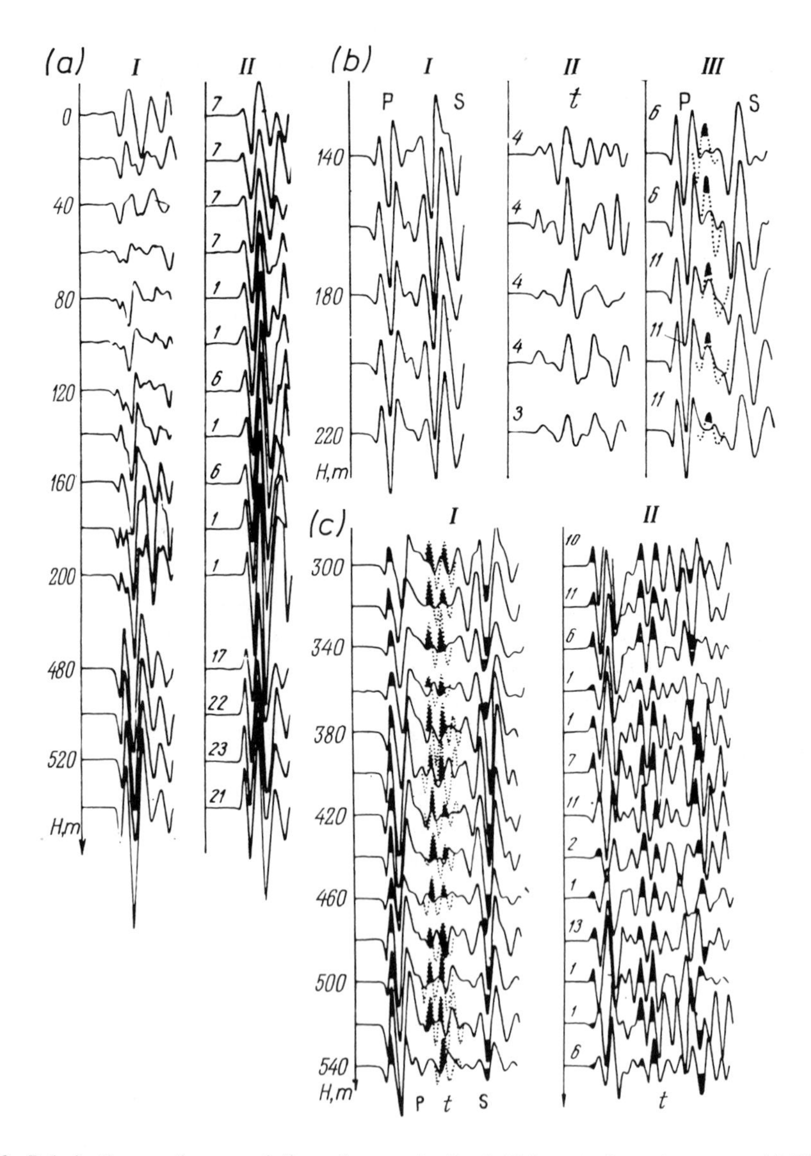

Fig. 58. Polarization-psotion correlation of waves in the initial part of a seismogram. (a) DH 188, l = 390 m, SP4; (b) DH 188, SP2, l = 210 m; (c) DH 198, SP4, l = 440 m.

(0–60 m) are those with ψ = 30°, whereas in the interval 80–200 m they are the components for which ψ = 0 (i.e., horizontal components), and at lower depths they are those for which ψ = 60° and 90°. In this way PPC positively identifies the direct wave all along the profile and at the same time shows that the direct wave is a complex pulse produced by superposition of several pulses.

If the particles in the interfering waves vibrate in different planes, the vibration planes are shifted in time or differ in shape, PPC is capable in some cases of separating and establishing continuity for the individual interfering waves.

The Z-component seismogram at I in Figure 58b shows direct P and S waves following in succession. Because the waves arrive along an inclined direction, the vertical components of both waves are practically the same. Yet, despite the seeming simplicity of the P wave, its tail part shows one more wave, t, which is separated and for which continuity is established in a PPC seismogram (at II in Figure 58b). The t wave arrives with some delay after the P wave, this delay being variable in magnitude. With a sufficiently large delay, the t wave may be recorded outside the area where it would interfere with the P wave. Figure 58c shows seismograms for the Z component (I) and the tracking component (II) of the t wave. In the latter continuity is positively established for the t wave all along the profile. In each case the t wave is horizontally polarized.

Extensive VSP work has shown [36] that the conditions of excitation strongly affect the shape of each wave, the structure of the entire seismograms, and as a consequence, the value of the observation. This explains the interest in the conditions of wave excitation and especially so in relation to the polarization method, which is based on separating and establishing continuity for each of several types of waves. The observer can no longer be restricted to the direct P wave; account must also be taken of the direct S wave. The importance of either wave may be different, depending on the objective sought and, especially, the structure of the upper part of a section. So, the conditions of excitation must be handled in a special way in each particular case. According to VSP evidence, the best way to analyse the conditions of excitation is to refer to the direct waves recorded at interior points of the medium. This is where the capabilities of polarization-position correlation may prove of special value.

Let us consider the record for a direct wave in the late arrivals. Figure 59 shows seismograms for an offset VSP profile with l = 1500 m. Although all along the profile the direct wave is recorded outside the area of likely interference with the refracted wave recorded in the early arrivals, no continuity can be established for the direct wave on the Z-component record (Figure 59a). In contrast, continuity can be established and the true difference in intensity between the waves can be ascertained on the inclined-component records (Figure 59b). Refering to the latter, the direct wave is so much stronger than the first refracted wave that the first amplification level proves insufficient for their continuity to be established concurrently. To make the direct-wave record more readable, the sensitivity of the respective channels has to be reduced by about an order of magnitude.

In addition to the velocities of P waves, drill-hole observations also yield data on the velocity ratios for P and S waves, which is a basic variable in describing the physical properties of rock that may be used in predicting the sequence of strata in a section. We shall consider the data collected at borehole 243 (Figure 60). The Z-component seismogram (Figure 60a) shows that the dominant waves are direct P and S waves. However, practically no continuity can be established for the P wave at the top of the profile. In effect, it may not even be recorded at some points. It is only with increase in depth that the P-wave record becomes more regular, and the record gradually gains in amplitude. In contrast, continuity is positively established for the P wave all along the profile on the tracking component seismogram (Figure 60b), and its shape is well preserved. This also applies to the S wave. On the Z-component seismogram, the S wave can only be separated at the top of the profile, and its trace progressively diminishes with increase in depth. On the other hand, the S wave can be reliably traced all along the profile

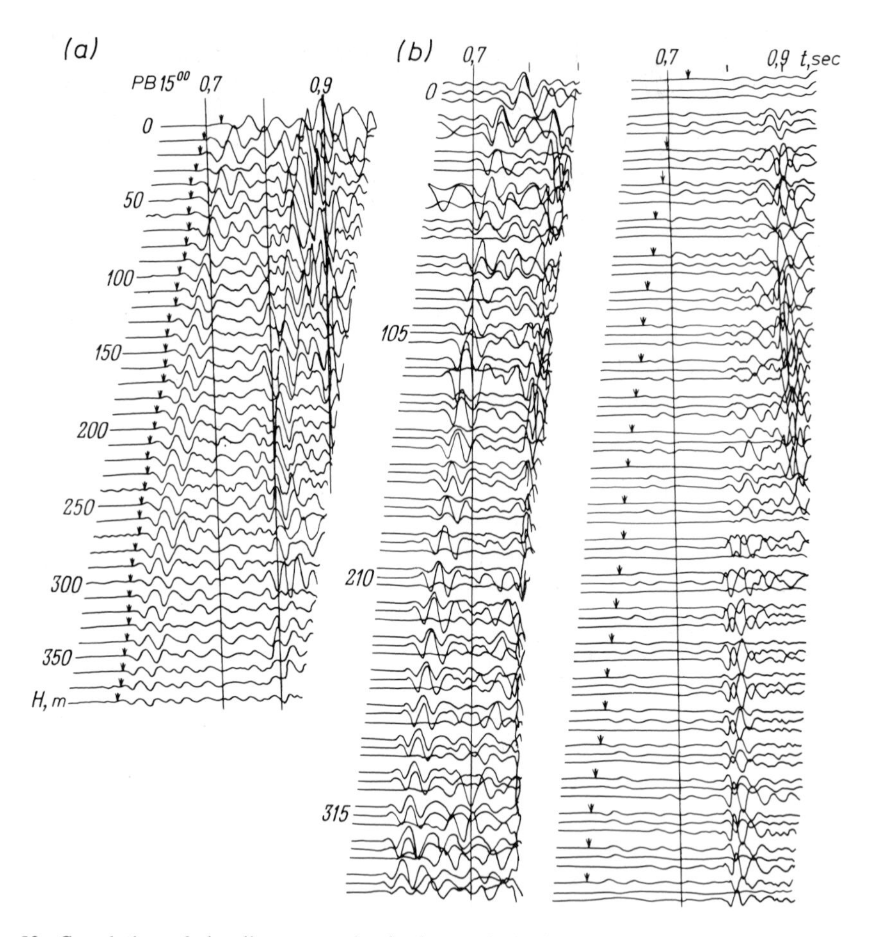

Fig. 59. Correlation of the direct wave in the late arrivals (l = 1500 m, DH 90a, Kudinovka area). (a) Z-component seismogram; (b) seismogram produced by a three-component symmetrical array at two levels of amplification.

on the tracking-component seismogram (Figure 60c). To sum up, the Z component in the present case is not optimal in regard to establishing continuity for the P and S waves. In contrast, the S wave is eliminated on the tracking-component seismogram for the P wave, and vice versa. To sum up, when applied to borehole observations, PPC extends the capabilities in section analysis appreciably.

Secondary Waves. As an illustration of PPC capabilities, we shall examine the observations made in the Zhairem deposit. Within the orebody area penetrated by drill-holes, the productive Upper Devonian Kanogenic sediments are buried under chert-carbonate rocks of the Lower Carboniferous series. The dip of the Devonian sediments varies from 30° to 50°.

The wave field was recorded at borehole 3110 from one shot-point located at a distance of 700 m from the drill-hole. On the fixed-component seismograms produced by a

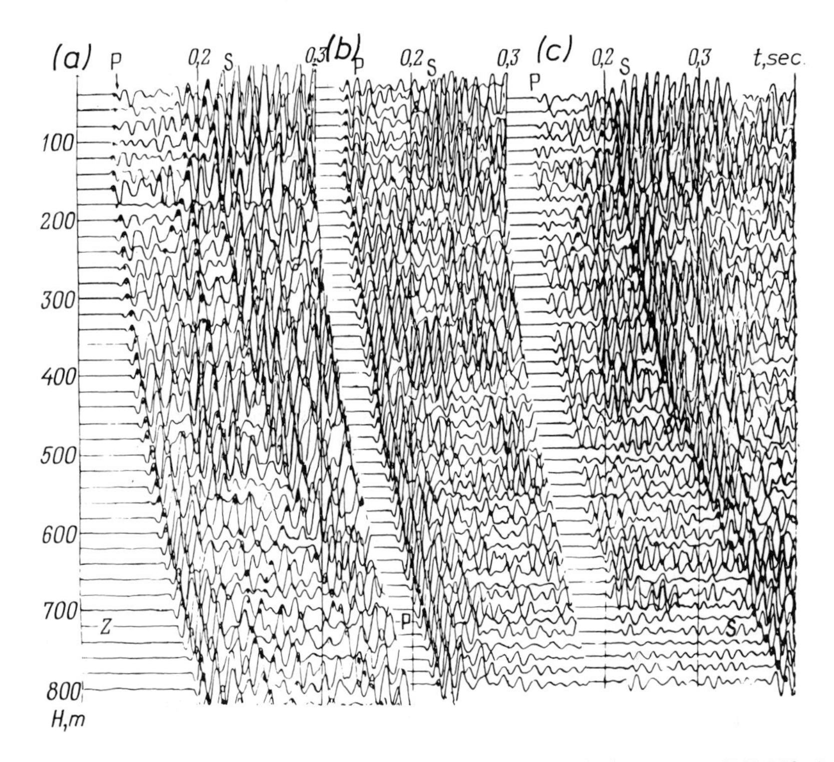

Fig. 60. Polarization-position correlation of direct compressional and shear waves (DH 243, Sayak deposit).

three-component *XYZ* cluster (Figure 61a), the individual waves can only be traced over some relatively short intervals of the profile. For none of the waves can continuity be established along the profile, using only one of the component seismograms. The *Z*-component record shows complex waves, with the first arrivals suddenly varying in intensity and polarity. It is only at lower depths ($H > 500$ m) that continuity can be established for the first wave more positively.

Figure 61b shows a set of seismograms for the fixed components variously oriented in space. The spatial alignment of the fixed components is shown in the stereogram. On the seismograms at $\psi = 60°$ (components 18 and 22) and $\psi = 30°$ (components 9, 11, 14, and 16), it is seen that the various waves can only be separated in some short portions of the vertical profile, and that in most cases components closely spaced together may show the same wave(s). For example, components 9 and 18 (see the stereogram) show a *t* wave at time 0.3 s at a depth of more than $H = 500$ m. In contrast, the components markedly differing from one another in direction show different waves. For example, the *t* wave at time 0.3 s is completely absent on traces 7, 14, 16, and 22, whereas trace 14 convincingly shows the wave at time 0.35 s. However, there may be cases where adjacent components will show the same waves in a drastically different manner or will display altogether different waves.

On all fixed-component seismograms, the waves are separated and their continuity established in terms of the well-known criteria of position correlation. The waves with

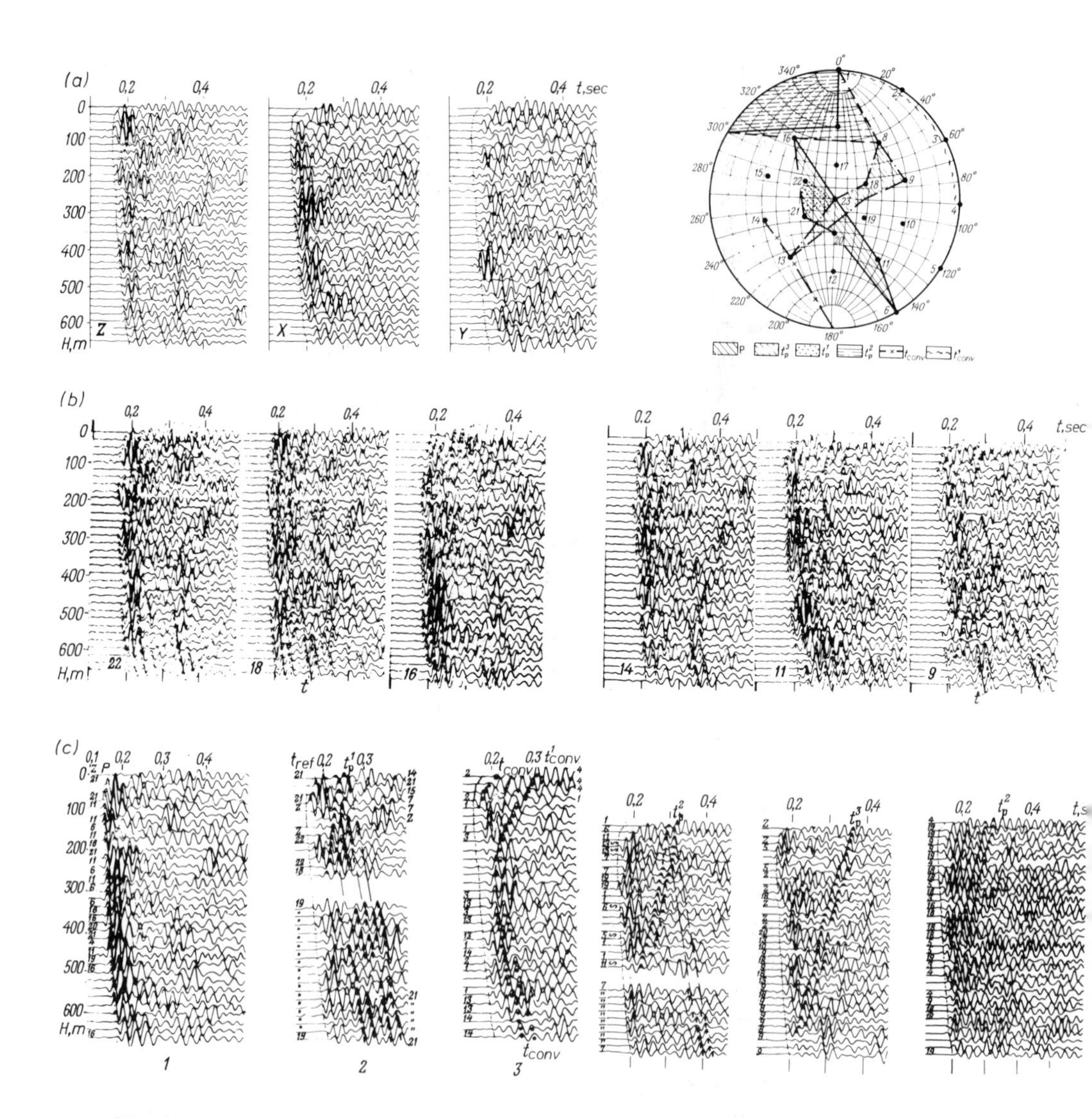

Fig. 61. Wave field observed in VSP (l = 700 m, DH 3110, Zhairem deposit). (a) *Z*-, *X*-, and *Y*-ponent seismograms and stereogram; (b) fixed-component seismograms (the orientation is shown on the stereogram); (c) tracking-component seismograms for some waves. The stereogram shows the spatial areas (fields) of tracking components for various waves.

a positive apparent velocity are marked by dots, and those with a negative relative velocity by dashes. For all the waves separated on the various components, travel-time curves are constructed from which the regular waves can readily be ascertained. From among the fixed-component seismograms for each wave, those components are chosen on which the respective wave stands out best. It is only within some parts of the seismogram, notably within the interference areas, that tracking components are chosen for every point on the polar seismograms.

Let us consider the combined tracking-component seismograms for the various waves, beginning with the first one. Because the first wave is recorded against a quiet background, its P tracking-component is in the direction of the wave vector at each point. On the combined seismogram (Figure 61c), it is seen that, although the first wave is complex in shape, it has about the same amplitude and its continuity is positively established along almost the entire profile. It is seen on the stereogram that the directions in which the tracking components of the first wave are oriented in space vary substantially, whereas the amplitude scarcely changes and corresponds to a direction of $\pm 180°$ to the source. At the same time, the angle with the vertical varies by as much as 150°: from horizontal ($H = 140$ m) to $\varphi = 30°$ ($H = 60$ to 660 m), with a reverse azimuth.

At the top of the section, the first recorded wave is the refracted wave t_{ref} associated with a boundary at a depth of about 100 m. Its tracking component (the one which is optimal with respect to its separation) is component 21 making an angle of 30° with the vertical. For the transmitted converted wave $PS_r(t_{\text{conv}})$ the tracking component is horizontal component 2 on which the t_{ref} wave does not appear, so that the t_{conv} wave is separated against a practically quiet background.

In the late arrivals the wave for which continuity is positively established as far as the ground surface is the reflected compressional wave t_p'. For it, the tracking components are components 21–23 (see the stereogram). Associated with the name reflecting boundary are the converted reflected wave t_{conv}' for which the tracking components are the horizontal components whose azimuth varies by as much as 60 degrees along the profile, and the converted downward-transmitted wave t_{conv} for which the tracking components are components 1 and 13. The compressional waves reflected from the ground surface are positively recorded at considerable depth on component 21, which makes an angle of 60° with the horizontal.

Among the waves reflected from an inclined boundary (at $H = 500$), a reflected compressional wave t_p^2 is most outstanding. Because the reflecting boundary is inclined, the tracking components for this wave run close to the horizontal and occur in both the upper half-space (7, 16) and the lower half-space (–11, –13), which can clearly be seen on the stereogram. Upon reflection from the ground surface, the t_p^2 wave can be followed to a considerable depth on component 23.

The wave associated with a boundary at a depth of 660 m is the reflected compressional wave t_p^3 whose tracking components vary appreciably along the profile. It is to be noted that the t_p^2 wave may be identified more reliably at higher frequencies, which is clearly shown in Figure 61c showing tracking-component seismograms produced for the t_p^2 and t_p^3 waves, using higher-frequency filters. This does not hold for the t_p^2 wave for which no correlation exists at depths of 200 to 400 m. Referring to the stereogram (see Figure 61a), for each individual wave the tracking components are concentrated in a relatively narrow sector. These sectors for different waves overlap little, if at all, and cover a good proportion of the half-space.

To sum up, polarization-position correlation offers a means of establishing continuity for various types of regular waves (reflected and refracted compressional, reflected, and transmitted converted waves), variously polarized and associated with four interfaces located in the depth interval from 0 to 600 m. Although it is of practical interest, this range of depths poses most problems in ore prospecting because the wave fields at the top of the section show an extremely complicated pattern. The observations carried out and the experience gained to date confirm the efficiency of PPC in analysing wave fields recorded by VSP.

SURFACE OBSERVATIONS

The applicability of PPC to surface observations in ore prospecting is of special interest because orebodies are known for their high wave velocities and complex structure. Consider the observations carried out in the Sayak copper orefield (the Northern Balkhash area). The objective of the observations was to investigate the steeply dipping contours of intrusive granodiorite bodies controlling the location of mineralization in the skarned rock of a volcanogenic – mentary sequence where the boundaries dip at 30° to 50°. In seeking this objective, resort was made to the polar method. The three-component clusters were spaced 25 m apart. Records were produced by a commercial Poisk–I–48–OV station. The spread length was 350 m, the shot-point spacing was 700 m, and the travel-time curve was 1400 m.

Consider the wave field produced from one shot-point within a profile interval where it appears very complex on the Z-component seismograms (Figure 62). No regular waves appear on the fixed-component seismograms produced by a three-component XYZ cluster (Figure 62a), except a shear wave on the X component. It is only within some portions of the profile and at different times that some waves may be noted, but they are unstable, and no continuity can be established. In contrast, PPC applied to analog materials based on fixed-component seismograms detected a multiplicity of regular waves being propagated in various directions and at various velocities. Figure 62b, c shows combined tracking-component seismograms for each wave.[1] The waves have been isolated by applying discrimination in terms of displacement direction at a point.

Wave identification may be further improved by raising the signal-to-noise ratio, which can be done by applying the traditional methods based on space-time discrimination to the composite tracking-component seismograms. In the present case, however, this was not necessary.

The travel-time curves of all waves observed in the time interval up to 0.5 s are correlated at common points (Figure 63), and these points serve as a basis for the construction of the section. Because the observations were made by the 'point-line-plane' scheme, the section is constructed in a vertical plane, so that space analysis can only be done in terms of wave polarization. Analysis in terms of displacement directions shows that most of the boundaries intersecting the intrusion are associated with transverse waves (shown by the dashed lines). Referring to the section, it is seen that the medium in the near-contact zone has a very complex structure. This is related to the complex contact of the intrusion dipping northwards and a layered-sedimentary complex dipping in the reverse direction, the presence of diorite-porphyrite dykes and faults. The boundaries in the sedimentary sequence are more extensive and more conformable than the intrusive contact.

The presence of a large number of waves in the wave field is confirmed by PPC carried out on the VSP data obtained at drill-hole 535. From the combined tracking-component

[1] The surface observations quoted here were processed by T. G. Chastnaya and R. N. Khairutdinov, at the KazVIRG.

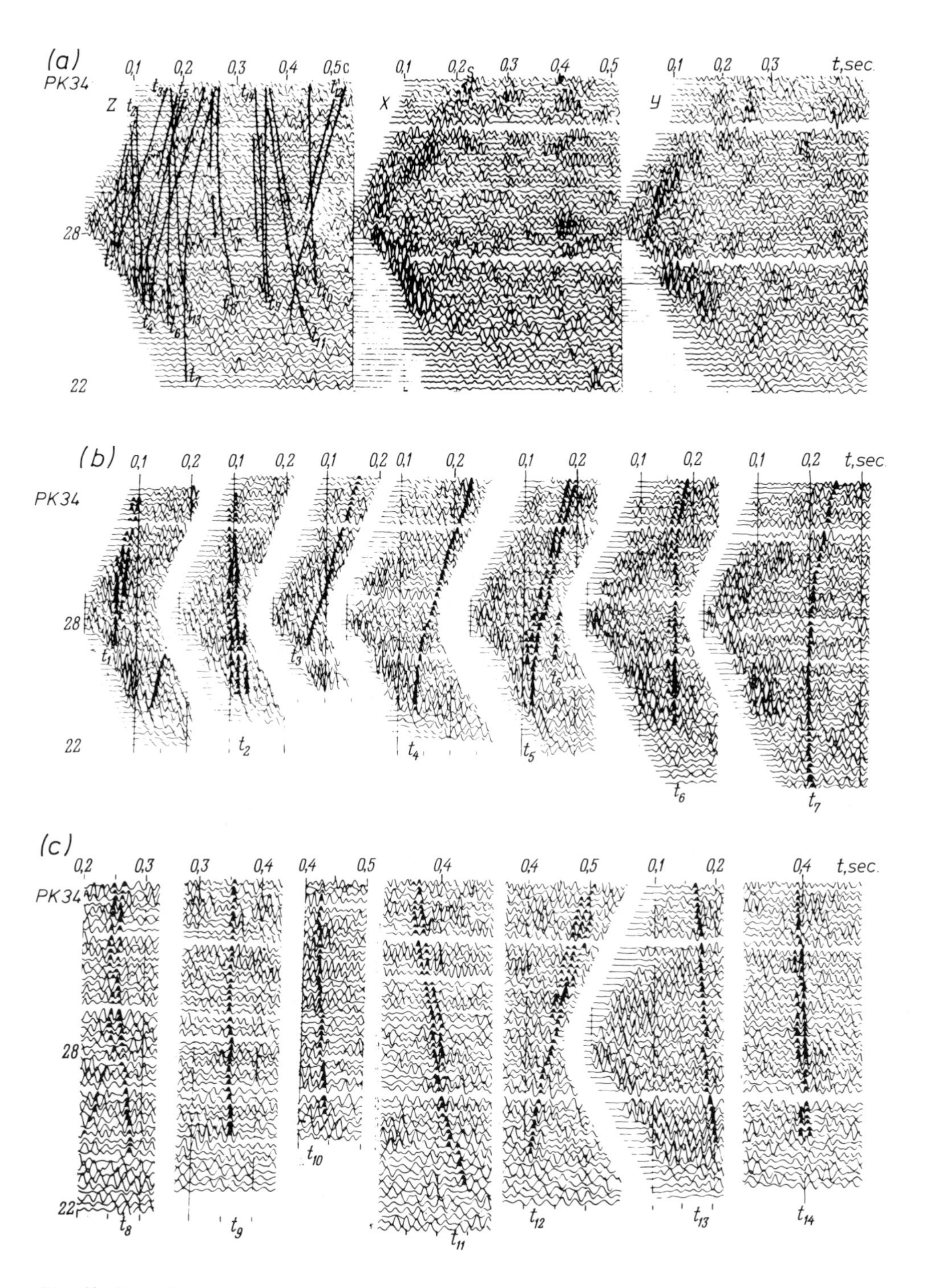

Fig. 62. Wave field plotted in surface observations. (a) Z, X, and Y fixed-component seismograms; (b, c) tracking-component seismograms for the detected waves.

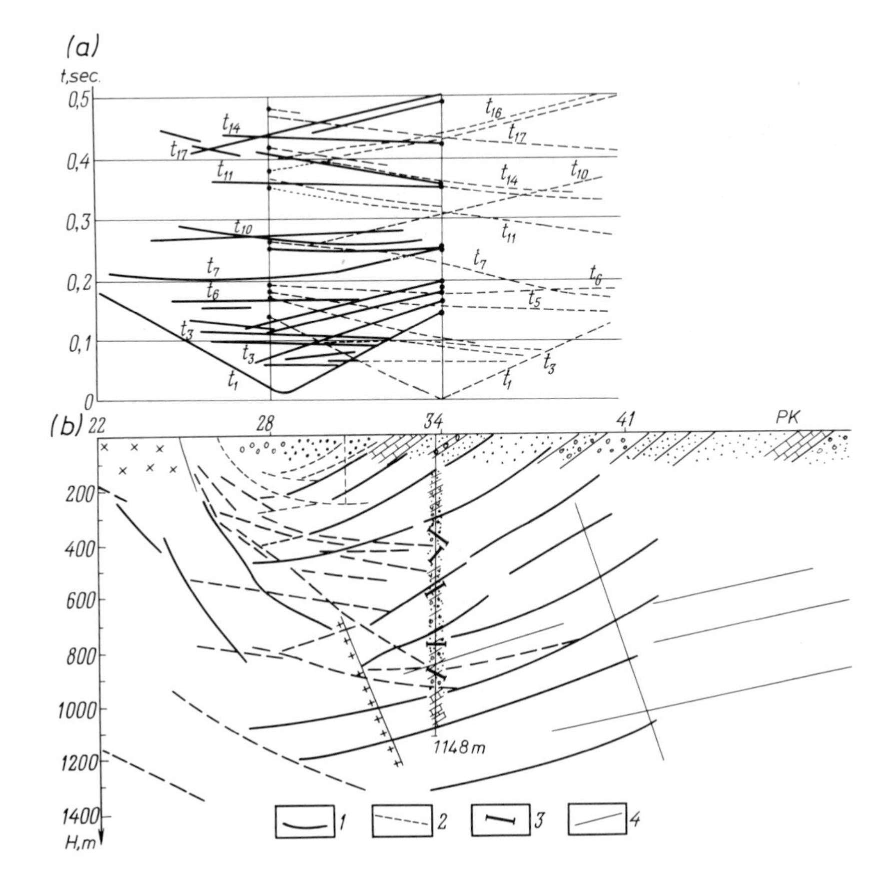

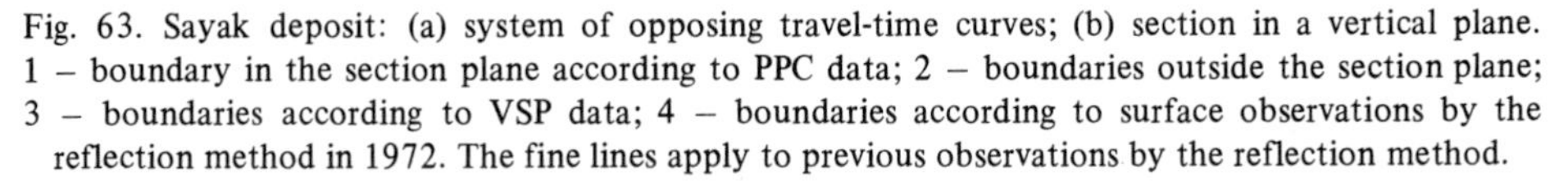

Fig. 63. Sayak deposit: (a) system of opposing travel-time curves; (b) section in a vertical plane. 1 – boundary in the section plane according to PPC data; 2 – boundaries outside the section plane; 3 – boundaries according to VSP data; 4 – boundaries according to surface observations by the reflection method in 1972. The fine lines apply to previous observations by the reflection method.

seismogram in Figure 64b, it is seen that the ground surface is reached by a large number of regular waves, which would be recorded on a horizontal profile. On the other hand, the Z-component seismogram (Figure 64a) shows no regular waves except for a direct S wave, as in surface observations.

Use of PPC in Seismic Prospecting for Oil

Experience with and the results of PPC as applied to seismic prospecting for ores have warranted the application of PPC to the more complex oil- and gas-prospective regions. To this end, acting in cooperation with KazVIRG and the ‘Krasnodarneftegeofizika’ Trust, IEP AS USSR has, since 1973, been carrying out large-scale observations, both on the surface and in drill-holes, in regions widely varying in geological structure. By far the largest effort has been undertaken on the Taman Peninsula where sharp diapir

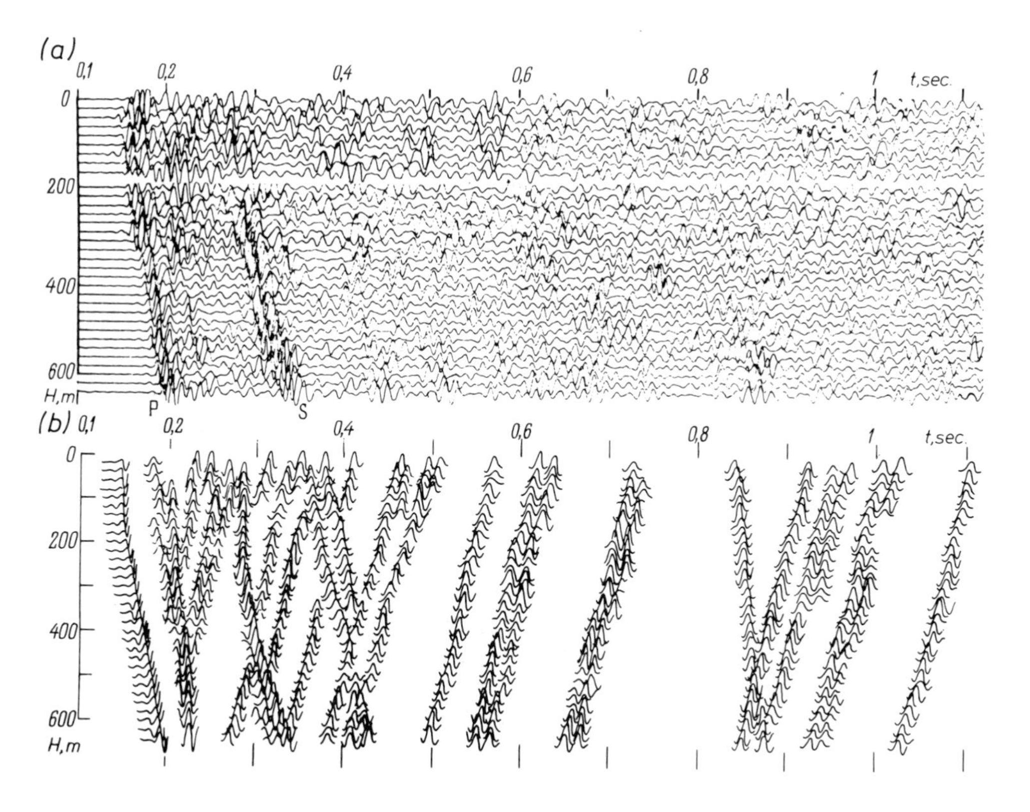

Fig. 64. VSP seismograms (DH 535, l = 760 m, Sayak deposit). (a) Z-component seismogram; (b) tracking-component seismograms.

folds are abundant and where complex spatial seismic-survey systems have proved ineffective.

We shall demonstrate the effectiveness of polarization-position correlation, using as examples surface and drill-hole observations under relatively simple platform conditions and in complex geologic structures (diapir folds).

PLATEAU CONDITIONS

Let us consider the VSP observations carried out in the Kushchevsk area, Krasnodar Region (USSR), characterized by a relatively simple horizontally-layered structure and relatively thin sediments (about 1400 m). Figure 65 shows VSP seismograms recorded in borehole 17 from two shot-points located at 1200 m and 1500 m from the drill-hole.

We shall first examine seismograms (Figure 65a) for the spatially fixed Z, X, and Y components and also two seismograms for components in a local coordinate system.

Since the shot-point lies 1200 m distant from the line of traverse, the first compressional wave on the Z component is poorly defined. On the seismogram for the P component, which is the tracking one for the first wave, its continuity is positively established. On the other hand, the reflected waves within the initial portion of the

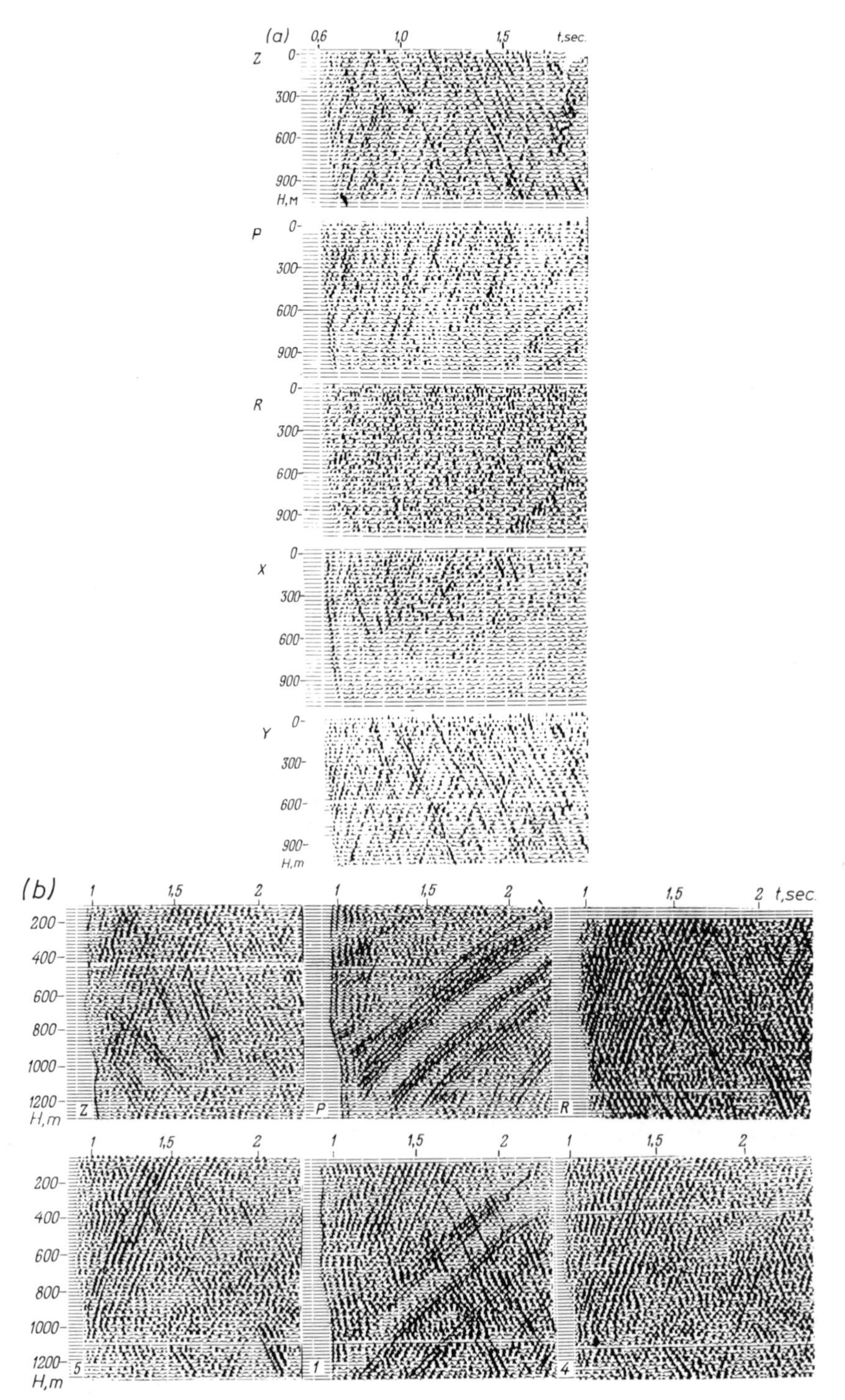

Fig. 65. VSP seismograms for various components. (a) Z, Y, X, P, and R components (from top to bottom) at l = 1200 m; (b) same, at l = 1500 m.

seismogram can only be detected in the topmost part of the profile. For the reflected waves, the tracking components are the R components, because their records display no incident waves. On the combined R-component seismogram, the reflected waves are detected positively both in the initial and in the later parts of the seismogram. It must be noted, however, that the R components can be used as tracking components only for the reflected waves in the first arrivals. For the waves in the later part of the record, other tracking components may be found.

The X-component seismogram mainly displays low-velocity waves in the later part of the record. The Y-component seismogram shows no regular waves, which can be explained by the axial symmetry of the section.

Figure 65b shows seismograms for a vertical profile intersecting almost all of the sedimentary sequence. The profile was shot in order to detect a transmitted converted wave from the basement under conditions close to the appearance of a basement-refracted wave in the first arrivals. With l = 1200 m, the wave recorded within the profile portion adjacent to the basement is still the incident wave. At l = 1500 m, the first recorded wave within the same profile portion is already a wave refracted by the basement.

On the Z-component seismogram, the first wave appears feeble and poorly defined. It is seen that the first arrivals along the profile are formed by different wave types differing in apparent velocity; these waves alternate in such a manner that the travel-time curve has a complex shape with sudden inflections. Obviously, under such conditions the fixed Z component cannot serve as a tracking component for the first waves. With more confidence, the first wave can be detected on the P-component seismogram. Because the R-component seismogram does not record the P wave, the reflected waves can be traced practically directly from the reflecting boundaries. The initial part of this component seismogram displays a multiplicity of singly reflected compressional waves being propagated towards the ground surface, and also multiple waves reflected from the ground surface and being propagated inwards. The compressional wave critically reflected from the basement (at the surface, it is recorded at about 1.7 s) can most positively be detected on component 5. The converted waves are most positively detected on components P and l, which show no reflected compressional waves.

Thus, despite the relative simplicity of the medium, the wave fields are characterized by a large number of wave types. None of the fixed components (Z, X, or Y) excited from any shot-point can be used as a tracking component. The first compressional wave and the reflected compressional waves in the initial portion of the seismogram can best be detected and their continuity best established on seismograms for tracking components which are the respective local coordinate axes (p and r) referenced to the direction of propagation of the P wave. Various waves recorded in the later part of the seismogram are detected and tracked on components variously oriented in space. Joint analysis of the various components with the use of PPC makes it possible to dismember the wave field and to detect and trace waves widely differing in nature, but associated with the same discontinuities in the section, so as to obtain a still better knowledge of the medium.

The use of PPC in tracking reflected waves already in the first arrivals, as demonstrated in the above examples, is of special importance in VSP work, because it substantially improves the accuracy of stratigraphic correlation of the reflected waves. It is to be recalled that with Z-component records alone, a reflected wave cannot be tracked near the reflecting boundary, because of the presence of a long and complex train of incident waves characterized by high intensity and occupying practically all of the initial portion of the seismogram.

OBSERVATIONS UNDER DIAPIR FOLD CONDITIONS

The Taman Peninsula is abundant in diapir and crypto-diapir folds, associated with a thick sequence of plastic clays, which form piercement cores and crop out on the surface. The Neogene deposits deformed into diapir folds form extensive sublatitudinal ridges alternating with synclinal troughs. The folds have steep slopes, and the dips of the limbs range from 40° to 80°. Because of this, the upper portion of the section displays steeply-dipping interfaces with sharply differing spatial orientation.

In the circumstances, different wave fields are excited not only in different areas, but also in different parts of the same area. Within each site, the pattern of the wave field varies according to the relative position of the line of traverse, the source, and the diapir structures. Among the basic factors responsible for the complex wave-field pattern is the superposition of a large number of waves polarized in different directions and differing in both velocity and direction of propagation. On the basis of the wave-field pattern, three zones may be singled out schematically: a zone of interstructural troughs, a pre-diapir zone, and a diapir zone. The seismograms recorded in the zone of interstructural troughs usually show reflected waves at arrival times up to 2.5 s, associated with the top part of the section (as high up as the roof of the Maikop deposits). At later arrival times, a complex interference-affected record is produced.

In the pre-diapir zone, the range of arrival times for clear reflections from the boundaries in the superincumbent Maikop strata is reduced to 1–1.5 s, and the record strength is somewhat lower than in the zone of interstructural troughs. The diapir zone is characterized by random records with durations of 3.5 to 4 s and a relatively low intensity. Practically no regular waves can be noted on the seismograms. The wave-field pattern becomes still more complicated in the near-arch and arch areas.

Under such conditions, despite all the effort expended for nearly 20 years (with short interruptions) in developing seismic-surveying techniques specifically adapted to the diapir tectonics of the Taman Peninsula, reflected waves associated with deep-seated boundaries were not detected either in surface observations or in VSP work. Because of this, seismic prospecting in the Taman Peninsula, so promising in terms of oil and gas, has all but been discontinued in recent years.

Trial runs using the polar method were carried out in the Fontalovskaya and Starotitarovskaya areas. Basically, the technique used was a combination of three-component observations at a point and three-dimensional observations in space. To this end, spatial spreads were used in each area.

Observations in the Fontalovskaya Area. The spatial spread employed in the Fontalovskaya area had its northern end located in an interstructural trough and its southern end in the near-arch and arch parts of a diapir fold. The spread consisted of a number of horizontal and vertical profiles. At the intersections of meridional and latitudinal profiles, drill-holes 500 m and 200 m deep were put down. All profiles (both surface and drill-hole) of the spread were shot from three shot-points. One shot-point (SP1) was located in a trough west of the spread, and the remaining two south of the spread in the diapir zone. The general aspect of the seismograms varied with the position of the profiles and shot-points relative to the diapir structures.

The wave-field pattern as a whole is illustrated on the seismograms recorded from SP1 along an in-line profile (Figure 66) and an off-line profile (Figure 67). The seismograms display reflected compressional (*PP*) waves, reflected and transmitted converted (*PS*) waves, direct shear (*S*) waves and reflected shear (*SS*) waves, and also low-velocity diffracted (*PS**) waves.

The initial part (1–1.5 s) of the seismogram recorded on the in-line profile shows a series of high-velocity compressional waves reflected from shallow horizontal boundaries in the supra-diapir sequence. For some of them, continuity is positively established all along the profile. The later part of the seismogram (2–3.5 s) shows reflected compressional (*PP*) waves associated with depths of prospecting interest. The converted reflected (*PS*) waves differ from the *PP* waves in somewhat reduced apparent velocity and lower frequencies. They form an extended group of waves recorded over a wide time interval (from 0.5–3 s). The direct shear (*S*) wave near the shot-point interferes with surface waves to produce an extended group of strong dispersive waves. On moving away from the shot-point, the shear waves overtake the surface waves, so that two groups of shear waves, S_1 and S_2 may be noted, differing in propagation velocity and polarization (*SH* and *SV*, respectively).

As a rule, the reflected shear (*SS*) waves are associated with boundaries in the supra-diapir bed.

The seismogram recorded on the offset profile (see Figure 67) shows the same groups of waves as the seismogram for the in-line profile. The wave-field pattern in this case differs, because the southern end of the profile (PKO–5) intersected diapir structures. Because of this, the *PS* waves in the initial portion of the seismograms have negative velocities, and their presence is detected already in the first arrivals. These waves are mainly recorded on horizontal and inclined ($\psi = 30°$) components.

The low-velocity diffracted waves of the *PS** type are associated with the wedging-out areas of the diapir flanks.

Sketchy as it is, the description of the wave field given above shows that the field is the product of interference between a large number of waves widely differing in nature. The presence of a low-velocity layer is responsible for the fact that the waves are naturally grouped around different components – the compressional waves around the vertical, and the shear waves around the horizontal traces. This does not, however, mean that the vertical or horizontal traces may serve as tracking components for the respective waves.

Now we shall demonstrate the capabilities of PPC within some selected parts of the profiles. Notably, we shall show that different waves may be recorded as differently on components only slightly differing in their direction in space. For example, compare seismogram 23 with seismograms 17, 18, 21, and 22 (Figure 68) recorded for the profile segment extending from PK16^{80} to PK17^{95} (see Figure 66).

At times up to 1.5 s, the records look all very much alike. At time $t = 1.5$ s, the t_1 wave can be positively identified on seismogram 21, less positively on seismogram 23, and practically not at all on seismograms 17, 18, and 22.

At time $t \approx 1.6$ s, a strong group of t_2 waves is noted. The record of this group on

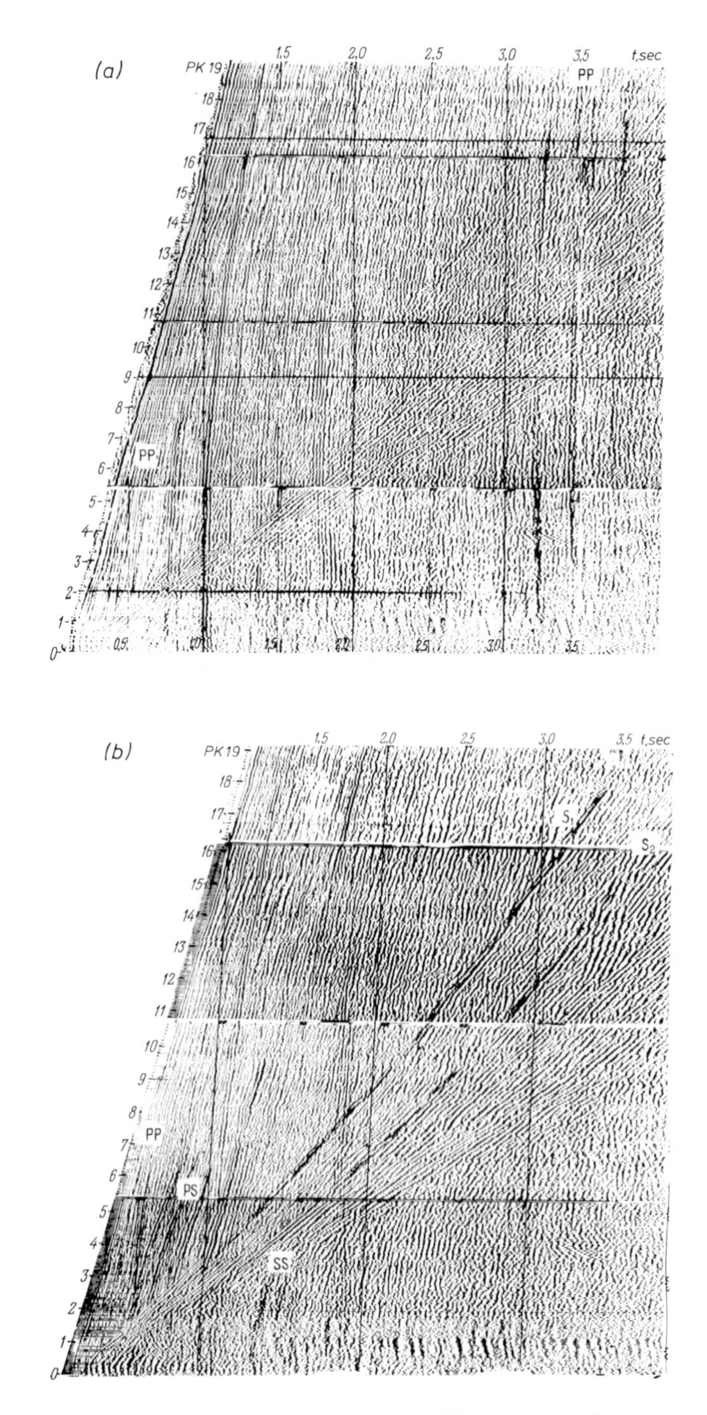

Fig. 66. Fontalovka area: (a) Z-component seismogram and (b) component III seismogram along an in-line profile.

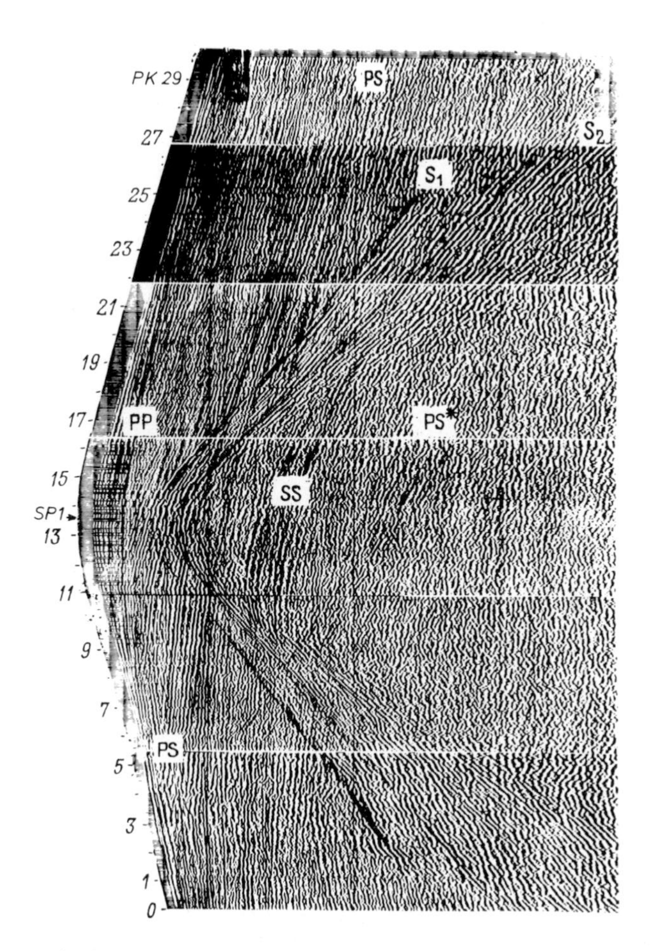

Fig. 67. Fontalovka area: seismogram along an off-line profile.

seismograms 17, 22, and 21 substantially differs from the records on seismograms 18 and 23. The latter display two waves, t_2' and t_2'', the former having a somewhat higher v_a and frequency. On seismograms 23 and 18, the t_2 group cannot be resolved.

At time $t \approx 1.7$ s, the t_3 wave can best be detected on seismogram 22, but it scarcely appears on seismograms 18 and 23; on seismogram 17 its correlation is dubious.

The group of t_4 waves (at time $t \approx 1.8$ s) can be divided into two waves, t_4' optimally recorded on seismogram 18, and t_4'' optimally recorded on seismogram 21. On seismograms 17 and 22 the two waves appear unresolved. It is interesting to note the record of this wave group on seismogram 23 where the t_4'' wave appears in the initial part of the section, whereas the t_4' wave is not to be seen at the early positions.

The t_5 wave ($t \approx 1.9$ s) can readily be identified on its tracking component record 22, but it does not appear on seismogram 21 and is detected with difficulty on seismogram 23. As regards the t_6 wave ($t \approx 2$ s), its phase can be identified on component record 23 where it causes the travel-time curve to bend. On component records 17 and 22 it

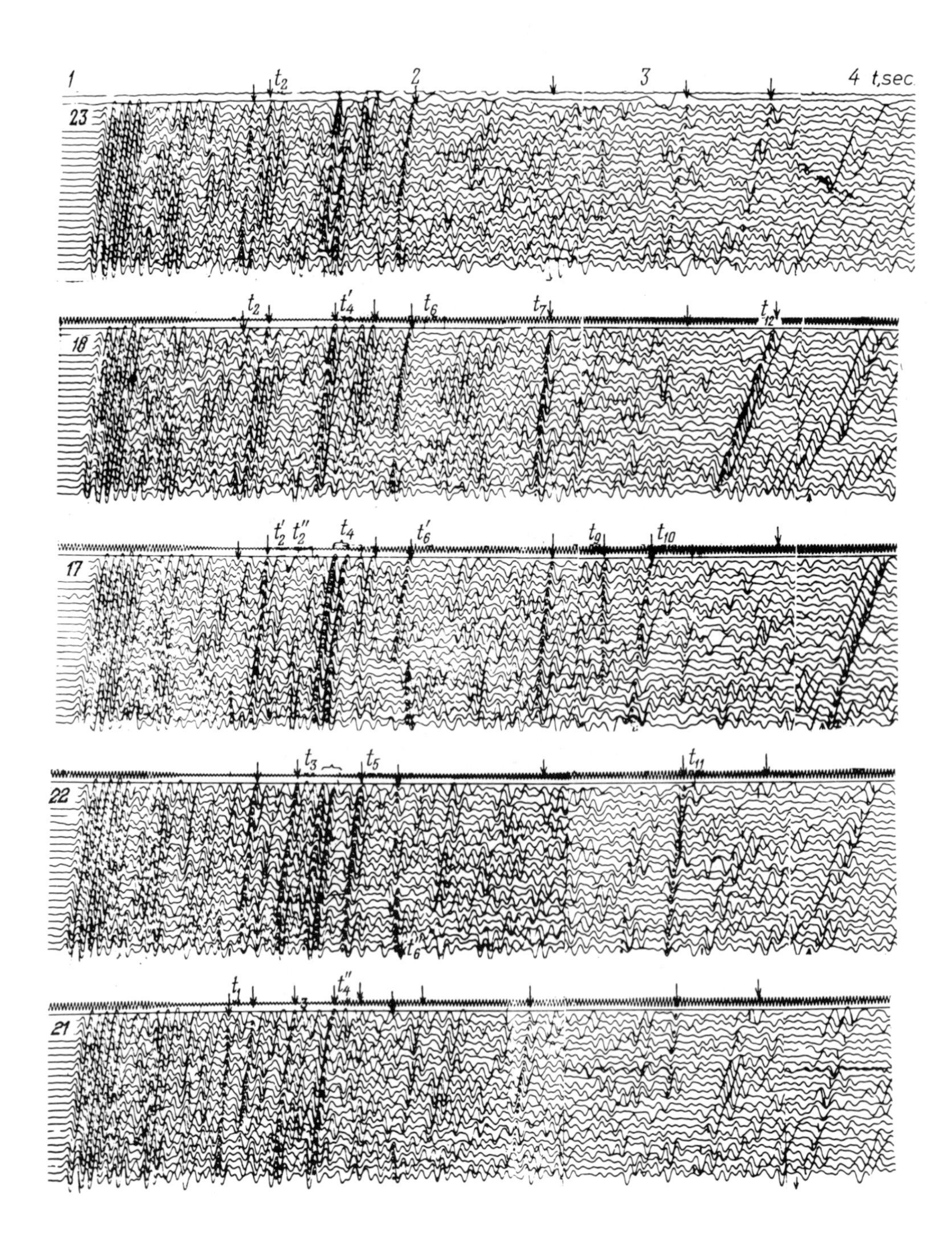

Fig. 68. Fixed-component seismograms over an in-line profile ($PK16^{80}$–$PK17^{95}$).

is seen that the bending is caused by the superposition of waves differing in the sign of apparent velocity: for t_6' the apparent velocity is positive, and for t_6'' it is negative. The t_6' wave may be reliably tracked on component record 18, but it is not seen on component record 17. At later arrival times (t = 2.4 s), the t_7 wave can readily be noted on component records 17 and 18, whereas on vertical component record 23 it is barely seen. The t_9 wave ($t \approx 2.9$ s) is optimally displayed on component record 17, the t_{10} wave ($t \approx 3$ s) on component records 17 and 22, the t_{11} wave ($t \approx 3.2$ s) on component record 22, the t_{12} wave ($t \approx 3.4$ s) on component record 18, and the t_{13} wave ($t \approx 4$ s) on component record 17. Although their intensity is high on component records 17 and 18, the t_{12} and t_{13} waves are scarcely seen on component record 23. According to their v_a, the t_1 wave (component record 21) and the t_0'' wave (component record 22) may well be reflected compressional *PP* waves. At a distance of $l \approx 2$ km from the shot-point, the *PP* waves differ very little in v_a from the dominant spurious waves mainly represented by the refracted multiple waves *PPP* and reflected converted waves *PS*. Because of this, their detection with the aid of polarization-position correlation is of obvious interest.

To sum up, for a sizeable proportion of waves, the tracking components are not verical, but inclined, lying in two directions roughly perpendicular to each other.

The substantial difference between the records of adjacent components can be seen also within the profile segment between PK3[60] and PK4[75] (Figure 69). Component

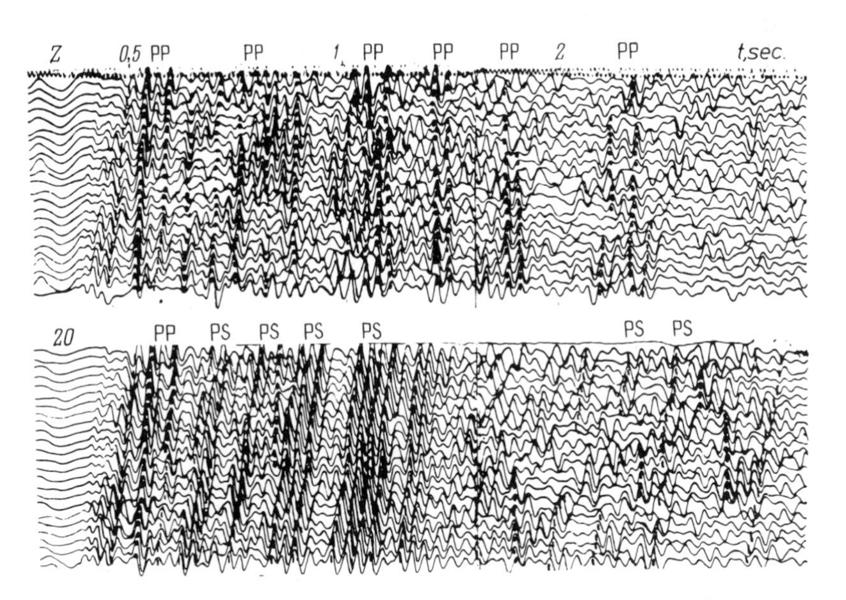

Fig. 69. Fixed-component seismograms over a longitudinal profile (PK3[60]–PK4[75]).

record 23 mainly shows reflected compressional *PP* waves, at times up to 2.2 s, whereas component record 20 predominantly displays reflected converted *PS* waves from shallow boundaries. No compressional waves appear on component record 20, despite the fact that tracking components 21–23 are close in direction to component 20. The *PP* and *PS* waves, dominant on adjacent component seismograms, scarcely upset each other's correlation, except for the strong group of *PS* waves at time t = 1.1 s, which somewhat distorts the *PP* wave on component record 23 at the same instant of time.

Consider a segment of the offset profile between $PK10^{80}$ and $PK11^{95}$. At times up to 1 s, component record 7 displays a complex, almost unresolved wave field (Figure 70). No dominant regular waves are to be noted here. At the same time, component record 8 highlights groups of compressional waves t_1 and t_2 associated with shallow reflecting boundaries. On the adjacent component records 7 and 9, these waves are barely noticeable. In the same time interval one can readily identify converted waves t_3, t_4, and t_5 which appear clearly on component record 7 and not at all on component record 8. Their tracking component is on record 6.

The t_6 wave may be reliably tracked on component record 9, whereas on component record 10, it does not appear at all, nor does it appear on component record 23. In contrast to the t_6 wave, the t_7 wave may be reliably detected on several component records, including component record 23. For the low-frequency *SS* waves ($t \approx$ 1.3–1.7 s), the tracking component is on record 6 where the wave group poorly resolved on other component records appears clearly divided into two dominant waves t_8 and t_9. The low-velocity t_{11} and t_{12} waves ($t \approx$ 1.9–2.2 s), securely traceable on component records 7 through 9, scarcely stand out on component record 10. The reflected compressional wave t_{10} (t = 1.8 s) may be reliably singled out on component record 23 and is almost unnoticeable on the other component records listed. For a group of compressional waves reflected from deep boundaries ($t \approx$ 3–3.5 s), the tracking component is on record 8, whereas on the other component records these waves are hardly seen.

Consider several component seismograms for the profile segment between $PK9^{60}$ and $PK10^{75}$ (Figure 71). The initial portion of component record 1 shows three converted waves (t_1, t_2, and t_3); and that of component record 2 only one such wave (t_2). The tracking component for these waves is on record 16. In the same time interval, component record 23 shows other waves. The shear t_4 wave is optimally displayed on component record 16, whereas it scarcely appears on component record 2. Reflected shear waves are recorded in about the same manner on all the component records listed. The low-frequency t_5 waves are mainly registered on component record 2.

The above examples show that in surface observations the tracking components, even for compressional waves, are inclined and not vertical components, and that wave correlation on the basis of tracking components may substantially improve wave-field analysis and help identify various wave types.

The fixed-component seismograms recorded for various waves in the Fontalovka area have served as a basis for stereograms, which show the spatial disposition of tracking-component areas (Figure 72). On the in- and off-line profiles, the areas of tracking components for the same waves are disposed identically relative to the ray plane Q_P of the first compressional wave (with allowance for the difference in profile orientation). The area of tracking components for reflected compressional waves lies to the right of the ray plane, which is another way of saying that the azimuthal deviation from the ray plane is $\Delta\omega = 90°$. The shear waves near the compressional-wave front are polarized in two mutually perpendicular directions, namely, nearly along the ray plane (S_2) and at right angles to the ray plane (S_1). The reflected shear waves *SS* and the reflected converted waves *PS* are oriented horizontally in the ray plane (the *SV* components). The low-velocity waves of the *PS** type are polarized in horizontal directions perpendicular to

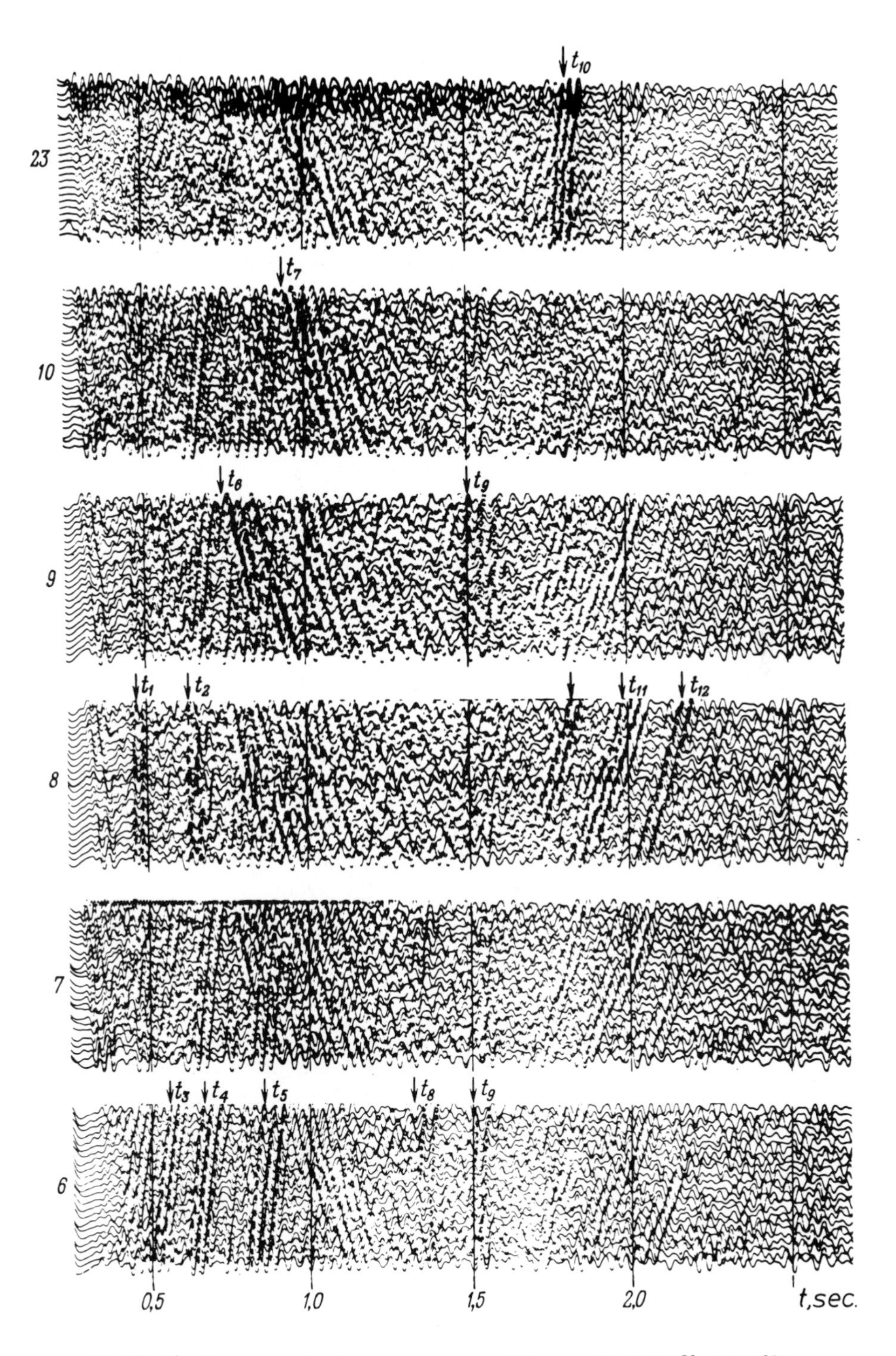

Fig. 70. Fixed-component seismograms over an off-line profile (PK10[50]–PK11[95]).

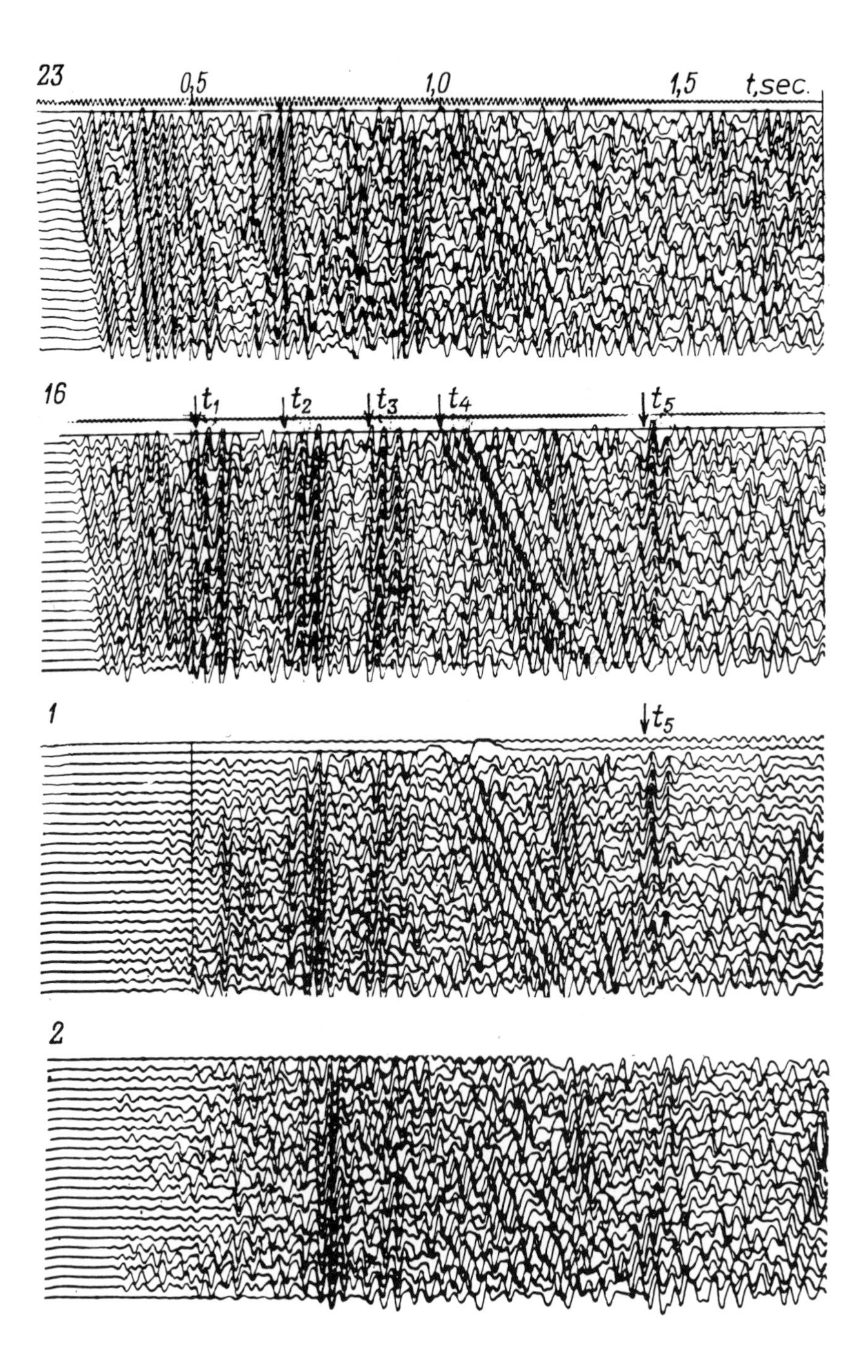

Fig. 71. Fixed-component seismograms over an in-line profile ($PK9^{60}$–$PK10^{75}$).

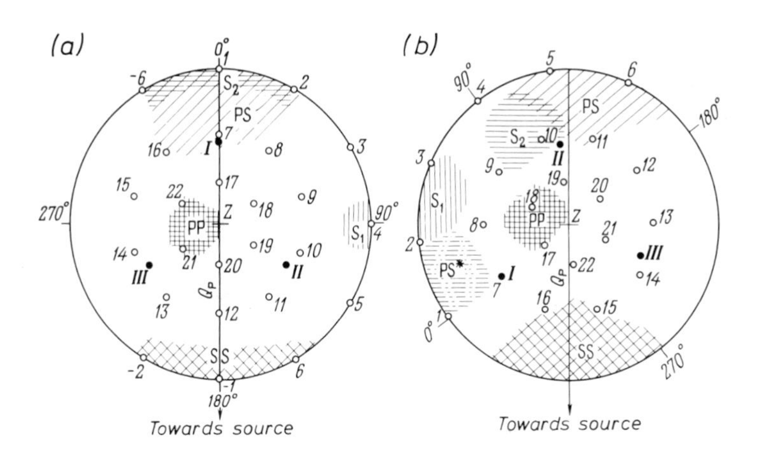

Fig. 72. Stereograms showing areas (fields) of tracking components for (a) an in-line profile and (b) an off-line profile.

to the ray plane. The accompanying stereogram shows that the polarization method can identify waves whose displacements cover practically all space.

Observations in the Starotitarovka Area. This area is part of the Karabet anticlinal zone comprising three uplifts whose domes contain Maikop clays. According to reconnaissance observations carried out in deep drill-hole 2, the vertical profiles are dominated by shear waves in addition to a direct compressional wave. The vertical profiles have been used to determine the velocity gradient. Component record 23 shows between the P and S waves a large number of waves being propagated downwards with the same velocity as the first compressional wave. These are various multiple waves associated with both the ground surface and the intermediate interfaces. The direct shear (S) wave produces a complex record varying along the profile and from record to record. Because of the strong background noise, no regular waves associated with deep boundaries and being propagated towards the ground surface are detected on the Z-component seismograms. Polarization-position correlation has shown [40] that the first-arrival record retains its general aspect all along the profile, and that of the large number of waves reaching the ground surface at least some are associated with deep-lying boundaries. An especially large number of waves (incident, reflected, compressional, shear, and converted) can be discerned in the top part of the profile (500–1500 m). At the time of field work, the fact that the reflected waves were recorded at later times (3.5–4 s) was interpreted as pointing to the existence of deep reflecting horizons, and warranted further exploration.

In order to ascertain how reliably reflected waves could be detected, resort was made to a spatial spread made up of a multiplicity of surface and vertical (drill-hole) profiles (Figure 73a). The drill-holes were put down at the intersections of the surface profiles. The drill-hole at the intersection of profiles 11, 14, 15, and 16 was 1600 m deep; those at the other intersections were 500 m deep. The spatial spread may be visualized as a set of individual three-dimensional elements each of which consists of three profiles (two horizontal and one vertical) intersecting at one point. All the profiles of the spread were

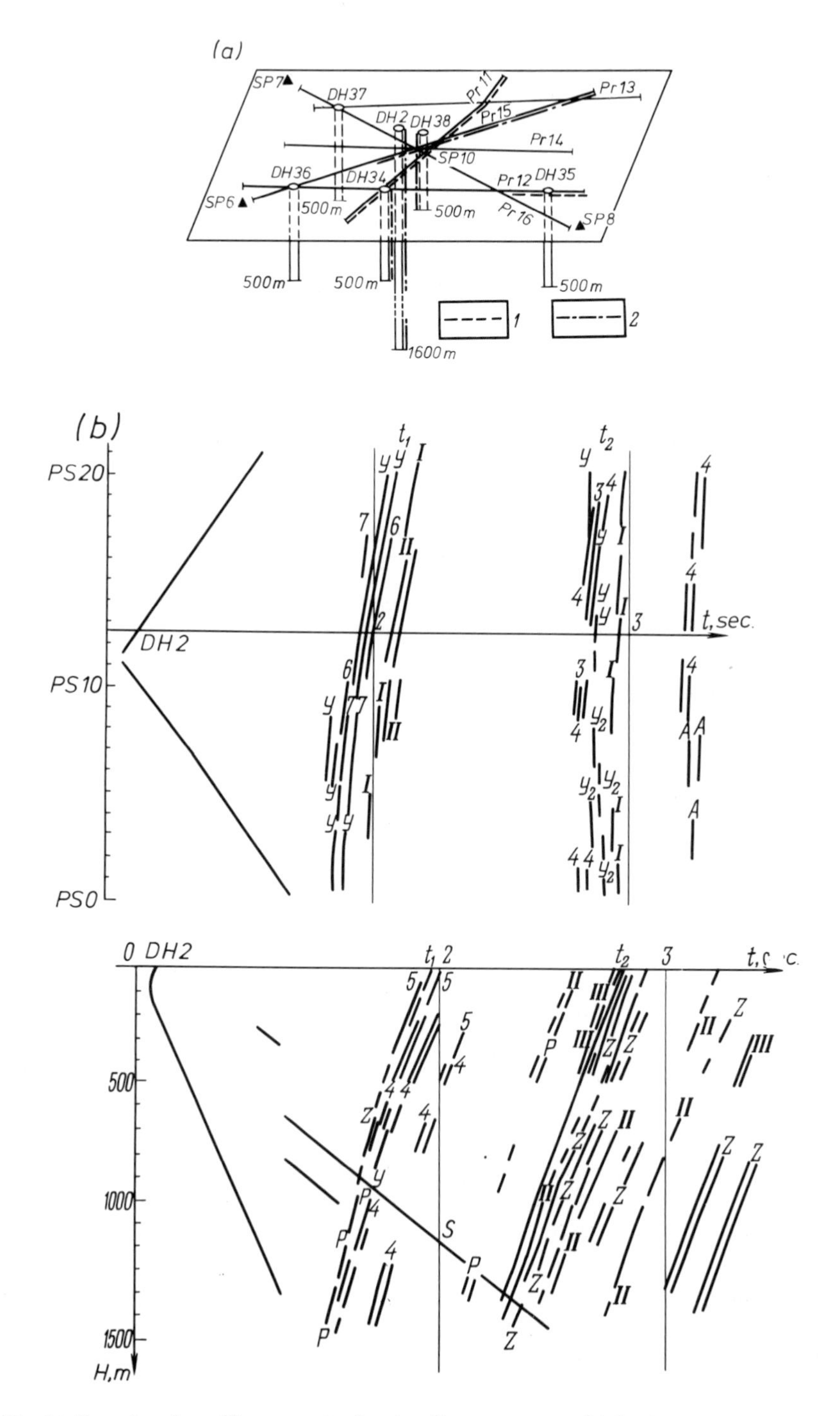

Fig. 73. (a) Spread and profile segments showing Z components. (b) horizontal and vertical travel time curves for waves t_1 and t_2: 1 – for t_1 wave; 2 – for t_2 wave.

shot from four shot-points (6, 7, 8, and 10), with SP10 located in the centre of the spread and the remaining at the ends of the surface profiles.

The wave field recorded in the Starotitarovka area substantially differs from that observed in the Fontalovka area in that the surface Z-component seismograms show practically no regular waves.

The observation data were interpreted on the basis of seismograms for spatially-fixed components and components in a local coördinate system referenced to the direction of first arrivals at each point. The data were processed, and the respective seismograms were prepared by the 'Polyarizatsiya' automated system developed by Starodvorsky of the 'Krasnodarneftegeofizika' Trust.

At times up to 2 s, the seismograms show a large number of strong spurious waves. These are mainly converted waves reflected from high-dip boundaries and also direct shear and associated secondary waves. The waves in the top part of the section were detected and analysed from a comparison of a large number of seismograms and combined travel-time curves based on records for tracking components. It is relevant to note that the vertical component was also not always a tracking component, either in VSP work or in surface observations. On the longitudinal profiles, correlation is optimal for waves on the components other than those lying in a vertical plane. Correlation deteriorates as the arrival (recording) time increases.

Practically no useful wave could be detected all along the profile on a seismogram for any one wave component in the time interval 2–3.5 s. Because the noise level in surface observations is usually very high, and there are no rigorous criteria for wave discrimination, the technique basically reduced to noting reflected waves on vertical profiles on the basis of apparent velocity. As the next step, such waves were identified with those on surface profiles, and their continuity established on surface profiles by means of polarization-position correlation. In this way, the same waves were identified and followed on all the profiles of the spatial spread. As an example of observations in the time interval from 2–3.5 s, we shall take up observations made from SP10.

Within some short portions of the profiles, the waves form very short line-up patterns and continually interfere with spurious waves. Since, however, the conditions for interference differ from wave to wave component, it is possible to establish continuity for the waves all along the profile on different component records. Figure 73b shows combined horizontal and vertical travel-time curves plotted on profile 16 for two groups of waves recorded at the surface at t_1 = 1.9 s and t_2 = 2.8 s. The travel-time curves are intermittent because they are compiled on the basis of seismograms for different components. The travel-time curve for the t_1 wave is more intermittent because the wave is recorded in an area of spurious waves being propagated downwards. The line-up patterns for the t_2 wave are more extended. Of course, more wave groups were noted in addition to the t_1 and t_2 waves, but they were less consistent. Therefore, we shall dwell only on the t_1 and t_2 waves.

The t_1 and t_2 waves were identified and their continuity established on the surface and vertical profiles of the spread by means of polarization-position correlation. They were found to be associated with boundaries in the Maikop deposits at depths from 2500 m to 3000 m ($H_9 \approx 3000$ m). The tracking components for the t_1 and t_2 waves were chosen from an analysis of seismograms for both the fixed components and the components in a local coordinate system. Since the source records had been obtained by the variable-density method, the tracking components were chosen for the more stable and extended waves with a specified velocity. It is to be noted that for most profiles the tracking components of the t_1 and t_2 waves were off-vertical components differently oriented in space, and it was only for some segments that the tracking components were the Z components. For the t_1 wave, the tracking component was component 23 on profile 11 and, partly, profile 12. For the t_2 wave, it was likewise component 23 on profiles 12 and 15 and also on the vertical profile at DH–2 (Figure 73a). For the most part, however, the t_1 and t_2 waves could be reliably detected only on off-vertical components. Leaving out wave correlation with reference to surface profiles, it is to be noted that on most profiles, the t_1 and t_2 waves were detected discretely on various components.

Kinematic data for the spatial spread served as a basis for constructing a map of isochrone lines (Figure 74) and for determining the angle of inclination of the boundaries

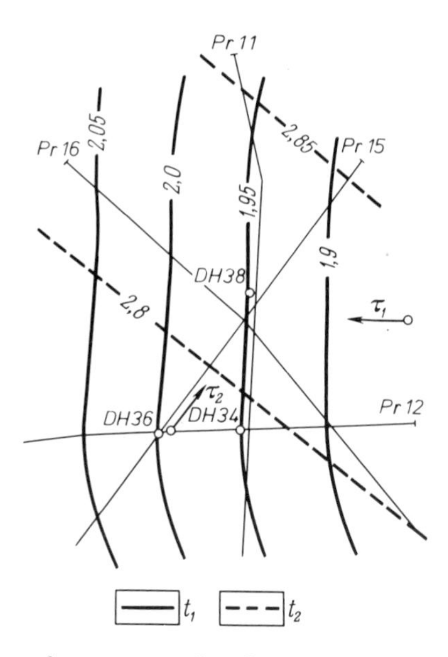

Fig. 74. Map of isochrones for waves t_1 and t_2 (in seconds). τ_1, τ_2 are the field-gradient vectors.

for the t_1 and t_2 waves (they were found to be 20° and 12° respectively.) With allowance for the angles of inclination, the depths of the boundaries under the shot-point were estimated to be $H_1 = 2100$ m and $H_2 = 2900$ m, which agrees with well-logging data. According to the latter, the Maikop sequence has a boundary at 2200 m where the formation velocity jumps from 1940 to 2170 m s^{-1}, and another boundary at 3050 m where the formation velocity jumps from 2170 to 2740 m s^{-1}.

An analysis of the spatial orientation for the tracking components is of special interest. Figure 75 shows the tracking-component areas and their directional fields established on the basis of such an analysis. For the surface profiles, the stereograms are shown at the ends of the profiles. The shaded strips along the profiles mark the position that the tracking sector would take up, if the stereograms were plotted at each point. For the vertical profiles the stereograms are shown at the drill-hole entrances. At the top of each

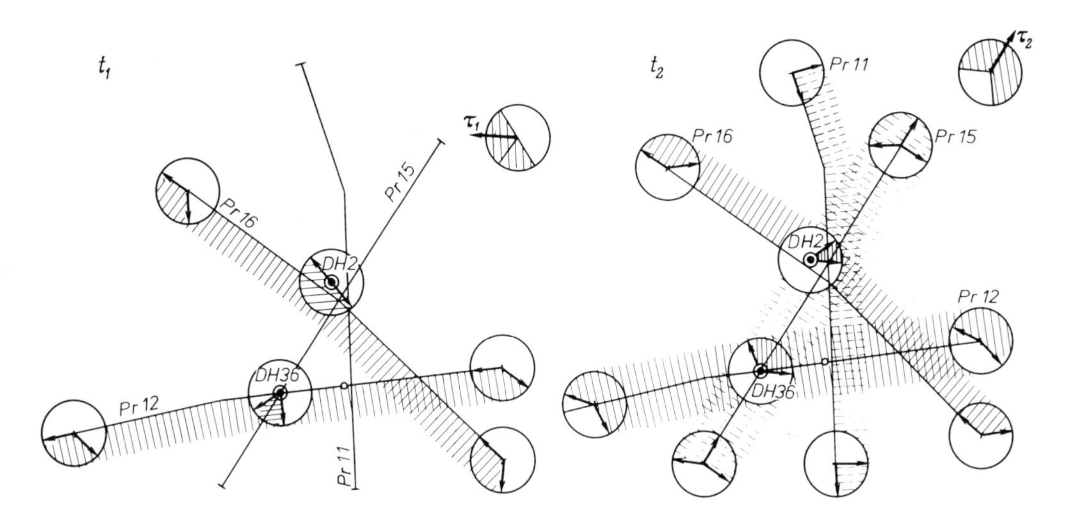

Fig. 75. Direction fields for the tracking components of waves t_1 and t_2 observed on a spread of horizontal and vertical profiles.

field is shown an overall stereogram and the direction of the field-gradient vector. A comparison of data has shown that on the whole the orientation of the tracking components agrees with the kinematic data. This implies, above all, that the orientation of the tracking components is mainly decided by the direction of wave propagation. This is so also because the tracking components were evaluated on the basis of correlation over an extended segment of each profile rather than at a point.

Using polarization-position correlation, the investigators have been able to detect and establish continuity for eight reflecting horizons at depths from 500 to 2100 m. It has also been found that in surface observations the Z components may be used as tracking components only for waves associated with shallow boundaries (not over 1200 m).

In the depth range from 2000 to 3000 m, the investigators have been able to detect and establish continuity for waves over all profiles of the spatial spread. For deep waves, the tracking components were, as a rule, off-vertical rather than vertical components, variously oriented in space.

Use of PPC in Establishing Continuity for Converted *PS* Waves Resulting from Distant Earthquakes

Regional work utilizing transmitted converted waves induced by distant earthquakes and explosions runs into difficulties both when detecting the waves and when establishing their continuity along the line of traverse. To obviate the difficulties, attempts were made to apply PPC to *PS* waves. The observations were carried out on a line profile. The spread, 100 km long, had 35 stations (points 108–140) spaced 3 km apart, which recorded waves simultaneously and were set up at the bedrock outcrops and in the sedimentary sequence. We shall limit ourselves to the observations carried out at the bedrock outcrops where the waves in the initial part of the seismograms arrive in a direction not rigorously vertical, and where various types of waves are recorded by all geophones, including the *XYZ* cluster. Emphasis will mainly be placed on CDR–I, which enables *PS* waves to be analysed in the zero-displacement plane for the *P* wave.

Referring to the position seismogram of a distant earthquake recorded at 0556 hrs. on 27 November 1974, it is seen that the initial portion of the *Z*-component record (Figure 76a) is formed by a group of compressional waves. The aspect of the wave record varies little along the line of traverse. Three waves may arbitrarily be singled out in Figure

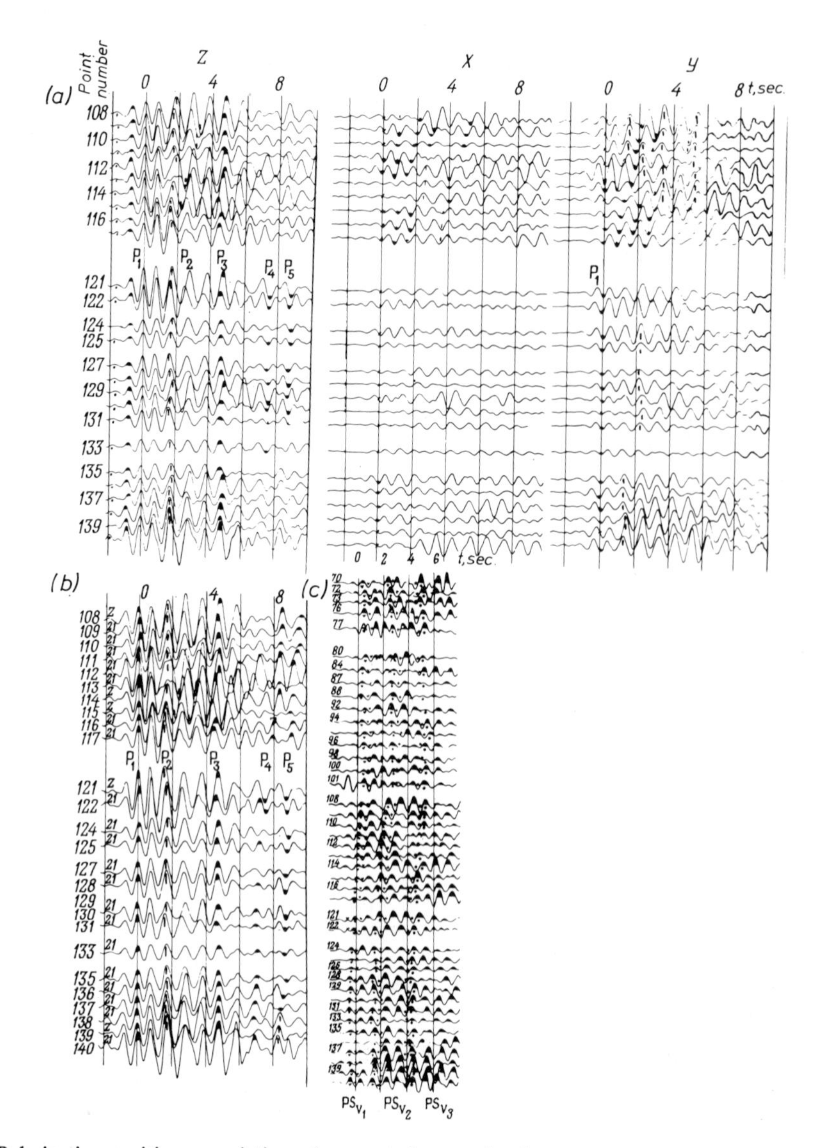

Fig. 76. Polarization-position correlation of converted transmitted waves excited by a distant earthquake. (a) fixed *Z*-, *X*-, and *Y*-component seismograms; (b, c) tracking-component seismograms for *P* and *PS* waves.

76a, namely, P_1, P_2, and P_3. The travel-time curves for all compressional waves in the initial part are parallel to that of the first arrival. The waves forming the initial part of the record are comparable in intensity, which decreases with time at a fairly low rate.

Whether transmitted converted waves can successfully be detected and their continuity established (see Chapter 4) depends to a large extent on the form of polarization of compressional waves. As careful analysis has shown, compressional waves recorded outside the area of interference with other wave types are usually polarized linearly, and that for the waves making up the initial group the directions of particle displacement lie all in the same plane. In practice, however, compressional waves are continually interfering with converted waves. Within the profile segment under consideration, the first converted wave PS_1 arrives 0.4 s after the first compressional wave, and this fact leads to the elliptical polarization of the wave and to errors in determining the direction of particle displacements of the P wave.

The interference of regular waves produces a wave whose characteristics remain fairly constant along the profile. In the case on hand, the waves are polarized in a vertical plane whose orientation varies along the profile within a relatively narrow range of angles. The components that can be used for tracking converted waves, that is, those least distorted by projections of P waves, are chosen by means of polar correlation at each point in the zero-displacement plane for the P wave. On seismograms for the traditional fixed components (Figure 76a) the amplitudes on the Y-component record are substantially greater than those on the X-component record, which is decided by the azimuth of wave arrival. These seismograms show noticeable projections of compressional waves on both the Y and X components. On the other hand, the X- and Y-component seismograms display regular waves, which differ in apparent velocity from compressional waves. These waves were analysed by means of polarization-position correlation. To this end, seismograms for the original components fixed in space and recorded by an ZXY cluster were supplemented by those for the tracking components of P waves (Figure 76b), horizontal components X_P in the ray plane and horizontal components Y_P in a plane perpendicular to the ray plane, and also seismograms of tracking components for PS waves. In the seismogram of Figure 76, phases of converted waves recorded outside the area of interference with P waves are shown filled-in and those inside the area of of interference by dashes. The directions of tracking components for PS waves lie in the Q_{0P} plane (components 8, 9, and 10) and in the azimuth of the vertical ray plane (components 2, 3, and 5). The waves that are best followed on the tracking-component seismograms are PS_1 and PS_2 (points 108–117 and 125–140). The PS_1 wave, converted at a shallow boundary, is linearly polarized and appears as an SV wave. The PS_2 wave can be tracked positively along the profile with Δt = 1.4 to 2 s relative to the first arrivals. In the direction of points 108–117, its move-out time gradually increases to $\Delta t \approx 3.5$ s (point 108). Outside the area of interference with other waves, it is recorded as an SV wave (points 108, 114, 115, and 139).

The PS_3 wave, trackable in the interval between points 108 and 117, is likewise polarizaed as an SV wave. The detected waves basically agree with data for a large number of earthquakes. The most important difference is that PPC applied to the interval between points 108 and 117 detects PS_2 and PS_3 waves, converted at inclined boundaries dipping in the direction of point 108. The presence of inclined boundaries is likewise responsible for the azimuthal deviations of the first compressional wave and the PS waves.

It should be noted that the scope of PPC work carried out with a view to detecting transmitted converted waves is still insufficient for any generalization to be made. Nevertheless, available evidence is sufficient for one to presume that PPC may serve as an efficient tool in analysing the character of waves forming the initial portion of records for distant earthquakes, which fact is vitally essential to further progress in the physics of the transmitted converted-wave method. Of special value in this respect may be planar three-component spreads and selective polarization filter systems. At present, work in this direction is already under way on a sufficiently large scale.

CHAPTER 6

STUDY OF MEDIA ON THE BASIS OF SEISMIC WAVE POLARIZATION

Polar and polarization-position correlation have substantially extended the possibilites of analysing complex wave fields. Still another source of information about the medium is seismic-wave polarization, a very sensitive variable strongly dependent on the structure of the medium. Unfortunately, in surface observations, wave polarization is strongly affected by low-velocity layers and the top part of the section, so the usefulness of the polarization method in the circumstances is very limited, and a good proportion of data thus obtained is of a qualitative character.

Effect of Low-Velocity Layers (LVL) on Wave Polarization

Early work on wave polarization [24, 51–56] showed that in surface observations the directions of particle displacement may well differ from the direction of wave arrival and that this discripancy is mainly due to the strong effect that the top part of the section and low-velocity layers have on wave polarization [30, 34, 35]. The top part of the section and, especially, the low-velocity layer are the most non-uniform parts of the section, whereas the bottom of a low-velocity layer is usually the most sharp boundary at which the velocity of compressional waves may change by a factor of three or four. The effect of a low-velocity layer is suddently to change the direction of particle displacement on passing across the layer. In some cases, the linear polarization of the waves may also be disturbed.

Apart from reflections in the case of surface observations, the direction of particle displacement is substantially affected by discontinuities in the top most differentiated part of the section. The presence of one or more layers close to the ground surface brings about wave interference and changes the direction of particle displacement. The extent of such changes depends on the velocity of elastic waves in the strata and underlying rock, the wavelength λ, and the layer depth h.

Reference [84] gives the apparent angles of incidence of a wave on to a free surface as a function of the ratio λ/h, as derived from theoretical seismograms for interference-produced waves. So long as the ratio λ/h is low, the particles in a P wave are displaced in the same direction as that of the wave in the layer; that is, the final result depends on the top layer alone. If the ratio λ/h is high, the particles in the wave are displaced in the direction in which the wave is incident on the lower boundary, which is another way of saying that the low-velocity layer has no effect on the direction of particle displacement. According to the angle of incidence, the transition from the low to the high values of the λ/h ratio lies in the interval from $\lambda/h = 2$ to $\lambda/h = 4.5$. For one low-velocity layer, the values of λ/h close to the asymptote are observed at $\lambda/h = 10$ to $\lambda/h = 15$. Following Gamburtsev's suggestion, the frequency dependence of the apparent angle of emergence of seismic radiation has been used to trace variations in velocity with depth in the top part of the section.

With advances in vertical seismic profiling, theoretical data about the effect of low-velocity layers on the direction of particle displacement have been supplemented by experimental evidence [36]. Yet, despite the extensive work in this direction, what was learned about the effect of the low-velocity layer in ordinary VSP studies has proved insufficient, first, because the observations were lacking in detail and, second, because the material usually obtained from the top part of the section was of low quality, mainly for technical reasons (elaborate drill-holes, poor quality of grouping etc.). Therefore, special-purpose experiments were carried out in 1975 in cooperation with 'Krasnodarneftegeofizika' Trust, having the objective of establishing the effect of the low-velocity layer on the wave-field pattern. These experiments have shown that under the low-velocity layer the particles are displaced in the same direction as the arrival of waves. It has also been found that the low-velocity layer can critically affect the direction of particle displacement and complicate the wave-field pattern.

OBSERVATION TECHNIQUE

The observations were carried out in two stages. During the first stage, the observations were made along a vertical profile intersecting the low-velocity layer. To make up for the effect of various technical factors, the observations were carried out in uncased drill-holes, all at the same time (from the same explosion) at all depths. To this end, a series of shallow drill-holes differing in depth and lying next to one another were put down at the intersection of profiles 2 and 5 in the Fontalovka area (see Chapter 5). At the bottom of the drill-hole, a three-component symmetrical seismometer cluster was set up, specifically designed for sub-LVL observations. The seismometers were fitted out so that they could be lowered in uncased drill-holes with the aid of a drilling tool, aligned and clamped at the bottom, and conveniently removed upon recording. The observations were carried out at 12 points at a time over a depth range of 0 to 78 m (Figure 77a). The shot-points were arranged on two mutually perpendicular profiles, spaced 100 m apart, the maximum distance being $l = 2000$ m on profile 5 and $l = 1300$ m on profile 2.

During the second stage, work was carried out on a segment of a horizontal profile (Figure 77 b), with observations made simultaneously at 12 points on the ground surface and under the low-velocity layer at a depth of 54 m (which was 2.5 times the depth of the LVL bottom). During both the first and second stage, all shots were fired at a depth of 42 m, that is, under the low-velocity layer. Owing to the specially designed sub-LVL seismometers and simultaneous recording of the same shot over the entire vertical profile on the ground surface and under the low-velocity level, the data obtained are free from any incidental noise and throw undistorted light on the influence of the low-velocity layer on wave polarization.

Now we shall examine the results of observations, beginning with the first arrivals to be followed by secondary waves recorded in the later part of the seismograms. For the first arrivals, the effect of the low-velocity layer on the direction of particle displacement is analysed quantiatively, and for the succeeding waves by comparing the tracking-component regions on the surface and below the low-velocity layer.

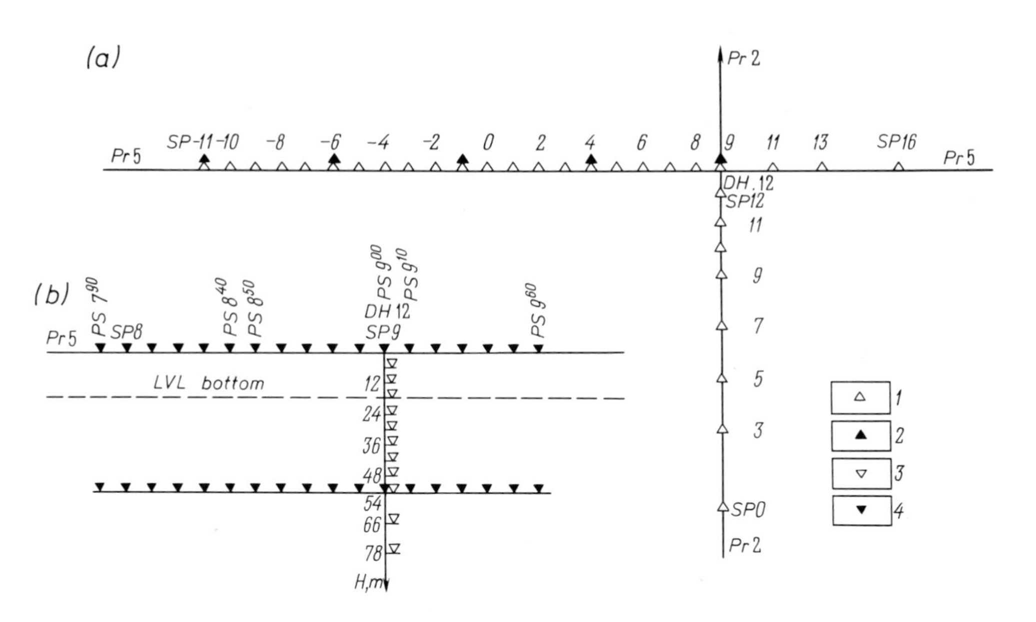

Fig. 77. Spread used in the study of effects of a low-velocity layer on polarizations. (a) layout of shot-points; (b) layout of reception points (in a vertical plane); 1, 2 – excitation points in stages I and II, respectively; 3, 4 – observation points in stages I and II, respectively.

FIRST ARRIVALS

The manner in which the low-velocity layer affects the direction of particle displacement in the first wave can be seen on the seismograms recorded for the vertical profile (Figure 78). On the Z-component record, the arrivals have the same sign along the entire profile. On the other hand, the three-component seismograms show that at the points lying in the low-velocity layer (at depths from 0 to 18 m), the traces of the first arrivals are in phase, whereas at the points under the low-velocity layer they show a phase reversal. More graphically, the effect of the low-velocity layer is demonstrated by the field of particle displacements in a vertical plane (Figure 79). The directions of particle displacement observed in drill-holes at various depths are shown at these depths under the respective shot-points. The fields of displacement directions are practically identical in any direction away from the drill-hole in and directly below the low-velocity layer. Because the shots were fired at a depth of 42 m, the directions of particle motion in the depth interval from 30 to 48 m are close to the horizontal. A distinctive feature of the observed wave field is that the direction of particle motion suddenly diverts towards the vertical and depends little on distance as the waves pass through and are propagated in the low-velocity layer. From an analysis of the isochrone field, it may be concluded that under the low-velocity layer the directions of particle displacement correspond on the whole to the direction of propagation for the first arrivals. This seems to be corroborated by the fact that the same true velocity has been computed from the directions of particle displacement and from the in-line time-versus-depth curve. It should, however, be noted that the directions of particle displacement show regular, although insignificant (10–15°),

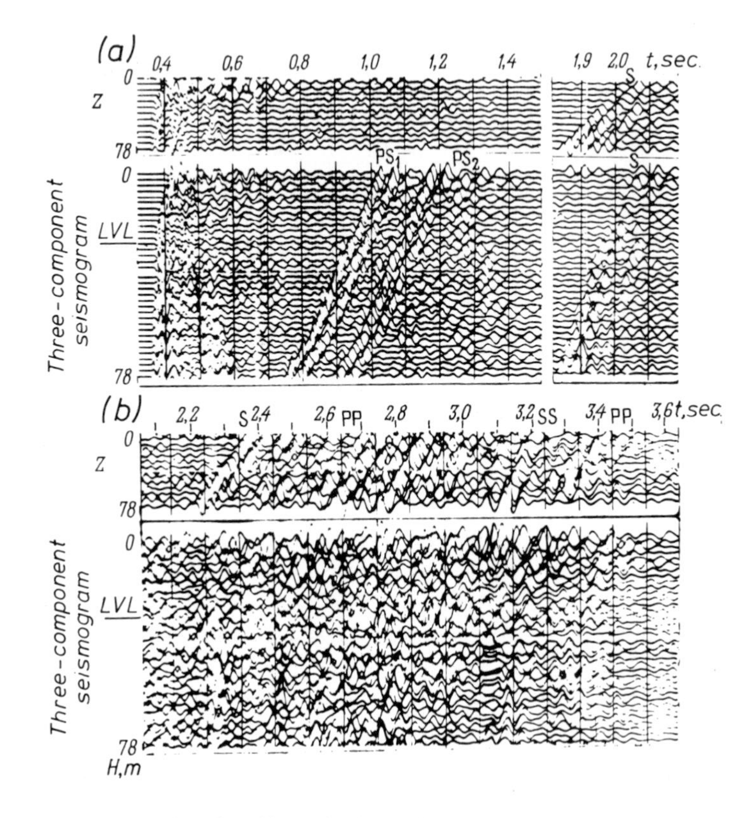

Fig. 78. Seismograms illustrating the effect of a low-velocity layer on the direction of particle displacement. (a) SP5; (b) SP2.

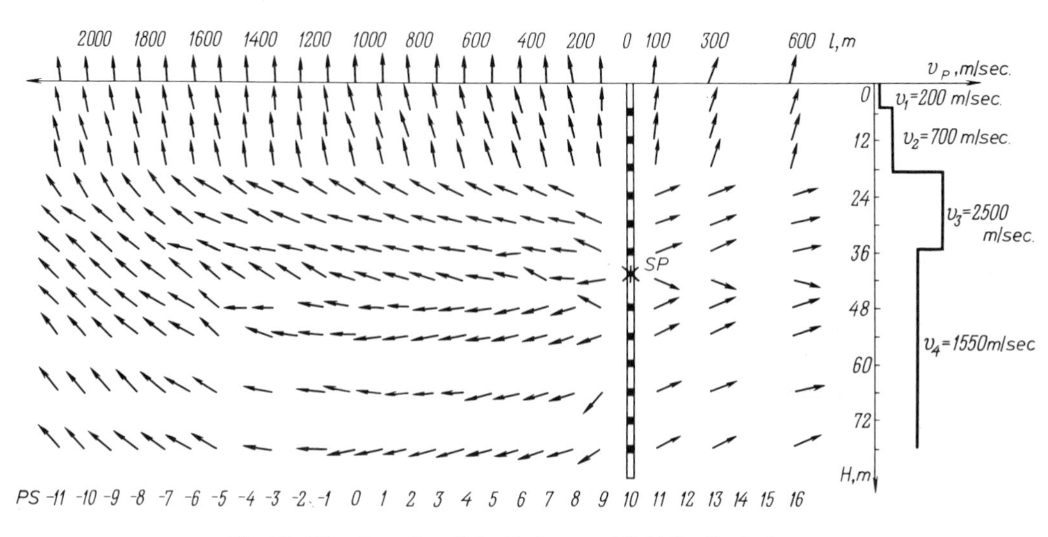

Fig. 79. Direction of particle-displacement field for the first wave.

deviations from the direction of wave motion in both the vertical and the horizontal planes. These deviations are associated with the layered structure and anisotropy of the medium. This fact is interesting in itself, because it may yield further information about the medium, but its discussion is outside the scope of this book.

LATER ARRIVALS

The later arrivals are of special interest, and a more detailed discussion is in order. Observations on a vertical profile (stage I) detected and established continuity for compressional (*P*) and shear (*S*) waves, multiple shear (S_m) waves, reflected compressional (*PP*) waves, reflected shear (*SS*) waves, and converted (*PS*) waves. For illustration, some of them are shown on seismograms recorded by a three-component symmetrical cluster (Figure 78a). As is seen, two converted waves, PS_1 and PS_2, are dominant; they are associated with shallow and steeply-dipping boundaries in a supra-diapir sequence. On the *Z*-component record, these waves can scarcely be tracked either under or within the low-velocity layer. The seismograms recorded by the three-component cluster show reflections of these waves from the bottom of the low-velocity layer and from the ground surface.

In contrast to the converted waves, the direct shear (*S*) wave (t = 2 s) may clearly be seen on the *Z*-component record where its intensity suddenly decreases as the wave passes through the bottom of the low-velocity layer. The same applies to the direct *S* wave (Figure 78b) recorded at greater distances (t = 2.4 s). Here, too, the record shows a single-mode wave reflected from the bottom of the low-velocity layer. The reflected compressional (*PP*) wave (t = 2.6 s) is clearly seen within the low-velocity layer on the *Z*-component record, but under the low-velocity layer the same *Z*-component record shows none of it. A second reflected compressional wave (recorded at t = 3.4–3.5 s) can be followed on the *Z*-component record both in and under the low-velocity layer.

In between the above compressional waves, the seismogram shows several reflected shear (*SS*) waves associated with shallow and high-dipping boundaries in the supra-diapir sequence. These waves can be positively traced on the *Z*-component seismograms both under and in the low-velocity layer.

From analysis of a large number of seismograms, it may safely be concluded that records for waves within the low-velocity layer are far more complex than those for waves under the low-velocity layer. This complexity appears to be caused by the presence of an extended train of multiple waves being propagated in the low-velocity layer. Such waves can be induced by both the direct wave and all strong secondary waves reflected from the ground surface. In some cases, if the instruments have a sufficiently broad dynamic range, it is possible to detect quadruple waves. In passing through the bottom of the low-velocity layer, confident correlation of *PP* waves is ordinarily upset, and the record appears substantially less regular.

To estimate the effect of the low-velocity layer on the wave field, we shall examine profile observations carried out by the polarization method simultaneously on the ground surface and under the low-velocity layer (stage II). From a comparison of records for the various fixed component waves on the ground surface and under the low-velocity layer, it is a relatively easy matter to outline tracking-component areas for the various wave types and establish the general pattern in their variations. In Figure 80, such areas are shown for two shot-points (l = 500 m and l = 1500 m) for sub-LVL and surface

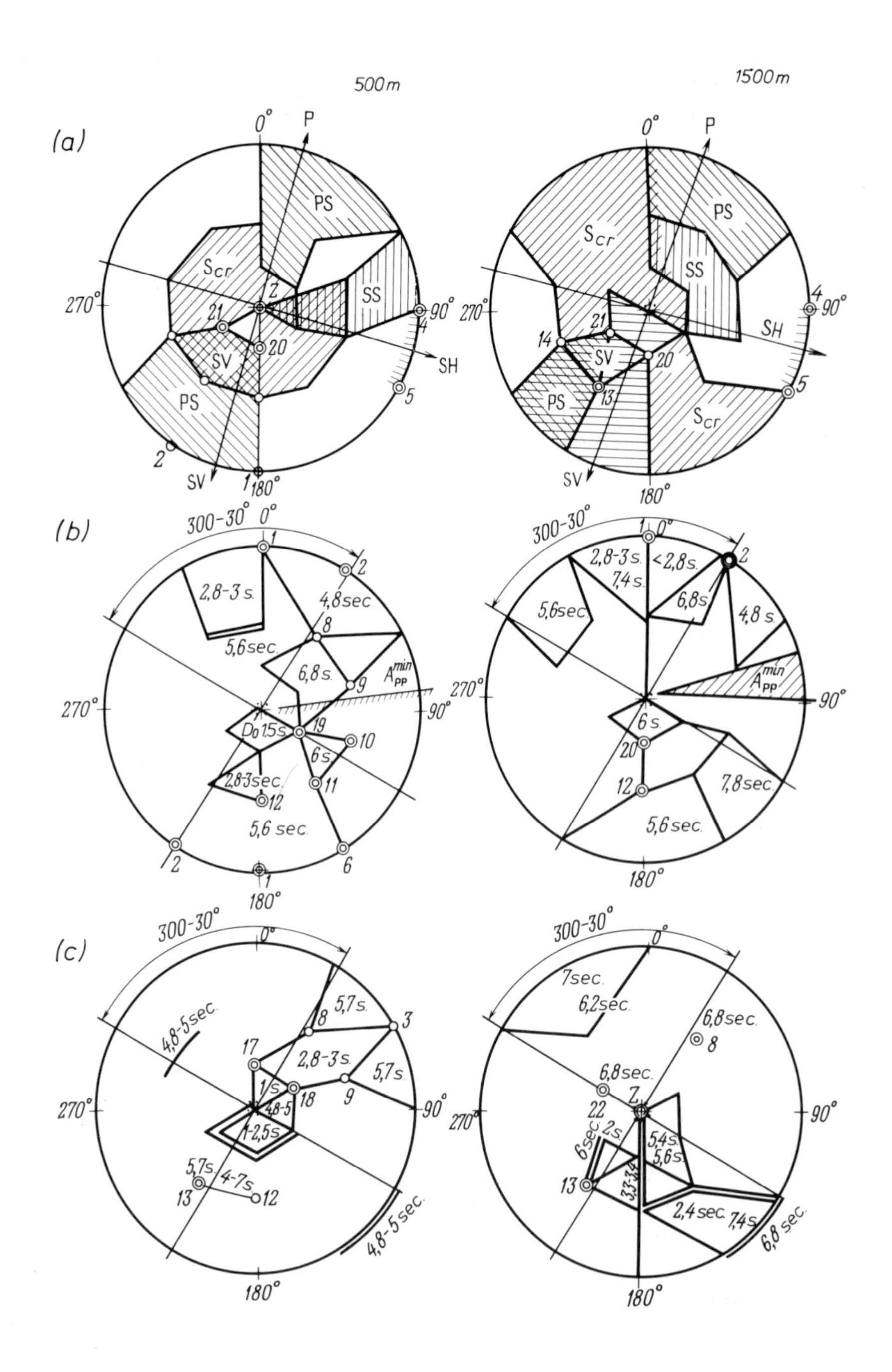

Fig. 80. Optimal detection areas (fields) for dominant waves at l = 50 m and l = 1500 m. (a) shear and converted waves under a low-velocity layer; (b) reflected compressional waves under a low-velocity layer; (c) reflected compressional waves on the ground surface.

observations. Within such an area, it is possible to find for each wave a tracking component for which the signal-to-noise ratio is maximum. The tracking-component areas for all waves are stable on all profiles. For the direct compressional (*P*) wave, the stereograms show only displacement azimuth.

Sub-LVL Observations. In sub-LVL observations, polarization-position correlation can detect and establish continuity for the dominant waves, which arrive at seismometers as compressional and as shear waves (Figure 81b). Whatever the position of the shot point, the tracking-component areas for the various waves show reasonable stability in space orientation. The small azimuthal deviations of the displacement directions in the *P* wave are accompanied by as small deviations of the tracking-component areas for the *SH* and *SV* waves.

With increasing distance to the shot-point, the direction of wave arrival changes, and so in a regular manner does the orientation of the tracking-component areas for the various waves. As an example, for the *SV* waves and l = 500 m the tracking-component area is located between *Z*-component records 21 and 22, whereas at l = 1500 m it occurs closer to horizontal component records 13 and 14 and moves closer to component records 1 and 2 as the shot-point distance is increased still more. In contrast to the *SV* waves, the optimal components for the *SH* waves at any shot-point distance are component records 4, 5, 9 and 10. For the *SS* waves, the tracking components usually are at right angles to the direction of displacement in other strong waves (*P*, *SV*, and *PS*). The multiple shear waves S_m occupy a sizeable proportion of the seismogram, following the direct *S* waves. For them, the tracking-component areas are markedly wider than for the other waves, because the direction of particle displacement in this group approaches the vertical with increasing time. Thus, the tracking-component areas for the shear waves take up an appreciable portion of the half-space.

From an analysis of the tracking-component areas for the most regular reflected compressional waves (Figure 80b) trackable under the low-velocity zone over the entire observation interval (the recording times are marked on the stereograms), it follows that:

(1) for all shot-points, the tracking-component areas for the *PP* waves under the low-velocity layer and also for the shear waves cover a sizeable proportion of the half-space;

(2) for all *PP* waves, except the most shallow ones ($t \approx 1.5$ s), the tracking component is the *Z* component;

(3) for many waves (e.g., those with arrival times of 2.8 and 4.8 s), the tracking components are the components approaching the horizontal; their number increases with increasing shot-point distance l;

(4) there is an area of minimal amplitudes for the *PP* waves, perpendicular to the ray planes of compressional waves.

Analysis also shows that in the case of sub-LVL observations the records are remarkably regular, and the orientation of recording areas for the dominant reflected compressional waves is decided above all by the direction of wave arrival and, to a lesser degree, by the superposition of spurious waves.

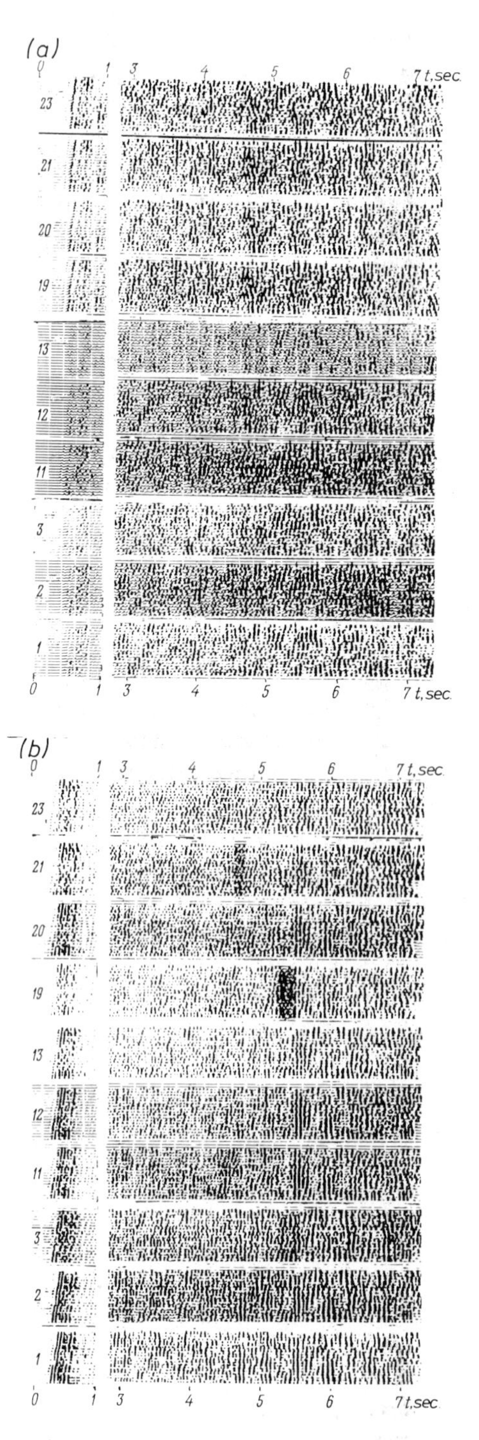

Fig. 81. Fixed-component seismograms produced by a computer for: (a) horizontal profiles on the ground surface; and (b) under a low-velocity layer.

Surface Observations. The records obtained for the surface profile show general irregularity. Practically no reflected compressional waves are trackable. The orientation of the tracking component areas for the *SH*, *SV*, *PS*, and *SS* waves is very close to that of the respective areas in the sub-LVL observations, except that the optimal *SV* components run closer to the horizontal.

Passage through the low-velocity layer brings about far more noticeable changes in the orientation of the tracking-component areas for the reflected compressional waves (Figure 80c). For most waves, the tracking-component areas approach the vertical. Yet, even on the ground surface, the tracking components for most waves are off-vertical rather than vertical. As in the sub-LVL observations, the orientation of these areas is above all decided by the direction of wave arrivals, although the superposition of spurious waves now plays a greater role.

As an example, consider the fixed-component seismograms recorded on surface profiles. No regular waves can be detected on the *Z*-component seismogram of the surface profile (Figure 81a). With the aid of PPC, it is possible to detect several reflected compressional waves only within some segments of the profile, mainly on the components making an angle of 30° with the vertical. The *Z* component may serve as a tracking one only for the first arrival.

The *Z*-component seismograms on a sub-LVL profile differ from those on surface profiles in that they are simpler in general aspect and display a markedly larger number of regular waves. But again, the *Z* component may serve as a tracking one only for the first arrival. The inclined components in the later part of the record show strong reflected compressional waves over the entire interval of recording times (up to 7.0 s). Since most of these waves are not recorded on surface seismograms at all, there is little ground for comparing the sub-LVL and surface seismograms.

It may be concluded from the foregoing that the low-velocity level strongly affects both the direction of particle displacement and the general aspect of the seismograms. In surface observations, this factor limits the possibilities of quantitative interpretation of polarization data and impairs the efficiency with which polarization discrimination can be applied to the compressional waves. Comparison of PPC data for surface and sub-LVL observations seems to indicate that sub-LVL work renders polarization discrimination an effective tool even under conditions of a thick low-velocity zone. As a matter of record, it should be added that the value of sub-LVL observations has been noted earlier even where the *Z* component was used alone [13, 64]. With the polarization method, sub-LVL observations are still more valuable, because they enable this method to be combined with the multiple coverage technique without any increase in effort.

In the light of available sub-LVL data, it appears that the polarization method will prove its worth also in the seismic exploration of offshore areas. The point is that the pressure sensors used in offshore seismic exploration cannot discriminate among waves on the basis of particle displacement. Also, there are some difficulties in using pressure sensors in shallow waters. These limitations can be circumvented by using three-component displacement sensors in bottom observations, so that the capabilities of the polarization method may be used in much the same manner as in sub-LVL observations on land.

Little has so far been learned about the effect that the low-velocity layer and the top part of the section have on the polarization of seismic waves. However, since they affect the applicability of the polarization method, further effort is in order, both theoretical and experimental, in this field. On the other hand, the wealth of VSP evidence and special investigations show that in observations at inner points of the medium the direction of particle displacement corresponds to the direction of wave arrival and varies in a predictable manner. This provides a basis on which wave polarization may be used as a further source of knowledge about the medium.

Study of Particle Displacement in a Vertical Plane

The direction in which the particles vibrate (are displaced) in a wave may throw additional light on the velocity characteristics of the medium, the character of the waves, and the location of areas where wave interference and conversion take place. The direction of particle displacement is far more sensitive to wave conversion than kinematic variables. It is a fact, for example, that it is not always possible to locate the point of wave conversion from a vertical travel curve, whereas this can be done easily from the direction of particle displacement. Thus, when a direct wave is converted to a head or multiply-refracted (curved-path) wave (the head wave emerges in the region of first arrivals), the direction of the particle-displacement curve suddenly passes into the region of negative angles. A sudden change in the direction of particle displacement inside the medium (not at an interface) may be an indication of wave interference or conversion.

A major source of information about the propagation velocity of seismic waves in the seismic-exploration frequency range is seismic well-logging. Among the well-logging data, those of primary importance are the travel-time curves for the first arrivals along an in-line profile, because they provide the starting point from which to compute the horizon and average velocity curves necessary for constructing a section. Further information about the medium can be gleaned from the polarization of seismic waves, by noting the direction of particle displacement in the first arrivals, which are recorded against a quiet background. It is to be noted that in anisotropic media the direction of particle displacement may be different from the direction of wave propagation, and this may impair the accuracy of qualitative data reduction.

PATTERNS OF VARIATION IN THE DIRECTION OF PARTICLE DISPLACEMENT ALONG A VERTICAL PROFILE

Early three-component observations in boreholes demonstrated that variations in the direction of particle displacement in the first arrivals along a vertical profile could be caused by discontinuities in the medium, superposition or conversion of waves. It has been established that there is a regular relationship between the direction of particle displacement in the first arrivals and the velocity section [30, 36, 43], and that this relationship may be used to glean more information about the medium.

In axisymmetrical media, where variations in the direction of particle displacement occur in a vertical plane, these variations along the line of traverse may conveniently be presented in the form of curves. Consider the basic features of these variations for some of the most commonly encountered structures [36].

Homogeneous Medium. In a homogeneous medium the direction of particle displacement is independent of the propagation velocity and is solely determined by the relative position of the observation points and the source; for a specified depth, it is further determined by the distance to the source. In the circumstances, the direction of particle displacement in the first-arrival compressional wave will always tend downwards. Because real media are practically never homogeneous, this pattern of variations can only serve for a comparison with the pattern of variations observed in inhomogeneous media.

In media with a vertical velocity gradient the direction of particle displacement in the first arrivals strongly depends not only on the relative position of the source and the detector, but also on the rate of rise in velocity with depth β. If the velocity increases with depth ($\beta > 0$), which is usually the case in real media, the vertical component record will always show regions where the direction of particle displacement alternatively takes a '+' or '−' sign. Given a fixed excitation source, it is characteristic of any vertical profile (except for an in-line one) that at the top of the section the first movement is always upwards (the wave arrives from below), which corresponds to a negative apparent velocity. The region of negative directions increases with increasing rate of change of velocity β, and with increasing distance between the source and the drill-hole, l. As the depth H increases, the direction of particle displacement tends more and more towards the horizontal until it changes sign, and the wave moves downwards. Naturally, this is accompanied by a change of sign on the vertical component record. The direction of particle-displacement curves change little in shape, but given the same β, the increase in l will cause them to shift along the abscissa. From the above pattern of variations in the direction of particle displacement, the observer may conveniently note the rise in velocity and quantitatively determine the rate of rise of velocity. A computed set of direction curves for particle displacement may well serve as a chart for finding the gradient β simply and quickly [36].

Layered Medium. In layered media, the shape of the particle displacement curves follows variations in wave velocity. At interfaces where the wave velocity suddenly changes, the curves display discontinuities. If the velocity changes from a lower to a higher value, the direction of particle displacement deviates away from the vertical. Conversely, if the velocity changes from a higher to a lower value, it deviates towards the vertical. At interfaces of the second kind, the variation in the direction of particle displacement will be gradual. It is specially to be noted that in areas adjacent to interfaces where secondary reflected waves are superimposed on the incident waves, the direction of particle displacement may deviate in some other manner.

EXPERIMENTAL OBSERVATIONS OF VARIATIONS IN THE DIRECTION OF PARTICLE DISPLACEMENT [30, 36]

Let us compare the direction of the particle-displacement curves, the vertical travel-time curves for the first arrivals, and the velocity section recorded for the Russian platform (Kuibyshev Region). Drill-hole 501 put down for VSP work traverses a Tertiary terrigenous sequence of sands and clays 133 m thick. The terrigenous sequence is underlain by gypsum and dolomite (133–189 m) of the Sosnovo Group, limestones (189–274 m) of the Kalinovo Group, and gypsiferous marls (274–282 m) of the Buguruslan Group, Lower Permian anhydrites and dolomites appear at a depth of 283 m.

Observations were carried out from shot-points located respectively at l = 140 m,

l = 240 m, and l = 800 m. The records made at l = 140 m mainly show a direct and a transmitted wave. The records made at l = 240 m show the conversion of a direct wave into a refracted wave. Those made at l = 800 m show a refracted wave (Figure 82a). Analysis of the wave field produced by both horizontal and vertical profiling unerringly identifies the nature of the recorded waves.

Direct Wave. At l = 140 m, the recorded multiply-refracted (curved-path) wave is trackable in the region of first arrivals as far as the interface (H = 133 m). At l = 240 m, the direct wave is seen to convert at a depth of 115 m into a refracted wave associated with the surface of the Sosnovo Group. The vertical travel-time curves for both shot-points (see Figure 82a) are hyperbolic in shape. On the records produced by a symmetrical three-component drill-hole array, the direct wave produces strong, sharply defined arrivals. The direction of particle displacement in the first wave near the ground surface approaches the vertical (the wave arrives from below), which is attributed to the effect of a low-velocity layer.

The direction of particle displacement curves for the direct wave in the top part of the profile lie in the region of negative angles ($\varphi < 90°$). As H increases, they cross the zero-value axis ($\varphi = 90°$) at a depth of 20 m for the shot-point at l = 140 m and at a depth of 40 m for the shot-point at l = 240 m, and smoothly pass into the region of positive angles. These curves are typical of a multiply-refracted (curved-path) wave. As already noted, they may be used to determine the rate of change of velocity with depth in the top part of the section. For the curve recorded from the shot-point at l = 240 m, the value of β is 0.0015 m^{-1}, which agrees well with ultrasonic well-logging data (Figure 82b).

Conversion of a Direct to a Refracted Wave. The vertical travel time (time-depth) curve recorded for the shot-point at l = 140 m shows a region of negative apparent velocity in the depth interval 123–128 m. However, it is difficult to identify a refracted wave in that interval, because the region is small, and the number of observation points is insufficient. Recourse to the direction of particle displacement shows that the conversion occurs in close proximity to the interface (within about 10 m of the boundary).

More reliably, the conversion of a direct wave to a head wave is seen on the records made from the shot-point at l = 240 m. The seismograms for a depth of 113 m show a weak arrival nearly one-tenth to one-fifteenth as strong as the direct wave. It can only be detected by a highly sensitive instrumental set-up. On three-component seismograms, this wave, in contrast to the direct wave, produces a line-up pattern and appears as a refracted wave associated with the surface of the Sosnovo Group. The refracted-wave arrival is immediately followed by a strong direct wave, which is also corroborated by the direction of particle displacement. The direction of particle displacement for the direct wave is shown next to that for the first arrivals. The conversion of a direct to a refracted wave is clearly seen on the direction of particle displacement curve which shows a discontinuity and passes into the region of negative angles.

Refracted Wave. On the records produced for the shot-point at l = 800 m, the first to be recorded is a refracted wave associated with a boundary not traversed by the drill-hole. The direction of particle displacement curve is practically parallel to the depth

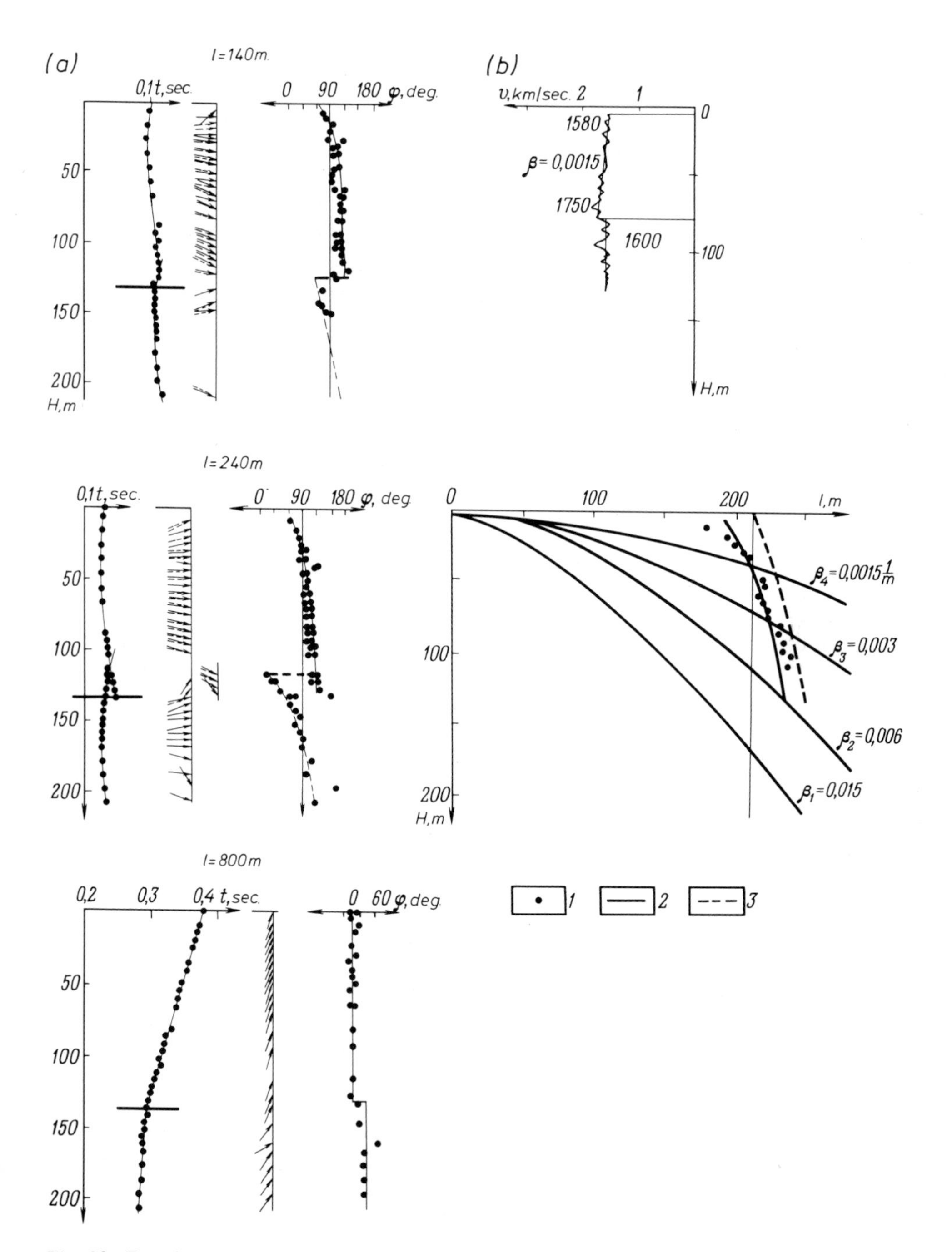

Fig. 82. Experiment study of variations in the direction of particle displacement in the first waves. (a) vertical travel-time curves, directions of particle motion and displacement curves for various values of l; (b) determination of β from directions of particle displacement; 1 – observed values; 2, 3 – computed curves for $\beta = 0.0015\ \text{m}^{-1}$ and $\beta = 0$.

axis. Under the boundary lying at a depth of 133 m, the direction of particle displacement makes an angle of about 45° with the horizontal. After refraction on that boundary, the direction of particle displacement approaches the vertical and makes an angle of 80° with the horizontal. From the difference in the direction of particle displacement before and after refraction, it is found that the velocity ratio of the compressional waves on either side of the boundary is 2.9 : 1, which agrees well with velocity-section data.

Conversion of a Refracted to a Transmitted Wave. At depths below 133 m, that is, under the interface, the records for the shot-points at l = 140 m and l = 240 m show a substantial difference in the general aspect. The weak in-phase arrivals of the refracted wave give way to stronger arrivals of a transmitted wave with phase reversals. From the direction of particle displacement curve for l = 240 m (see Figure 82), it is seen that immediately under the interface the direction of particle displacement for the transmitted wave departs from that for grazing incidence and is close to that of the refracted wave. It must be noted, however, that in some cases of refracted-to-transmitted wave conversion the direction of particle displacement in the transmitted wave at grazing incidence is close to the direction of the grazing ray.

DIRECTION OF PARTICLE DISPLACEMENT IN THE TRANSMITTED WAVE AT GRAZING INCIDENCE

Of special interest are cases where wave polarization is abnormal or where the direction of particle displacement in the first wave is different from that of its propagation. Such cases include downward transmitted waves within the portions of the profile adjacent to a refracting boundary (grazing rays).

Consider the computed direction of particle displacement for the transmitted wave at grazing incidence in a homogeneous layer lying in a homogeneous half-space. As the wave crosses the boundary, the transmitted wave on the vertical profile is converted to a head wave. As will be recalled, in a homogeneous medium a head wave is linearly polarized, and the direction of particle displacement does not vary from point to point along the vertical profile. When it is at nearly grazing incidence, the transmitted wave is elliptically polarized. Thus, conversion of a head to a transmitted wave upsets linear polarization.

As an illustration, consider the directions of particle displacement computed for two layers, using Podllyapolsky's formulae [98]. The computations have been made for three distances from the point of emergence of the head wave, a refracting-boundary depth of H = 250 m, ν_1/ν_2 = 0.5 (Figure 83a), and ν_1/ν_2 = 0.9 (Figure 83b). In the figure, the direction of particle displacement at the head wave front is shown at the intersections between the vertical profile and the refracting boundary. The direction of particle displacement in the transmitted wave at grazing incidence departs from that of wave propagation and can be resolved into two components, a radial component in the direction of transmitted-wave propagation and a component close to the direction of particle displacement in the head wave. Immediately under the boundary, the radial component is small, and the ratio of the major to the minor axis of the displacement ellipse is so high that the phase shift on the seismograms does not exceed 0.002 s, which lies within the limit of reading accuracy. In practice, this magnitude of ellipticity cannot

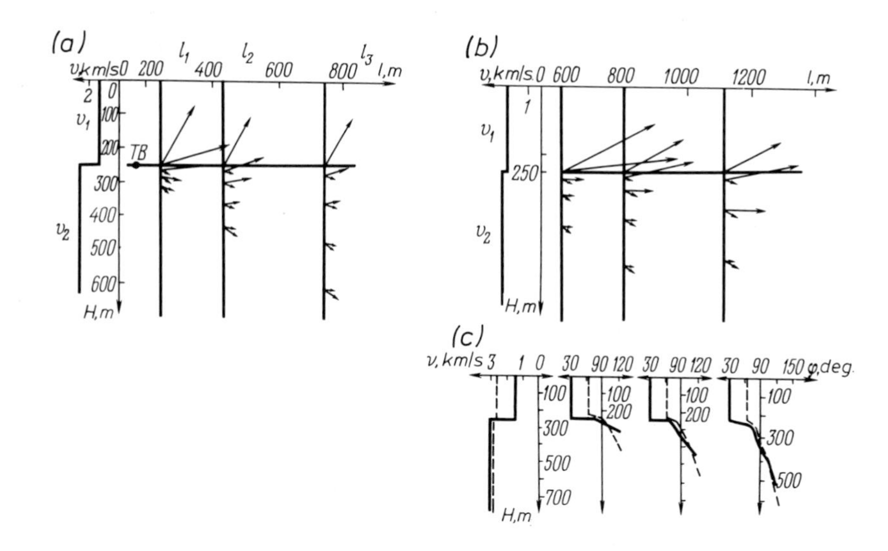

Fig. 83. (a) And (b): directions of particle-displacement and (c) displacement curves for a transmitted wave at grazing incidence (vector scale for $l_1 : l_2 : l_3 = 0.1 : 0.5 : 1$).

always be detected, and it appears that directly under the boundary the direction of particle displacement in the transmitted wave remains close to that in the head wave. On moving away from the boundary, the second component gradually approaches the radial one (the angle between them decreases), then rapidly falls in intensity until at a certain depth only the radial component remains, which is an indication that the direction of particle displacement is along the direction of propagation for the transmitted wave now linearly polarized. The direction of particle displacement lies on a straight line joining the point of emergence (PE) of the head wave and the observation point.

From a comparison of the computed directions of particle displacement for the head and transmitted waves, it is seen (Figure 83c) that conversion of the head of the transmitted wave is accompanied by a change in the direction of particle displacement, with the region of negative angles under the boundary increasing with growing difference between the velocities across the boundary. From analysis of variations in the direction of particle displacement for the transmitted wave, it is possible to identify the mechanism leading to the formation of the wave recorded above the boundary, and also to evaluate the rigidity of contact in real media. The latter consideration is especially valuable because it has a decisive bearing on the nature and intensity of waves.

IDENTIFICATION OF SECTION DETAILS

Seismic well logs do not give sufficient detail for relatively thin layers to be identified, although, as is known [36], the latter may markedly affect the wave field. In most cases, vertical travel time (time-versus-depth) curves tell very little. It appears that the detail of investigation could substantially be improved by using sonic and ultrasonic frequencies. Unfortunately, most drill-holes used for VSP work are encased and are therefore unavailable for ultrasonic logging.

The direction of particle displacement is more sensitive to velocity and may be used for identifying the section detail. Figure 84 illustrates the manner in which the direction of particle displacement in the first arrivals may be used to identify the details of a section at the Nura-Taldy deposit, which is an intrusion underlain by a relatively uniform shale sequence. The vertical time-travel (time-versus-depth) curves appear to vary gently. The depth at which the point of minimum occurs on the travel-time curves increases as the shot-point is moved away from the drill-hole entrance, which is characteristic of media with a continuous rise in velocity. On the other hand, reference to the curves of Figure 84b discloses the presence of layers varying in thickness and showing sudden changes in velocity. The $\varphi(H)$ plot displays several intervals of depth where discontinuities occur (100–200 m, 290–320 m, and 430 m). The discontinuity at 100–120 m

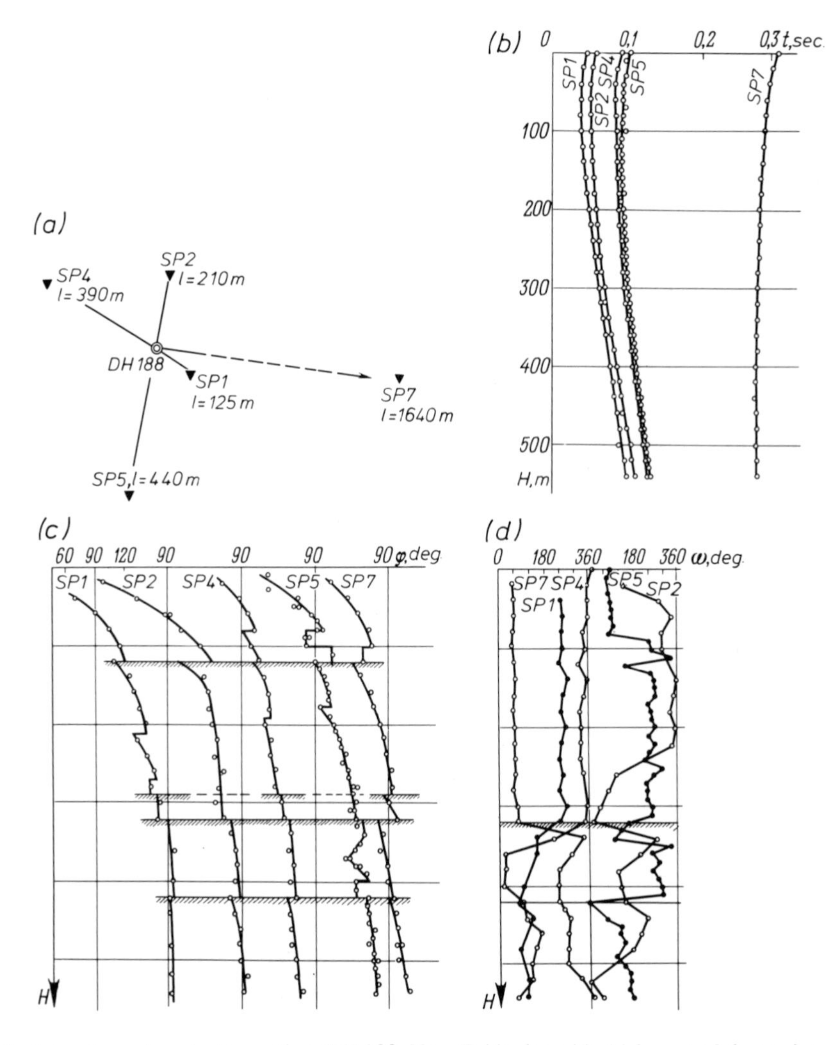

Fig. 84. Dissection of a velocity section (DH 188, Nura-Taldy deposit). (a) layout of shot-points and drill-hole; (b) vertical travel-time curves; (c) $\varphi(H)$ curves; (d) $\omega(H)$ curves.

is associated with the bottom of a gradient layer related to a weathered zone, and mainly appears on the $\varphi(H)$ plots. The interval of depth from 290 to 320 m is associated with the dome-shaped roof of the intrusion, so that it stands out more clearly on the $\omega(H)$ curves than on the $\varphi(H)$ curves. The feature at 430 m is presumed to be a boundary corresponding to facies changes within the intrusion. The identification of detail in the section based on analysis of particle displacement generally agrees with ultrasonic logging data.

DETERMINATION OF TRUE VELOCITIES FROM OFF-LINE VERTICAL PROFILES

Ordinarily, true velocities are found from vertical travel-time curves recorded in the course of observations on longitudinal vertical profiles, with the shot point located at the mouth of the drill-hole used. In many cases, an in-line profile cannot be shot for technical reasons, and recourse has to be made to off-line vertical profiles. To obtain true velocities, the data obtained from off-line vertical travel-time curves are usually adjusted in order to correlated them with in-line travel-time curves. This involves applying corrections for the departure of the seismic rays from the vertical. Since, however, the path of these rays is not known, simplifying assumptions have to be made as regards the medium, and this appreciably devaluates the data thus obtained. In contrast, the direction of particle displacement observed on off-line profiles leads to true-velocity curves directly based on an off-line travel-time curve without any preliminary calculation. This purpose may also be served by apparent velocities of propagation.

The apparent velocity v_a and the true velocity v are related in a known manner to the direction in which waves arrive at an observation point: $v = v_a/\cos\varphi$, where φ is the angle between the raypath of the wave and the vertical. The value of v_a can be found from a vertical travel time curve, and the angle φ can be determined from the direction of particle-displacement curves. As an illustration, we shall use the data gathered during the observations in the Starotitarovka area. The source data include vertical (off-line) travel-time curves, plots of apparent velocity as a function of depth, and direction of particle dispalcement curves, $\varphi(H)$.

The vertical travel-time curves (Figure 85a) are gently varying curves. The $v_a(H)$ plots (Figure 85b) display a portion associated with the top of the section at a depth of about 700 m, where v_a suddenly increases. This may be due to the approach to a minimum point on the vertical travel-time curve where the apparent velocity tends to infinity and can usually be determined with an appreciable error. The $v_a(H)$ plot for SP11 has a somewhat different shape. Here, changes in v_a are observed at other depths, and the curve has several points of inflection. The latter may be associated with the changes in reception conditions on moving away from the diapir dome towards its edge. It should be noted that there is practically no difference between the v_a curves recorded under approximately the same subsurface conditions (in the dome part of the structure), but from shot-points differing in azimuth – one across and the other along the strike (SP6 and SP8).

The directions of the particle-displacement curves $\varphi(H)$ (Figure 85c) are more differentiated and ragged. As a rule, a sudden change in the direction of particle displacement is associated with lithofacies changes, which control the elastic properties of the medium. Conversely, the absence of sudden changes in the direction of particle displacement curve, $\varphi(H)$, points to the stability of the elastic properties of the medium.

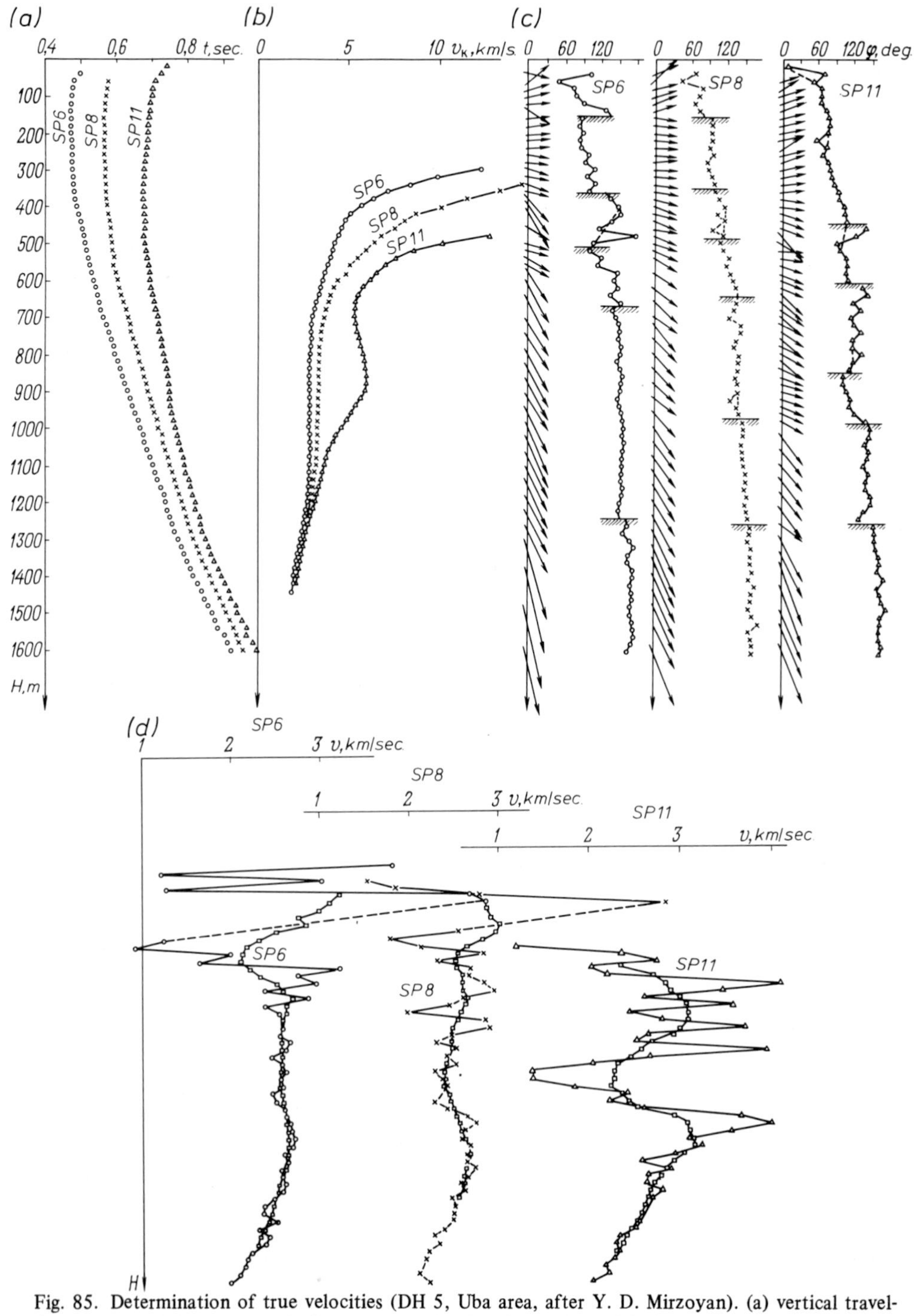

Fig. 85. Determination of true velocities (DH 5, Uba area, after Y. D. Mirzoyan). (a) vertical travel-time curves; (b) apparent-velocity curves; (c) directions of particle-displacement and curves of displacement directions; (d) true velocity curves.

The change in $\varphi(H)$ is most noticeable at a depth of 420 m (SP6) and at a depth of 670 m (SP11). At greater depths, the $\varphi(H)$ curve stabilizes. From a comparison of $\varphi(H)$ curves, several intervals can be noted where they all show a discontinuity. At the top part of the section, such a discontinuity is observed in the depth interval 150–160 m; other discontinuities occur at 670 m, 970 m, and 1250–1300 m. In the lower part of the section, the sudden change in the direction of particle displacement is associated with the top of Maikop sediments. On the $\varphi(H)$ curves, this feature is noticeable for all the shot-points.

Correlation of the above data leads to true velocity curves which show several boundaries responsible for abrupt changes in velocity in the section in question. It should be noted that the averaging of the curves mainly brings out thick strata (tens or even hundreds of metres thick). When averaged, such strata appear to be uniform, although the shape of the $\varphi(H)$ curves suggests the existence of thinner strata differing in properties from the average one. In the interval 700–1350 m where the section consists of a monotonous clay member, there is a uniform constant-velocity layer where the velocity is fairly high (2700–2800 m/s) and which is stratigraphically related to a undifferentiated N_1^{2+3} sequence. Despite some differences, data recorded from the distant shot-point (SP11) agree on the whole with the values of v determined on the other vertical profiles.

To sum up, the velocity section based on vertical travel-time curves has the following features. At the top part of the section down to depths of 600–700 m, the section traverses minor strata 100–1500 m thick where a lower and a higher velocity is observed. Next comes a layer in which waves are proagated at a constant velocity (2700–2800 m/s). The Maikop sediments (H = 1300 m) show lower velocities (1900–2000 m/s), which is apparently related to the reduced rock density in the diapir dome.

FIELDS OF DISPLACEMENT DIRECTIONS

The possibilities of analysing the wave field and the medium may be substantially enhanced by reference to fields of displacement directions. From them, one can conveniently establish the pattern of variations in the direction of particle displacement continuously in a vertical plane. In principle, such a field can be constructed by making observations in several drill-holes arranged along a common line and using a single shot-point. Because of the amount of drilling involved, however, such spreads are only resorted to in studying the topmost part of sections. Another, and more normal approach is to make observations in one drill-hole and use a large number of shot-points arranged on the ground surface along the line of traverse (profile).

For the first time, the power of direction field analysis was convincingly demonstrated in a study of spurious waves ordinarily observed over large areas in Middle Asia [33, 36]. As an example, we shall examine displacement fields for media with low velocity differentiation, which pose certain difficulties. Among other things, the head and multiply-refracted (curved-raypath) waves excited in such media are difficult to resolve. This is because velocity gradients are small, and horizontal travel-time curves for multiply-refracted waves scarcely differ from those for head waves.

Variations in the direction of particle displacement in the first arrivals readily bring out wave refraction and the accompanying increase in velocity with depth. On the other hand, it is a fact that wave refraction is difficult to notice from kinematic data, and this calls for the use of a complex system of overlapping travel-time curves. Apparently, this is the reason why the nature of the waves recorded in the first arrivals was for a long time misconstrued. As already noted, variations in the direction of particle displacement not only bring out the very fact of refraction, but also enable the observer to determine the coefficient of velocity increase with depth.

COMPUTED DISPLACEMENT FIELDS

Let us compare computed displacement fields for two media in which the velocity varies in a different manner, namely, a two-layered medium where the velocity shows mild changes in magnitude, and a medium where the velocity rises linearly (Figure 86). In the case of an abrupt change in velocity produced near the shot-point, the wave

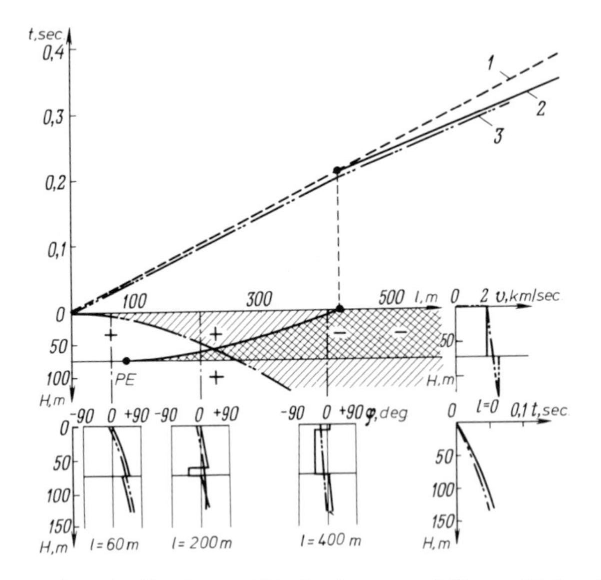

Fig. 86. Travel-time curves for (a) a direct wave, (2) a head wave, and (3) a multiply-refracted (curved-path) wave; change-of-sign fields, and displacement curves.

first observed on the horizontal profile is a direct wave, which then gives way to a head wave. At low values of *l*, direct waves arrive at points along the vertical profile from above (let this direction of wave arrival be designated by a '+' sign). Within the recording interval, the head waves arrive from below (so the direction of wave arrival takes a '–' sign). The regions of positive and negative directions of particle displacement in the first arrivals are separated by a change-of-sign line, which is the locus of interference between the direct and head waves. At the boundary, this line commences at the point of emergence (PE) for the head wave, whereas on the ground surface it lies at the intersection of the travel-time curves for the direct and head waves. The positive directions of particle displacement lie to the left and the negative directions of particle displacement to the right of this line.

So long as velocity varies linearly with depth, the first arrivals on the horizontal and vertical profiles are multiply-refracted waves. Although no wave conversion occurs, the vertical profile shows a change of sign owing to the curvature of the raypath. In this case, the change-of-sign line commences at the origin of the coordinates and passes downwards with increasing l. It should be noted that in the cases in question, the change of sign does not always occur in the same manner. For the vertical component of a multiply-reflected wave, the direction of particle displacement gradually passes from the negative into the positive region. The change-of-sign line follows the line of horizontal displacement. In the case of conversion from direct to head waves, there is an abrupt change in velocity, and the direction of particle-displacement curve shows a discontinuity.

Observed Displacement Fields. Consider the particle-displacement direction field, which was observed in Middle Asia in order to analyse the velocity behaviour of the top part of the section (Figure 87a). Curves for the direction of particle displacement are constructed along horizontal lines (Figure 87b). For depths less than 60 m, these curves lie completely in the field of negative angles and are practically parallel to the abscissa. At 60 m and deeper, near the shot-points, the curves lie in the positive field (the waves arrive from above). With depth, the size of this field and the value of angles increase. On moving away from the drill-hole the direction of particle displacement approaches the horizontal, and the curve for the direction of particle displacement passes into the field of negative angles.

The curve for the vertical profile (see Figure 87b) lies in the field of negative angles, starting with $l = 500$ m. In the depth interval 0–20 m the angle with the horizontal is 70°. At the boundary lying at $H = 18$ m, the curve changes abruptly by 27° and approaches the axis of horizontal displacements. At the boundaries lying at $H = 36$ m and $H = 55$ m, the direction of particle displacement changes abruptly by 9° and 7°, respectively. In each of the three layers, the curve is practically parallel with the depth axis. At 55 m and deeper, the direction of particle displacement gradually approaches the depth axis. The direction of particle displacement becomes horizontal between a depth of 150 m and 160 m. At greater depths, the curve passes into the region of positive angles, whereas at the maximum observed depth (210 m) the angle with the horizontal does not exceed 10°.

At $l = 1450$ m, the curve for the direction of particle displacement likewise experiences discontinuities at interfaces, but these abrupt changes are smaller than at $l = 500$ m. With increasing depth, the curve slowly approaches the axis of horizontal displacements. This pattern of variations in the direction of particle displacement corresponds to a medium with a velocity gradient. The observations agree well with the computed curve for $\beta = 0.001\ \mathrm{m}^{-1}$. The same goes for the direction of particle displacement curves computed for the same value of β, but along the horizontal directions and at different levels.

As follows from the experimental evidence cited above, quantitative analysis of the pattern of variations in the direction of particle displacement may yield further information about the recorded waves and the details of velocity sections.

ANISOTROPY OF VELOCITIES

The polarization of shear waves can be used to detect and investigate the anisotropy of the medium. The point is that shear waves are far more sensitive to anisotropy than compressional waves. This is explained by the fact that velocity anisotropy gives rise to

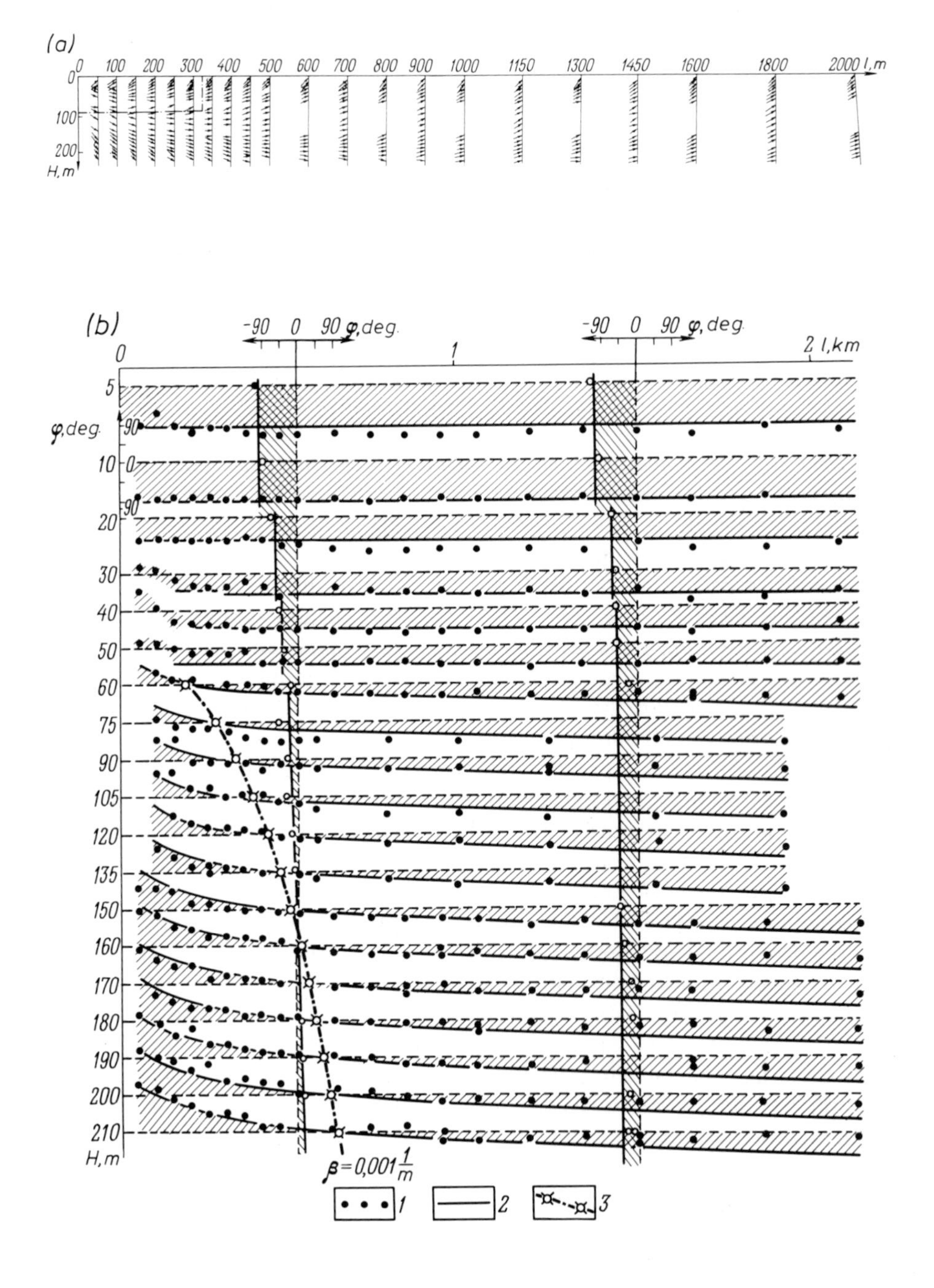

Fig. 87. (a) Particle-displacement field and (b) displacement curves in a vertical plane. 1 – observed values; 2 – computed curves; 3 – change-of-sign curves for direction of particle displacement.

two types of shear waves, *SV* and *SH*, polarized in different planes and being propagated at different velocities. In being propagated at different velocities, *SV* and *SH* waves are shifted relatively to one another in time and, interfering with one another, may form an elliptically-polarized wave. The parameters of the ellipse may vary according to the conditions in which the original waves interfere. In isotropic and axisymmetrical media, converted transmitted waves must be polarized precisely in the ray plane. If an isotropic medium contains inclined boundaries, *SH* components may appear in the shear wave. In such a case, however, there will be no phase shift between the *SV* and *SH* waves. Thus, the presence of a phase shift between *SV* and *SH* waves is an indication that the medium has anisotropic properties in the frequency range in question. This is the reason behind so much interest in the polarization of shear waves.

Anisotropy is especially noticeable in sedimentary sequences with a well-defined layered structure. In such media, the velocities of wave propagation along and across the layers may differ appreciably; for *P* waves the difference may be as large as 5 to 20%. For the crystalline crust, the coefficient of anisotropy v_Z/v_H is substantially lower (1.04 to 1.06). Laboratory and theoretical studies have shown that one-sided compression of cracked rock brings about a marked velocity anisotropy for shear waves. Since the anisotropy of the medium may have a marked effect on the polarization of shear waves and, as a result, on the detection of converted transmitted waves, conflicting views are currently held as regards their polarization.

We shall illustrate the uses to which the polarization of shear waves may be put, using field material gathered at the Sayak deposit. In terms of velocity properties, the observed complexes are little differentiated. In the near-contact zones, only limestones show low-velocity zones, which fact is attributed to their intense metamorphism and hydrothermal effects. The lithologic medium is only slightly differentiated and consists of interstratified sandstones and volcanic formations. VSP work was carried out in a drill-hole put down to a depth of 1140 m in the exocontact zone of a large intrusive massif. The drill-hole did not traverse the intrusion. The relative position of the shot-points and the observation drill-hole is shown in Figure 88a.

The in-line vertical travel-time curves for *P* and *S* waves (Figure 88b) can be approximated by practically straight lines. At the maximum depth of observation ($H = 1100$ m), it is found that $v_P/v_S = 5350/3160 = 1.69$. A careful comparison of the travel-time curves recorded for the various shot-points will show some kinematic differences associated with discontinuities in the section. For example, the hyperbolic $t(H)$ curve for the *P* wave recorded from SP5 is characteristic of an off-line profile in a homogeneous medium. On the other hand, the travel time curves recorded from SP2 and SP4 show a well-defined minimum and an apparent negative velocity above the point of minimum, that is, the first arrivals are direct multiply-refracted *P* waves. The difference in velocity for SP4 and SP5 equidistant from the observation borehole, but lying on mutually perpendicular azimuths, is likewise associated with discontinuities in the medium, v_{av} from SP4 being higher than v_{av} from SP5. Thus, the medium in question is axisymmetrical and it shows different velocity characteristics in mutually perpendicular directions. As is seen from Figure 88a, the shot-points marked SP2, SP3, and SP4 are arranged across, and the shot-point marked SP5 along the strike of the discontinuities in the volcanic-sedimentary sequence.

The travel-time curves for the *S* wave are similar to those for the *P* wave. They show a

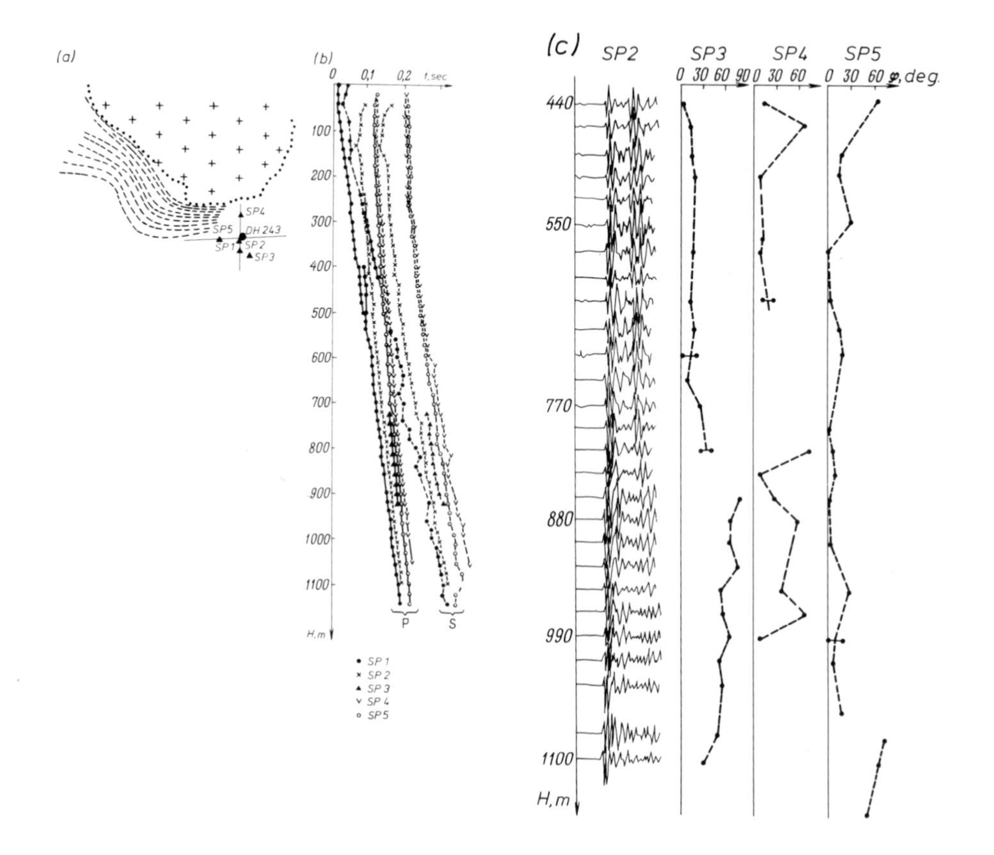

Fig. 88. Observations in DH 243, Sayak deposit. (a) layout of shot points and borehole; (b) vertical travel-time curves; (c) Z-component seismograms and displacement curves for S wave.

somewhat larger scatter in t values, because the waves in the late arrivals are less resolved, and different phases may well be chosen for correlation. This is the reason, among other things, why the minima on the travel-time curves for the S waves are less definite than on those for the P waves. The first phases of the P and S phases are about the same in frequency, which is approximately 100 Hz. Against the background of the low-frequency previous record, the first arrival of the S wave stands out clearly owing to its higher frequency.

Analysis of S waves for polarization shows that they are non-linearly polarized at nearly all of the points, except the head of the S-wave pulse, which retains linear polarization. Loss of linear polarization is related to the superposition of pulses polarized in various (often mutually perpendicular) planes and recorded with some time delay. Stereograms for direction of particle displacement for the first phases of S waves, where the vibrations are almost linearly polarized, are shown in Figure 44. It may be assumed that the non-linear polarization of S waves is related to the anisotropy of the medium, which is also confirmed by kinematic data.

Records of direct shear waves excited by explosions and earthquakes show that in all cases a relatively weak linearly-polarized phase is followed by strong nonlinearly-polarized phases of the *S* waves. In our opinion, this pattern is common and holds out special promise for anisotropy studies.

Let us examine the distribution of displacement directions with depth. Figure 88c shows plots of displacement directions for the first phase of the *S* wave. The angles between the vertical ray plane of the *P* wave and the direction of particle displacement for the first phase of the *S* wave are laid off as abscissae. The relation stands out especially clearly on records from SP2. In the depth interval 400–700 m, the directions of particle displacement lie in the vertical ray plane (they can be identified with the directions of particle displacement for *SV* waves). At greater depths, the directions of particle displacement change abruptly and approach those for *SH* waves. The *Z*-component seismogram (SP2) shows an abrupt change in the record shape for the *S* wave at a depth of 700 m as a result of a change in its polarization. The $\varphi(H)$ plot (SP4) shows a similar change in the polarization of the first extremum for the *S* wave with increasing depth; here, the greater scatter in values is attributed to the lower quality of the records obtained from SP4. In contrast to the records from SP and SP4, those from SP5 do not show such a relationship; at all depths, the directions of particle displacement in the *S* wave tend towards the ray plane of the *P* wave. It is only at some of the deepest points ($H > 1034$ m) that the directions of particle displacement make large angles with the vertical ray plane of the *P* wave.

To sum up, like the kinematic data, the observed pattern of variations in the polarization of the *S* wave may be related to a nonaxisymmetrical medium. As will be recalled, SP5 was located along the strike, and SP4 and SP2 along the dip of the discontinuities.

Undoubtedly, wave polarization may serve as a source of independent and detailed evidence about the medium. Before all the potentialities offered by this source can be fully utilized, however, a good deal of work, both experimental and theoretical, will have to be done. This task can be substantially simplified by recourse to polar correlation.

Location of the Boundary Position from Transmitted Waves in VSP

As seismic waves pass through inclined boundaries, their direction of propagation changes, and so does the direction of particle displacement in the first arrivals. These changes, whose nature and extent vary according to the spatial orientation of the boundaries and relative velocities, may be utilized to locate the position of the boundaries in the area adjacent to the drill-hole. In the general case of construction of the medium, axial symmetry is not maintained, the ray planes may be inclined, and the wave propagation is accompanied by changes not only in the angles in the vertical plane, but also in azimuths. In the circumstances, variations in the direction of particle displacement and wave propagation must be analysed in three dimensions. In regard to media characterized by complex boundary geometry and large angles of tilt, analysis of displacement azimuths is of special importance.

Consider the pattern of variations in $\varphi(H)$ and $\omega(H)$ for steeply-dipping boundaries, which are often encountered in orefields. For example, they have been discovered in drill-holes in the Nura-Taldy area (the top of an intrusion, $\varphi_n = 45°$) and the Zhairem area for which observation data have already been discussed.

Let the boundary be represented by a normal *n* whose coordinates are $\varphi_n = 45°$ and

$\omega_n = 180°$ (Figure 89a). The direction in which the waves incident on the boundary arrive varies with the distance l between the shot-point and the drill-hole, and the observa- depth H, but in all cases it will lie in the half-space bounded by the boundary plane Q_b.

Let us construct one of the conical surfaces making the same angle $\varphi_n = 20°$ with the normal, locate on this surface the incident ray with an arbitrary azimuth of incidence ω (point a), and pass the plane of incidence Q through the point a and the normal n. The refracted ray will likewise lie in this plane, and its direction will be determined by the ratio of velocities at the boundary. With $v_2/v_1 < 1$, the direction of the refracted

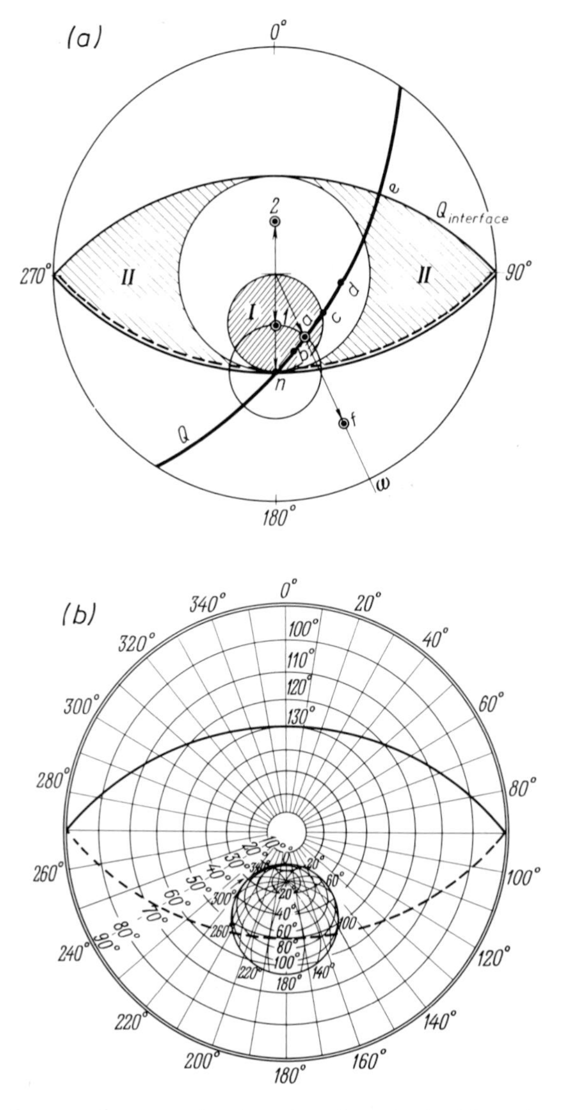

Fig. 89. Change of the raypath upon refraction at a planar inclined boundary and (b) a chart for determining the raypath upon refraction.

ray will be in the plane between the directions of *n* and *a* (point *b*, $v_2/v_1 = 0.61$), whereas with $v_2/v_1 \geqslant 1$, it will lie between directions of *a* and *e* (point *c*, $v_2/v_1 = 1.64$). Point *a* corresponds to $v_2/v_1 = 1$. As is seen, $\varphi_b > \varphi_a > \varphi_c$, which implies that as the velocity under the boundary increases ($v_2 > v_1$), the angle φ approaches the vertical; as the velocity decreases ($v_2 < v_1$), the angle φ approaches the horizontal. Naturally, if the angle φ is reckoned from the normal to the boundary rather than the vertical, the picture will be reversed. Since, however, the position of the normal relative to the boundary is unknown, and the analysis uses angles with the vertical and not with the normal, it is important to consider the general pattern of variations in these angles.

From all the likely directions of incidence of rays on the boundary, it is convenient to isolate fields I and II each of which shows a different pattern of variations in φ at the boundary with changes in the velocity ratio. In field I ($\varphi < \varphi_n$), the angle φ decreases with increasing v_2/v_1. In field II ($\varphi > \varphi_n$), the angle φ increases with increasing v_2/v_1. In all other directions of incidence, either case may occur on crossing the boundary. Similar fields can be delineated for any other angle of tilt of the boundary. In considering these fields, it is important to note two factors that are essential in the subsequent interpretation of experimental curves. First, the values of $\varphi(H)$ at the boundary obtained at any point on the vertical profile shot from equal distances but in opposite directions (points 1 and 2 on the stereogram of Figure 89a), decrease from 180° to 90° at point 1, and increase from 0° to 180° at point 2. Second, at the points of the profile shot from the same direction but different distances to the shot-point (points *a* and *f*), the angle φ increases at point *a* and decreases at point *f*.

Thus, if a section shows the presence of inclined boundaries, the consideration ought not to be limited to plots of $\varphi(H)$, as this would lead to an erroneous idea about the pattern of variations in velocity across the boundary; to avoid this, consideration must be extended to include variations of angles in space. Let us follow the pattern of these variations and examine the likely approaches to locating the position of the boundaries from the direction of particle displacement in the first arrivals.

CHANGE IN THE DIRECTION OF A SEISMIC RAYPATH UPON REFRACTION AT A PLANE BOUNDARY

Seismic waves are propagated in three dimensions. In surface observations, this feature manifests itself in so-called azimuth deviations of seismic rays. In [25], the author examines fields of azimuth deviations for reflected, refracted, and head waves in the case of a planar inclined interface, and also for waves reflected from dome-shaped and cylindrical boundaries, with allowance for refraction at an intermediate boundary.

Proper consideration of changes in the direction of the raypath in space is essential in the interpretation of VSP data where the direction in which the ray approaches a three-component cluster in a drill-hole serves as the point of departure for constructing an intermediate interface. This consideration is also important in earthquake seismology, especially in the case of weak local earthquakes, when some characteristics of the tremor have to be ascertained from data supplied by a single station [56, 57], and also in the technique based on converted transmitted waves excited by distant earthquakes. Undoubtedly, the number of cases where this consideration is essential will increase with further advances in seismic methods and with improvement in their accuracy.

Consider the pattern of changes in the direction of a seismic ray upon refraction on one plane boundary. It is to be noted that analytical methods yield unwieldy solutions for problems of geometric seismics, so it is more advantageous in many cases to have recourse to graphic and semi-analytical methods; they substantially simplify the solution and yield results sufficiently accurate for practical purposes [26, 86, 109]. In the treatment that follows, all constructions will be made in stereographic projection.

To make the presentation more graphic, we choose the angle that the inclined boundary makes with the horizontal as equal to 40°, the velocity ratio equal to $\nu_2/\nu_1 = 2.5$, and the azimuth of the boundary dip equal to 180°. If we mentally join the centre of the stereo-net to the source, point 0, then the angles locating the direction of the normal to the interface will be as follows: azimuth, $\omega_n = 180°$; angle with the vertical, $\varphi_n = 40°$. Upon refraction, the rays entering the second medium will only be those, which are incident on the boundary at an angle less than the critical angle. The rays incident on the boundary at the limiting angle form in the upper half-space a critical ray cone whose intersection with the interface is the critical contour defined as the locus of intersection between the critical rays and the boundary. On the stereogram of Figure 89b, this critical contour is shown as a circle in the lower hemisphere. In this case, the projection of the boundary on the stereogram cuts the critical contour, and it may appear that the critical contour partly lies outside the plane of the boundary. Actually, the projection of the boundary, the critical contour and the line of equal deviations lie in different hemispheres: the boundary line in the lower hemisphere (the dashed line), and the critical contour in the upper hemisphere (the solid line).

The observer imagined to be at the source would see all of the lower half-space in the ray cone whose axis is the normal to the boundary, and whose sides make the critical angle with the normal. Using stereographic projection, it is possible to ascertain, which of the rays emerging from the source may enter the second medium, and determine the angles at which they are refracted or graze the boundary. Since our further discussion will solely be concerned with transmitted waves, we shall consider only the rays lying inside the ray cone. The directions of the rays in the second medium upon refraction on the intermediate interface, may be ascertained from the stereo-net. Let us imagine these directions as lines of equal azimuths and of equal angles with the vertical. As is seen from Figure 89b, these lines form a stereo-net of their own, with a shifted pole of projection. Thus, inside the critical cone there appears one more stereo-net defining the direction of the rays in the second medium. Here, the meridional lines are the loci of the rays making equal angles with the vertical in the second medium. The lines of equal azimuths meet at one point corresponding to the ray directed vertically in the second medium. At this point, the concept of azimuth has no sense. The lines of equal azimuths and of equal angles of inclination are drawn at intervals of 20° on the stereo-net.

Thus the region of the critical contour encompasses the lines of the two stereo-nets. Those of the larger net define the directions of the rays in the first medium upon emergence from the source. Those of the smaller net, enclosed inside the critical cone, define the directions of the rays in the second medium upon refraction. Thus, each point of the area enclosed in the critical cone represents two directions – in the first medium prior to refraction, and in the second medium upon refraction.

From analysis of changes in the direction of rays, it is seen that for rays with the same azimuth in the first medium, the azimuth deviations from the original directions in the second medium decrease, and the deviations from the vertical increase as the angle with the vertical in the first medium increases. Azimuth deviations substantially increase with increasing change in velocity at the boundary and also with increasing angle of tilt of the boundary.

In illustrating the use of the above chart, we shall use the data of Table III as examples.

TABLE III
Changes in the Direction of the Ray (degrees)

Direction of ray prior to refraction		Direction of ray after refraction	
ω	φ	ω	φ
150	30	80	38
150	40	110	55
160	30	105	26
160	40	130	46
160	50	137	72
170	20	20	25
170	30	130	18
170	40	155	40
170	50	160	67

A change in the angle of tilt of the boundary leaves the chart pattern unchanged. When the boundary is horizontal ($\varphi_n = 0°$), the normal to the boundary is vertical and passes through the centre of the net. The critical contour is defined by the ray making an angle of 90° with the vertical upon refraction. In such a case, as will be recalled, all rays lie in the vertical planes, and there are no azimuth deviations; only the angles with the vertical undergo a change. In the circumstances, the pole of the inner stereo-net lies on the normal to the boundary. The lines of equal azimuths in the second medium (the radial lines) run along the respective lines of equal azimuths in the first medium. The lines of equal angles with the vertical appear on the stereo-net as concentric circles. Using this chart, it is an easy matter to determine the angles with the vertical for all rays in the second medium. Naturally, similar charts can be used to determine the direction of the rays incident on the refracting boundary from below and reaching the point of observation upon refraction. This technique is applicable to earthquake data recorded by individual stations and, above all, to work by the converted transmitted-wave method.

CHANGE IN THE DIRECTION OF RAYPATH ALONG A VERTICAL PROFILE

Let us trace variations in the direction along which the first arrivals approach the seismometers stacked along the drill-hole. In stereographic projection, this can most conveniently be done by use of a locus. Since to an imaginary observer moving along the drill-hole the first waves appear to arrive in the same direction along which he would see the source, the respective locus on the stereo-net will, for brevity, be referred to as the visibility line. Each point on this line corresponds to a certain drill-hole point at a particular depth. The visibility line for a drill-hole traversing two plane boundaries is described without any details of construction in [31]. For waves in which the direction of particle displacement is the same as the direction of wave motion, the visibility lines can be used to determine the direction of wave arrival and to interpret seismic data.

Let us construct the visibility line for a medium containing one plane interface the normal to which is such that $\omega_n = 180°$ and $\varphi_n = 20°$. We also assume that there are three boreholes and one shot-point (Figure 90b), and the velocity ratio is $v_2/v_1 = 0.5$. The azimuths of the directions from the shot-point to the respective boreholes are: $\omega_1 = 34°$, $\omega_2 = 234°$, and $\omega_3 = 320°$. The depths of observations are: $H_2 = 8.7$ reference

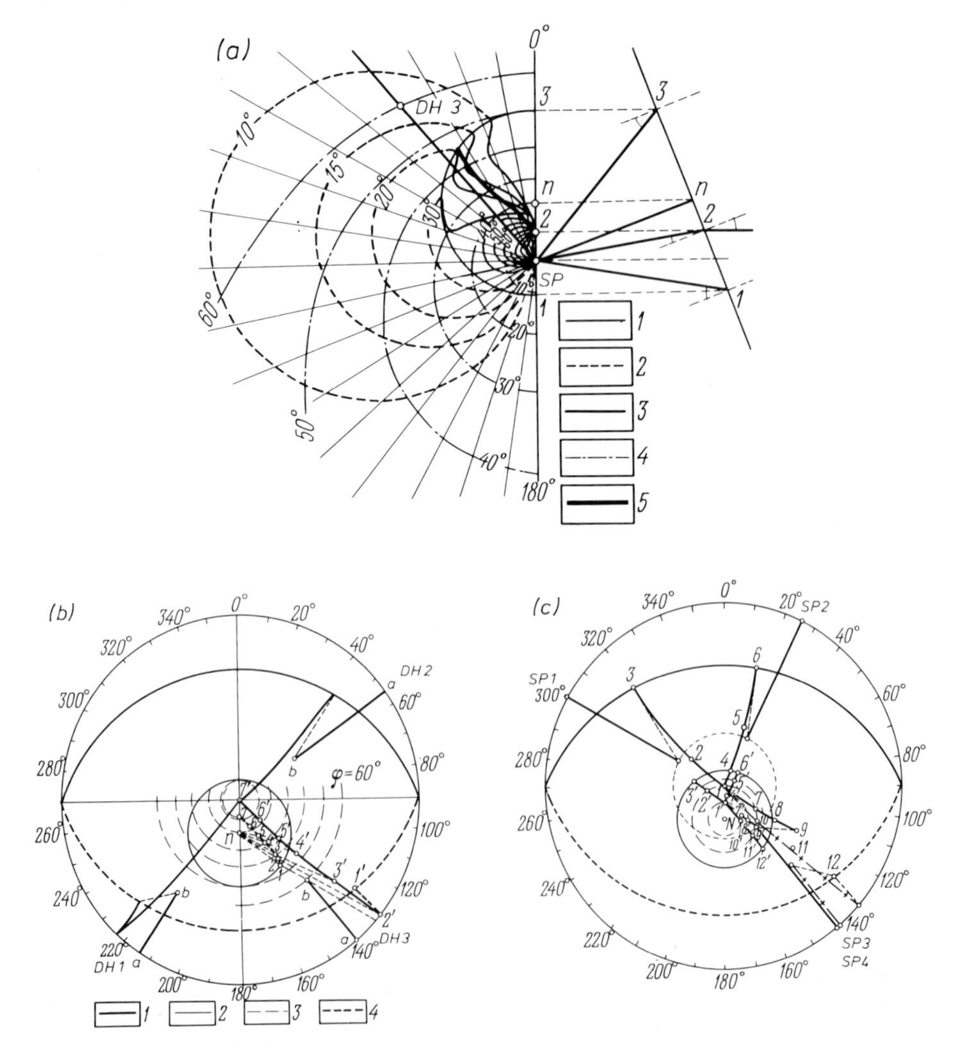

Fig. 90. Construction of visibility lines. (a) azimuth-deviation chart and section along the line of dip; 1 – azimuth-deviation lines for transmitted waves; 2 – azimuth-deviation lines for grazing rays; 3 – critical contour; 4 – projections of the line of intersection between the conical surfaces and the interface onto a horizontal plane; 5 – locus of projections for the intersections between the interface and the rays entering the drill-hole. (b) stereo-net and visibility lines for DH No. 3: 1 – visibility lines; 2 – critical contour; 3 – conical ray surfaces; 4 – surface bounded by rays entering the drill-hole upon refraction. (c) visibility lines for DH No. 3 and four shot-points.

units and H_3 = 4.77 reference units. To construct the visibility line, the first step is to identify the rays, which, on leaving the source and being refracted on the boundary, will enter the appropriate drill-hole. In plan, the projections of these rays will pass through the drill-hole entrance. The constructions may conveniently be made on the azimuth deviation chart (at *a* in Figure 90). The critical contour on the chart encloses the lines of azimuth deviations for the transmitted waves. Outside the contour are the lines of azimuth deviations for grazing rays. These lines are constructed on the basis of the conical ray surfaces emerging from the source and cutting the interface so that the resulting sections are circles. The projections of these circles on a horizontal surface appear as ellipses. Referring to the lines of azimuth deviations for transmitted waves, we may locate the locus of intersection between the interface and the rays, which will enter the borehole upon refraction. Under the interface, but in close proximity to it, the rays capable of entering the drill-holes are those for which the projections of intersections with the interface appear on the plan, but near the critical contour. As the depth is increased, the rays capable of entering the drill-hole will be those that leave the source in directions lying closer to the vertical.

The stereo-net of Figure 90b shows the construction of the visibility line for DH No. 3 and the visibility lines for DHs Nos 1 and 2. Let us go through the steps involved in the construction of the visibility line for DH No. 3. An observer on the ground surface would see the source along a horizontal line with an azimuth of 140° (point *a*). If the observer is imagined to be the interface (H = 4.77 reference units), the azimuth at the source will not change; only the angle with the vertical will change. At the boundary, the source will be seen at an angle of 60° (point *b*) – this is the last ray of the direct wave. To construct the visibility line under the boundary, the surface that will be formed by the rays, which enter DH No. 3 upon refraction, is transferred from the plan onto the stereo-net (Figure 90b). This line is enclosed in the critical contour. To obtain the critical contour, conical ray surfaces are plotted on the stereo-net. The directions of the rays in the second medium upon refraction may be found by reference to the net.

At the boundary the observer will see a grazing ray (point 1). On crossing the boundary, the visibility line shows a discontinuity. Immediately below the boundary, the observer will see the source beneath himself. In going farther down, the observer will finally be level with the source (point 2′). Descending still farther, the line of sight on the source will approach the vertical (points 3′–6′). At a greater depth, when the distance from the shot-point to the source may be neglected, the observer will see the source overhead (point 7′). For simplicity, the net does not show the visibility curves for head waves that should be recorded above the boundary. It is seen from Figure 90b that, on crossing the boundary, the rays undergo azimuth deviations up the dip of the boundary.

More valuable information may be gleaned from an analysis of several visibility lines constructed for several shot-points and a single observation drill-hole. Figure 90c shows visibility lines for DH No. 3 and four shot-points. Three of them (SP1, SP2, and SP3) are located at the same distance from (3 reference units), but on different azimuths (ω_1 = 300°, ω_2 = 25°, and ω_3 = 140°) relative to the drill-hole. The fourth shot-point (SP4) lies on the same azimuth with SP3 (ω_4 = 140°), but its distance from the drill-hole is 8.25 reference units. The velocity ratio at the boundary is taken as $v_2/v_1 = 1$.

The waves arriving from SP1, SP2, and SP4 travel at a grazing angle, because the directions of these shot-points to the intersection of the boundary with the borehole lie outside the critical contour. As a consequence, the waves in the first medium must be head waves over the entire interval of the vertical profile or near the boundary. At the boundary, the visibility lines show a discontinuity. Passage across the boundary changes both the azimuths of the rays and their angles with the vertical. The azimuth deviations for different shot-points take different signs: positive for SP1, and negative for SP2 and SP4. This agrees with the general rule: the rays deviate up the dip of the boundary.

For SP3 located at the same distance from the drill-hole as SP1 and SP2, but down the dip of the boundary, there are no grazing rays (the source is seen within the critical contour). The waves experience refraction, and the rays likewise deviate up the dip of the boundary.

The shape of the visibility curve varies with the kind of waves recorded. For better insight in to the matter, let us consider the field of direct and grazing waves for one plane interface Q ($\omega_n = 180°$, $\varphi_n = 20°$), according to the location of the shot-points on the observation plane. Let the shot-points lie on concentric circles, with the drill-hole as centre. Let each circle be called a 'source ring'. The parameter of such a circle may be the angle φ at which the ring is seen from the intersection of the drill-hole and the boundary. In our example, the sources are seen at angles φ equal to 5°, 10°, 15°, 23°, 30°, 36°, 45°, 48°, and 50°. The fields within which direct and grazing waves are recorded are shown in Figure 91a. All the sources whose directions are within the critical contour (points 1–8) give rise to transmitted waves in the second medium. All the shot-points whose directions are equal to the critical angle give rise to grazing waves in the second medium and to head waves in the first medium. For each ring source, there is a ray plane of its own on the ground surface.

For the sources with angles equal to 5° and 10°, the rings lie completely within the critical contour, and they correspond to transmitted waves. Under the interface, the source rings are deformed into more elaborate ray surfaces. For the source ring with $\varphi = 10°$ only one shot-point with $\omega = 0$ touches the critical contour and produces a grazing wave (point 1). Referring to the source ring with $\varphi = 15°$ inside the critical contour, we may determine the sources whose azimuths lie within the sector 055°–305°. These sources induce transmitted waves, which form a ray surface in the second medium touching the plane at points 2–2. The shot-points with azimuths of 055° and 305° lying in the critical contour (points 2–2) produce the first grazing waves (points 2′–2′) whose azimuths are 026° and 334°, respectively. The waves between these points on the Q plane are grazing waves associated with the shot-points whose azimuths lie between 055° and 305°.

As the angle φ (and, as a consequence, the distance to the shot-point) is decreased, the field of transmitted waves decreases, and that of grazing waves increases. For example, at $\varphi = 45°$ transmitted waves are only excited by the shot-points whose azimuths lie between 144° and 216° (points 6–6). The remaining sources on the same ring induce grazing waves occupying an interval on the Q plane bounded by azimuths 6′–5′ and 5′–6′. For the ring with $\varphi = 48'$, the region of transmitted waves is still smaller (azimuths 156°–224°), with most transmitted waves being practically horizontal. For the source ring with $\varphi = 50°$, only one shot-point with an azimuth of 180° (point 8) touches the critical contour. The remaining sources induce grazing waves, which cover the entire Q plane, and head waves are recorded in the first medium. As the angle of tilt of the

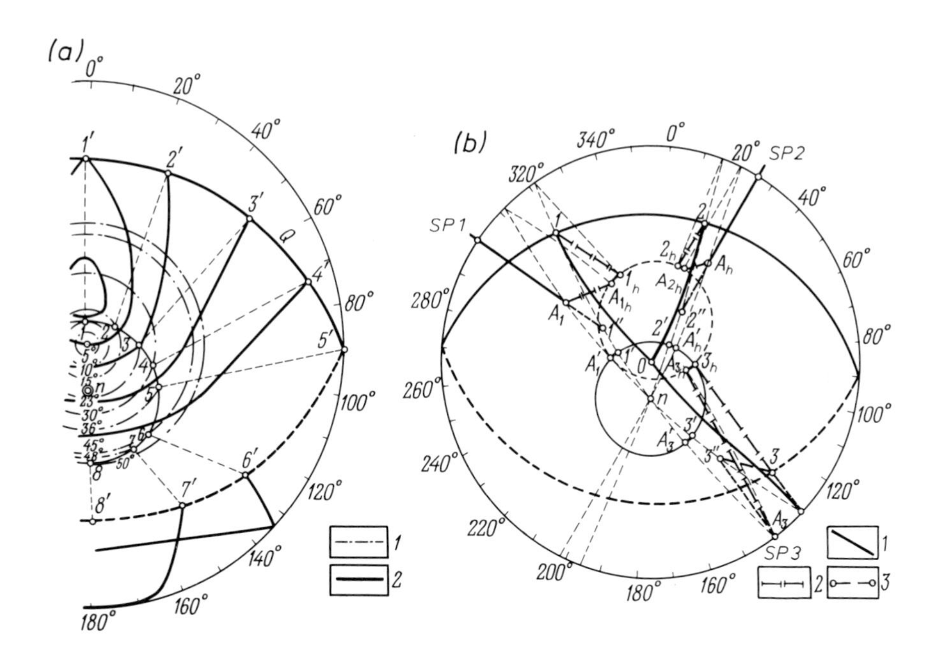

Fig. 91. Construction of a visibility line for head waves. (a) Regions of direct and grazing waves (owing to symmetry, the 0–180° azimuth region is shown): 1 – source lines; 2 – ray surface. (b) stereo-net and visibility lines: 1 – for first arrivals; 2 – same, for later arrivals; 3 – points corresponding to the same depth, for each shot-point.

interface and the velocity ratio increase, the field of transmitted waves decreases. In contrast, the field of grazing waves first increases, passes through a maximum, and decreases again. For the sources lying at the same distance from the drill-hole, the shape of the visibility line strongly depends on the azimuth at the source.

Wherever a grazing wave is observed at the interface, a head wave is recorded in the first medium along the vertical profile either over the entire interval above the boundary or only in the area adjacent to the boundary. Since, at present, the direction of particle displacement may be reliably determined only for the first arrivals, it is of special interest to study the direction of particle displacement in head waves. In contrast to direct waves, the direction in which head waves arrive markedly depends on the orientation of the boundary and the velocity ratio both above and below the interface. This somewhat complicates the task, but, as will be shown shortly, the same factors can be utilized in solving the inverse problem.

Consider the solution of the direct problem involving head waves induced at four shot-points (Figure 91b). Let us begin with SP1. To locate the incident ray, we construct on the stereo-net a plane passing through the normal n and the grazing ray (point 1). The direction of the incident ray lies at the intersection of this plane with the critical contour (point 1′). Laying off the critical angle from the normal in the direction opposite to the incident ray (point 1), we locate the reflected ray (the initial ray of a head wave). On the visibility line, this ray must be taken with the inverse azimuth, point 1_h. This point belongs to the visibility line when the observer is placed in the drill-hole level with the interface. The last (topmost) point on the visibility line for head waves (A_{1h}) is located by selecting on the plan and the stereo-net an incident ray such that the head-wave ray would pass through the drill-hole (point A'_1). Figure 91b shows the visibility lines for the head waves induced by all the three shot-points. For SP1 and SP2, the head wave appears in the first arrivals on the vertical profile at depths corresponding to points A_1 and A_2 on the visibility lines. Lower down on the vertical profile, as far the boundary, head waves are observed in the late arrivals (point 1″). Under the boundary, the first to be recorded is a transmitted wave.

For SP3, head waves are recorded in the first arrivals over the entire interval of the vertical profile. All visibility lines for head waves, as for transmitted waves, deviate up the dip of the boundary. The visibility lines for head waves are segments of the critical contour taken with inverse azimuths. For one plane boundary and several shot-points, they lie on one conical ray surface – the critical cone. Let us trace the visibility lines for the first arrivals from some of the shot-points. For SP1, the first arrivals on the vertical profile, starting with the ground surface, are formed by a direct wave. At point A_1, the observer sees the source likewise in the direction of the head wave (point A_{1h}). The portion of the profile from point A_1 to the interface is shown on the visibility line for the head wave (A_{1h}–1_h). At the interface, a transmitted wave appears, and as the depth increases SP1 will be seen along the transmitted rays (points 1–0).

Thus, conversion of waves along a vertical profile or passage across a boundary brings about a change in the direction of arrival for the waves. This change occurs in both the vertical plane and azimuth, and may be appreciable. Knowledge of the pattern in which such a change occurs may be utilized to locate the position of the boundary and to determine the velocity ratio at the boundary, which is the objective of inverse problems.

SOLUTION OF INVERSE PROBLEMS

An inverse problem, as we define it, refers to locating the position of the boundary (the direction of the normal) and finding the velocity ratio from the visibility lines. Let us take up the solution of an inverse problem first for transmitted, then for head waves. To locate the position of the boundary plane, it will suffice to locate its normal in space. As will be recalled, the incident and the refracted rays lie in the normal plane. In observations at the boundary, this plane will also contain the grazing ray. The normal to the interface is found as the trace of the intersection between the ray planes and may be constructed on the basis of th visibility lines. Such constructions for three shot-point (SP1, SP2, and SP3) are carried out in Figure 92. The ray planes that intersect along the normal are passed through the discontinuity points (1′–1, 2′–2, and 3′–3) of the visibility lines for the transmitted waves.

The ratio of the velocities for compressional waves being propagated in layers above and below the interface may be found from the ratio of the angles of incidence and refraction which are α_1 and α_2 for SP3, and γ_1 and γ_2 for SP1. For SP3, $\alpha_1 = 26°$, $\alpha_2 = 61°$, $v_1/v_2 = \sin\alpha_1/\sin\alpha_2 = 0.342/0.482 = 0.5$. For SP1, $\gamma_1 = 28°$, $\gamma_2 = 70°$, and $v_1/v_2 = 0.5$.

To locate the normal to the boundary, it will suffice to have visibility lines for two shot-points. For better accuracy, it is desirable that the azimuths of these points should

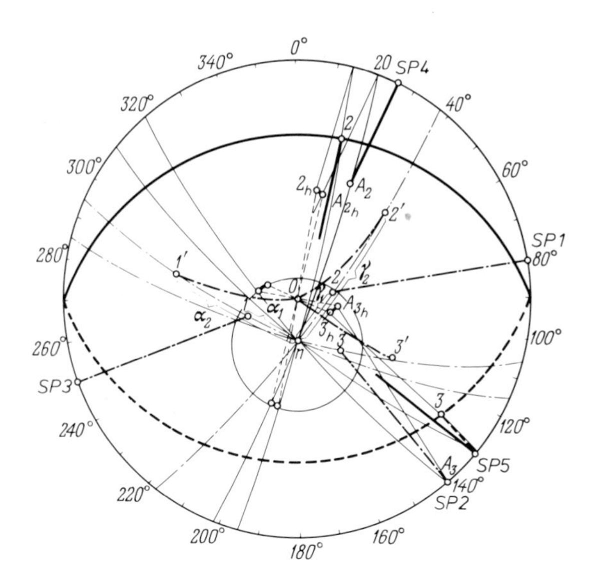

Fig. 92. Construction of a refracting boundary on the basis of the visibility line.

differ by at least 30°–40°. The third point provides a check on the results thus obtained. When the direction of the normal is known in advance, the velocity ratio may be found from one visibility line for transmitted waves.

By reference to the visibility lines for head waves, the normal to the boundary may be found as the trace of the intersection between the normal planes. A normal plane may be passed through every point on the visibility line for the head waves. This plane can be located by referring to the direction of the direct-wave ray. As an alternative, normal planes could be located by reference to the direction of particle displacement in grazing rays, in which case one visibility line would be sufficient to locate the position of the interface in space. Since, however, the direction of particle displacement in a grazing ray may differ from the direction of the ray [36], the visibility line in the grazing-ray field is rather fictitious. To solve an inverse problem on the basis of head waves, it is likewise desirable to have several visibility lines. The velocity ratio may be determined by reference to the whole or a part of the critical contour, which is re-arranged by using the visibility lines for head waves with inverse azimuths. By reference to the critical contour, it is possible to construct the normal to the boundary.

The foregoing appears to prove the usability of the direction of particle displacement in the first arrivals as a source of information about the position of the boundary in the areas traversed by a drill-hole, and also the velocity ratio for seismic waves above and below the interface. To make this usability a practical tool, however, requires above all skills in tracking variations in the direction of particle displacement.

To illustrate how the direction of particle displacement may be used to locate the position of a refracting boundary, we shall refer to the Zhairem ore deposit (Figure 93a). Data processing involves joint analysis of $\nu(H)$, $\varphi(H)$, and $\omega(H)$ curves for a direct transmitted wave. As an alternative, each may be analysed separately to yield quantitative data about the angles of tilt and azimuths of boundaries.

Should we presume the sharp changes in the $\varphi(H)$, $\omega(H)$ curves for transmitted rays to be associated with inhomogeneities of the earth featuring discontinuities in velocities

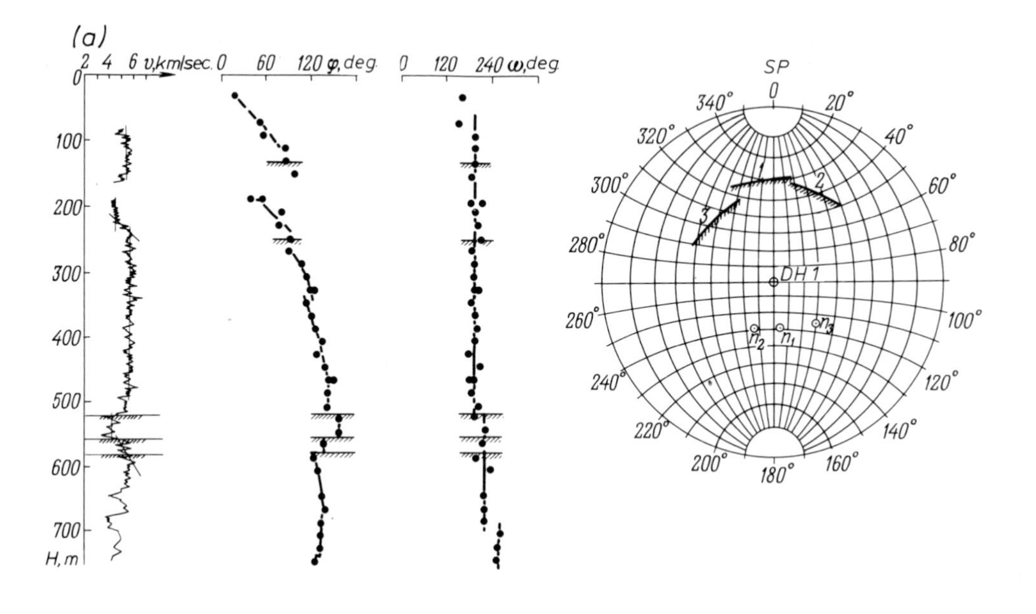

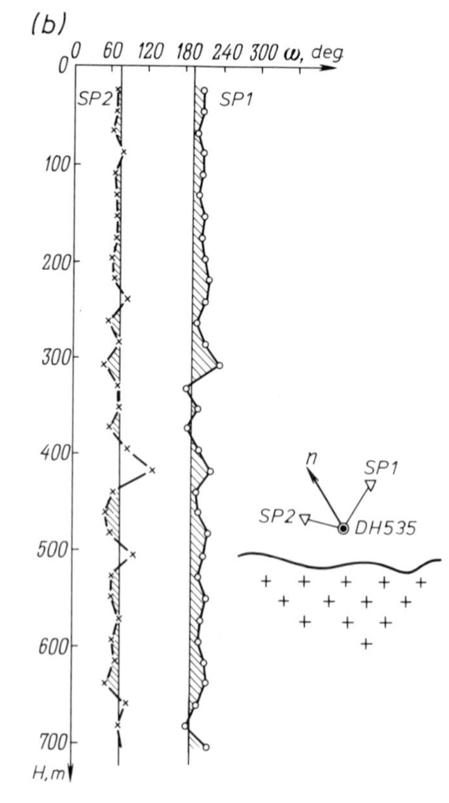

Fig. 93. Determining the orientation of a boundary at (a) Zhairem deposit and (b) Sayak deposit.

and different orientation in space, we would be in a position, resorting to certain simplifying assumptions, to locate the orientation of such interfaces. At every observation point, the direction of the ray incident on the boundary will be sufficiently close to the direction of the transmitted ray determined in the preceding interval of the section. Thus, each preceding point on the $\nu(H)$, $\varphi(H)$, and $\omega(H)$ curves may be treated as identifying the direction of the incident ray, and each succeeding point as identifying the direction of the transmitted ray. In this assumption, we neglect the changes in φ associated with changes in the relative position of the source and observation point, using as a basis the idea that such changes from point to point are insignificant if their spacing is sufficiently close (10–20 m). With such assumptions, we can establish at each point the direction of the normal to an element of the boundary from the directions of the incident and refracted rays, and also from the ratio of the velocities above and below the boundary. The presence of azimuth deviations in depth interval 490–570 m indicates that the bedding is not horizontal. The directions of the normals to the boundaries thus obtained are shown on the stereo-net.

If information about velocity distribution in the medium is lacking, the usability of the direction of particle displacement is substantially limited. Without going into detail, it may be noted that resort to the $\omega(H)$ curves and the averaged azimuth deviation related to the overall direction (averaged over the entire profile) in which the strata dip, makes it possible to determine the direction of dip quantitatively.

For example, the azimuth deviations for DHs 535 and SP1 in the Sayak area (Figure 93b) are seen to take a '+' sign ($+\Delta\omega$); this is an indication that the azimuth of the wave arrival exceeds the azimuth of the source. The azimuth deviations for SP2 in the same area and the same borehole are seen to take a '–' sign ($-\Delta\omega$). In view of the existing relative position of the shot-points and the drill-hole, the above azimuth deviations show uniquely, albeit qualitatively, that the stratification dips along the line passing between SP and SP2 in the direction opposite to the emergence of the intrusion.

Locating the Position of the Boundary from Converted Transmitted Waves in Surface Observations

Polarization is one of the criteria used in identifying converted transmitted (*PS*) waves on seismograms. In surface observations of horizontally-stratified media, *PS* waves must be polarized in the ray plane passing through the source or, in other words, the radial components of displacement must substantially exceed the tangential components. However, this requirement is not always met on real seismograms. Indeed, the tangential components are comparable with the radial components [2, 68–71, 100, 101].

Also, the directions of displacements for *PS* waves (azimuths) may differ from those for *P* waves, and the latter may differ from the azimuths on the source. The presence of strong tangential components has been ascribed to various factors. Of course, it is only too natural to ascribe their presence to the spatial position of the conversion boundary. This hypothesis has been advanced by Puzyrev [131]. Many authors believe that the principal factor is the anisotropy of the medium with respect to the elastic properties. As already noted, both factors may lead to the appearance of tangential components.

Taking a computed case as an example, we shall illustrate the effect of inclined boundaries on the polarization of *P* and *PS* waves and discuss the likely approaches to the inverse problem – that of locating the position of the conversion boundary from the polarization of transmitted converted waves. Let the medium be two-layered with a single plane inclined boundary (Figure 94). It is convenient to make all constructions in stereographic projection. Let Q be the horizontal plane of observation (the ground surface)

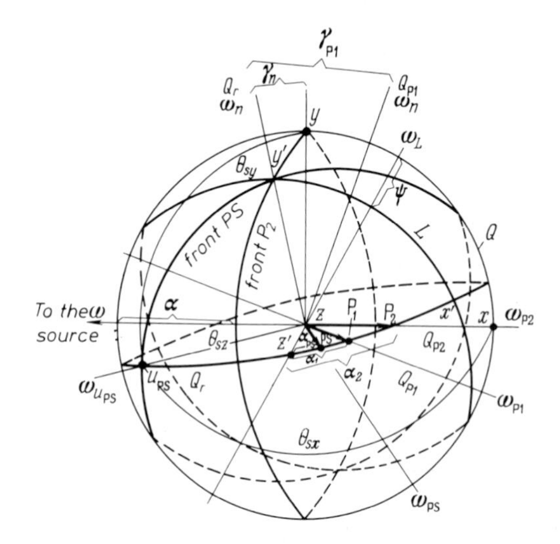

Fig. 94. Diagram for computation of converted wave components.

and L be the inclined plane boundary whose spatial position is such that the angle with the horizontal is $\psi = 20°$, and the azimuth of the line of rise is $\omega_L = 30°$. Also let the ratio of the velocities in the first and second media be $v_1/v_2 = 1/1.5 = 0.66$, and the velocity ratio v_P/v_S in the first layer be equal to 1.7 Suppose also that a distant source with an azimuth ω induces a plane compressional wave P_2 with an azimuth ω_{P_2} incident on the conversion boundary, and that the front of this wave makes an angle $\alpha = 50°$ with the horizontal.

To facilitate the computation of the components of the P_1 and PS waves upon refraction and conversion, we choose two left-hand rectangular coordinate systems: x, y, z referenced to the observation plane, and x', y', z' referenced to the conversion boundary. The z-axis is perpendicular to the Q plane, and the z' direction is perpendicular to the L plane. The x' axis runs along the line of intersection between the L plane and the Q_r ray plane, so that x' makes an acute angle (about 17° in our example) with the positive direction of the x-axis, whereas the y' axis runs along the line of intersection between the P_2 wave front and the boundary (L) plane.

As a result, in the first coordinate system the Q plane contains the x and y axes, whereas in the second coordinate system the L plane contains x' and y' axes. Upon refraction at the L boundary, two waves will be propagated in the upper medium – the compressional wave P_1 and the converted wave PS. The directions in which they are propagated lie in the ray plane passing through the normal to the boundary, which runs along z' axis and the direction of the incident P_2 wave. To locate these directions in the ray plane, it will suffice to determine the angles of refraction α_1 and α_2 for the P_1 and PS waves and to measure them from the normal in the direction of the incident rays. In our case, $\alpha_{PS} = 20°$, $\alpha_1 = 36°$, and $\alpha_2 = 63°$. In this way, we have located the direction of propagation for the P_1 and *PS* waves in the first medium. In an isotropic medium, the displacement vector U_{PS} of the *PS* wave lies in the same plane, at right angles to the direction of progation, and is found as the line of intersection between the ray plane Q_r and the plane of the *PS* wave front.

Thus, if the boundary were horizontal, then all the directions of wave propagation would lie in the same vertical plane coincident with the direction towards the source, and there would be no tangential components. The fact that the boundary is tilted changes the direction in which the waves are propagated in space (both in azimuth and in angle with the vertical). For a compressional wave, the azimuth deviation is $\omega_{P_1} - \omega_{P_2}$, and for a converted wave it is $\omega_{PS} - \omega_{P_2}$. The particle displacement vector U_{PS} for the *PS* wave makes with the coordinate axes the angles θ_{sy}, θ_{sx}, and θ_{sz}, respectively equal to 104°, 080°, and 017°. Now we can calculate the components of the displacement vector for the P_1 and *PS* waves according to to the amplitudes of displacement for the P_2 wave in the lower medium. Upon conversion at an inclined boundary, the P_1 and *PS* waves undergo azimuth deviations so that the directions of particle motion no longer lie in vertical planes passing through the source and the observation point.

The manner in which the components vary with the characteristics of the medium and the spatial position of one conversion boundary relative to the wavefront of the wave incident upon it from below is traced in reference [94]. Leaving out the computations, we shall recapitulate some of its findings. Of all the characteristics of the medium, the direction of particle displacement in *PS* waves is most affected by the position of the conversion boundary relative to the direction towards the source. The decisive factor in this respect is the angle of tilt of the boundary. At large angles of tilt ($\varphi > 30°$), the tangential components can be comparable with the radial components. Azimuth deviations for *PS* waves (relative to the direction towards the source) are almost insensitive to variations in the velocity ratio v_1/v_2 for compressional waves and to the direction of incidence of the wave on the interface from below. The strong dependence of the direction particle displacement in converted waves on the geologic structure serves as the basis for a technique, which utilizes the polarization of compressional and converted transmitted waves for determining the characteristics of the section, notably for locating the position of the conversion boundary (the angle of tilt and the direction of the dip). Currently, such problems are solved on a digital computer. The additional information yielded by the technique expands the capabilities of exploration (notably, it refines the position of the conversion boundary) and improves the accuracy with which the depth of boundaries can be determined with allowance for their inclination. This is of special value in cases where observations are carried out by scattered stations.

Locating the Position of a Refracting Boundary from the Polarization of Compressional Waves

As already noted, the direction of particle dispalcement in the first arrivals may differ from the direction at the epicentre, and this difference makes it possible to form an idea about the structure of the medium beneath the station. This is especially valuable where the waves are excited by weak local earthquakes and whose analysis entails a number of difficulties. One of them is that, owing to the strong attenuation of the high-frequency components of the waves (15–30 Hz), the intensity of the records falls off with distance. We shall discuss the manner in which earthquake records produced by one or two stations can be processed to circumvent the above difficulties.

The method at one time regarded as most promising in this respect is that proposed by Golitsyn and based on determining the azimuth of the epicentre from the direction of particle displacement in the first arrivals. This technique appreciably improves the accuracy with which the displacement vector can be located by means of correlation techniques, notably polar correlation. Unfortunately, it is not so efficient in determining the azimuths of the epicentre in localities with complex media. A major source of error in determining the azimuth of the epicentre is that it does not coincide with the raypath, although the error may be minimized by recording the high-frequency, rather than the low-frequency component. The error is especially appreciable in the topmost part of the section, and one has to take into account variations in the direction of the raypath as the wave is propagated from the focus to the station.

More conveniently, the position of a boundary beneath the station may be located from the direction of particle displacement in the first arrival. We shall illustrate this statement by reference to the records of weak local earthquakes in the Khait epicentral zone (Garm District, Tadzhikistan, USSR).

The exploration party used a network of stations capable of amplifying the signal 1×10^6 to 1.5×10^6 times. In fact, some of the stations were capable of boosting the signal several million times. Records were produced by eight-component polar arrays of inclined seismometers [55, 56]. In our illustration we shall use the observations carried out by two high-sensitivity stations: Nimichi (K–2) and Nimchak (K–1). Typical seismograms of weak local earthquakes recorded by these stations are given in Chapter 4.

SPATIAL INTERPRETATION

Spatial interpretation of weak earthquake records may be divided into two stages. One state involves analysis of any anomalies in the raypath. The other is concerned with the detection of the discontinuities that may have produced the above anomalies, so that appropriate corrections can be applied to the observed direction of the raypath and the true direction deduced.

Consider this technique of locating the refracting boundary on the assumption that the boundary is plane, from the azimuth deviations of the two stations. Data-processing shows that for the K–1 station most rays are concentrated within a narrow sector of azimuths from 250° to 290°. For the K–2 station, the dominant wave arrivals have azimuths from 330° to 060°.

A comparison of the observed directions of the raypaths with the azimuths of the epicentre determined by kinematic techniques from the data gathered by the network of stations reveals large azimuth deviations for each station (Figure 95a). These deviations are especially large for the K–1 station (Nimchak). The azimuth deviations for the K–2 and K–1 stations take different signs. Whereas for the K–1 station, the azimuths of pathrays are smaller than those of the epicentre (the azimuth deviations are negative), for the K–2 station the azimuths of the raypaths are greater than the azimuths of the epicentre (the azimuth deviations are positive).

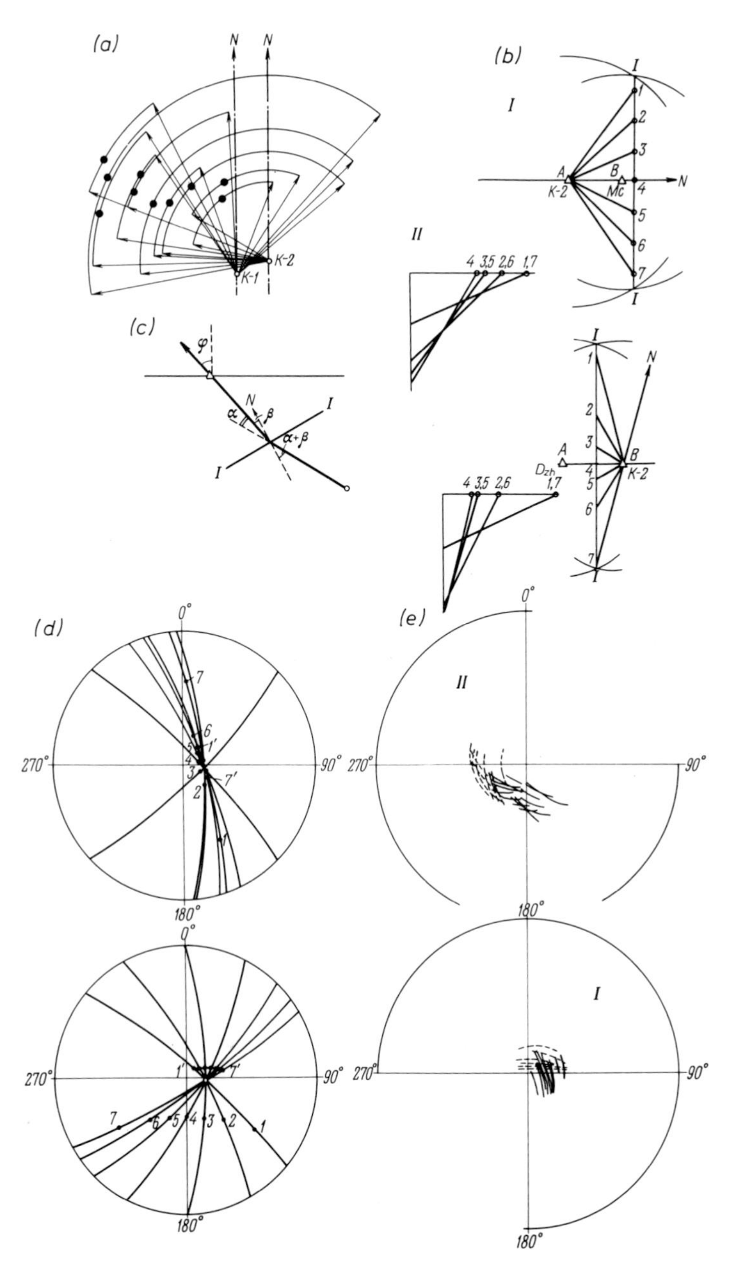

Fig. 95. Locating a refracting boundary. (a) Azimuth deviations in direction of particle displacement according to data for K–1 (Nimchak) and K–2 (Nimichi) stations (directions corresponding to the same earthquake are joined by an arc drawn through the epicentre of the earthquake, as located from kinematic data supplied by the station network). (b) I – construction of a line of likely epicentres for one earthquake recorded by two pairs of stations, namely K–2 and Ms, K–2 and Dzh. II – determination of the angle with the vertical. (c) positions of rays in a ray plane. (d) construction of lines of likely normals (1′–7′). (e) lines of likely normals (for Nimchak on the left and for Nimichi on the right of the figure).

The high values of azimuth deviations precluded the use of the azimuth method for locating the earthquake epicentres. An attempt was made to establish the pattern of variations in azimuth deviations so that appropriate corrections could be applied. To this end, it was necessary first to locate the interfaces in the medium, especially in the top part of the section, since they are decisive in determining the magnitude of azimuth deviations. Some evidence about the construction of the surface layers of the crust may be gleaned by observing natural earthquakes. The technique is as follows.

Suppose that a change in the direction of the raypath from the focus to the station is associated with one plane interface lying at a shallow depth. Assume also that the interface has practically no effect on the move-out time between the shear and compressional waves, Δt_{S-P}, and only changes the direction of the raypath. With such simplifying assumptions, the position of this boundary can be located by two pairs of stations.

If we describe spheres with stations A and B as centres, with radius $R = \Delta t_{S-P} v$, where v is the velocity of propagation of fictitious waves, then the line of intersection between the two spheres will be the line of likely hypocentres. The projection of this line on to a horizontal plane, the observation plane, is the locus of likely epicentres (line I–I). Each point on this line corresponds to a different hypocentre. On selecting any point on line I–I as the epicentre and using the known hypocentral distance, it is possible to locate the corresponding hypocentre and determine the angle between the vertical and the ray emerging from the hypocentre. This can conveniently be done graphically.

From the pre-refraction direction of the raypath thus found and the observed direction of the raypath after refraction, it is an easy matter to locate the normal to the interface. This can, for example, be done, using the Wulff stereo-net. We mentally place the centre of the stereo-net on the interface at the point of incidence of the ray and mark on a tracing-paper overlay the directions of the ray before and after refraction (i.e., the observed direction). Using these directions, we can construct a ray plane, which will contain the normal to the interface. To locate its position, it will suffice to determine the angle of refraction, using the equation $\sin \beta / [\sin (\alpha + \beta)] = v_1/v_2$, where α is the angle between the directions of the raypath before and after the refraction (Figure 95c).

If we measure the angle β on the stereo-net from the observed direction of the raypath away from the incident ray, we shall locate the likely normal to the plane interface in question. Each point on the line of likely epicentres corresponds to a likely direction of the normal. Proceeding in the same way at each point on the line of likely epicentres, we can find the respective directions of likely normals describing some surface. On the stereo-net, this surface will produce the line of likely normals. The line of likely normals (1′–7′) constructed on the basis of earthquake records is shown in Figure 95d.

The direction of the actual normal is located as the point of intersection between the lines of likely normals. If several stations are operating together, the direction of the normal may best be determined, if the lines on which the stations lie are mutually perpendicular.

The above technique was applied to locate a plane refracting interface beneath the K–1 (Nimichi) and K–2 (Nimchak) stations. Under the Nimichi station, the boundary was located on the basis of data supplied by the Dzhafr station, west of Nimichi, and also by the Mussafiron station, north of Nimichi. The two pairs of stations, Nimichi-Dzhafr and Nimichi-Mussafiron, lie on mutually perpendicular directions, which enhances the accuracy of determinations. Each pair was used to observe a group of earthquakes, so that a band of lines of likely normals was obtained for each pair. The likely direction of the normal is determined by the area of intersection between the two bands (Figure 95e).

In the case of Nimchak, the two pairs of stations used were Nimchak-Mussafiron and Nimchak-Dzhafr. The family of likely normals is shown in Figure 85e. It is seen that for the Nimchak station there is an appreciable scatter among the lines of likely normals as compared with the Nimichi station. Probably, the cause is a more complex medium.

EFFECT OF INTERMEDIATE REFRACTION

Using the located interfaces, it was possible to compute azimuth deviations and apply appropriate corrections to the observed azimuths. The corrected azimuths were compared with the azimuths of the epicentres as determined from kinematic data supplied by the entire network of stations (Figure 96). From the comparison it was found that the

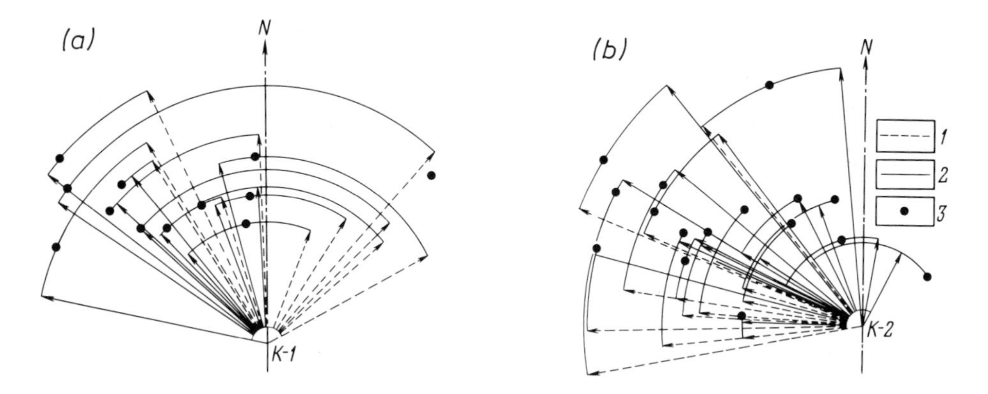

Fig. 96. Azimuth deviations for (a) K–1 station and (b) K–2 station. 1 – observed raypaths; 2 – bearings on the epicentre corrected for azimuth deviation; 3 – earthquake epicentres located on the basis of data supplied by the station network.

corrected azimuths showed a better agreement with the azimuths of the epicentres than the observed azimuths. For the K–1 station, the corrections were positive, so the final value of raypath azimuths increased. For the K–2 station, the corrections were negative. In this way, locating the interfaces on the basis of polarization data enables the azimuths of the epicentre to be corrected for intermediate refractions.

It may be added that the interfaces found in the above manner under each station have no geologic meaning and are only virtual boundaries related to changes in the raypath, whereas the changes might be caused by some more complex seismological discontinuities in the medium.

We have cited several examples illustrating the capabilities of wave polarization in getting more direct data about the medium. Not all of these capabilities have been implemented to the same degree. The technique based on analysis of the direction of particle displacement is in its infancy, and many matters related to procedures and apparatus are awaiting further elucidation. It is very likely, however, that as more knowledge is accumulated about the polarization of seismic waves in real media, new potentialities will be discovered, which can now only be guessed. The results already obtained give every reason to rate the technques described herein as very promising.

CONCLUSION

The efficiency of seismic exploration, and above all seismic prospecting, depends on the capabilities of wave-field analysis. To this end, present-day seismic exploration predominantly uses two wave field parameters: frequency and apparent wave velocity. It may be expected that the efficiency of seismic exploration will be materially improved not only through the more intense use of the traditional wave-field parameters, but also through the use of one more independent variable (wave polarization), a spatial and temporal characteristic of seismic waves directly related to the structure of the medium.

The polarization method is based on a joint analysis of wave polarization at a point and wave propagation in three dimensions, which is a combination of what we call CDR–I and CDR–II observation systems, Methodologically, this is equivalent to a combination of three-component observations at a point and three-dimensional observations in space. Wave analysis at a point resorts to CDR–I and polar correlation. Wave analysis along the line or plane of traverse uses polarization-position wave correlation (PPC).

With CDR–I, waves are discriminated at a point according to the direction of particle displacement, which is done by grouping of the first kind. Polar correlation of seismic waves is based on the analysis of records according to the orientation of wave components and is a further extension of correlation principles to seismic exploration.

Polarization-position correlation of seismic waves is based on wave discrimination at a point based on polarization combined with wave discrimination according to the direction of wave propagation. With this method, one follows what we call the tracking (or optimal) component for which the signal-to-noise ratio is a maximum. In the general case, the orientation of the tracking component is decided by the pattern of variations in the direction of the wave vector along the line of traverse and by the conditions under which spurious waves are superimposed at each observation point. Depending on the complexity of the wave field, various techniques of polarization analysis may be used differing in selectivity and completeness. Some of the them use polarization characteristics specified in advance, others operate with optimal characteristics determined by self-tuning filters directly from the wave field being analysed.

Practical application of the polarization method to seismic prospecting for petroleum and ores and in regional studies has confirmed the new potentialities that the method offers in both surface observations and in VSP work. In the light of this experience, the following may be stated.

The waves observed in real media both on the ground surface and at interior points of the medium are mostly non-linearly polarized. Most often, but not always, the first arrivals of direct compressional waves are linearly polarized. The polarization of direct waves may be used for quantitative analysis and monitoring of excitation conditions and, especially, the direction of the source.

As a rule, direct shear waves are non-linearly polarized. In most cases, the departure from linear polarization is caused by the superposition of waves polarized in various planes and shifted in time. The polarization of shear waves may serve as a source of further information about the medium.

Converted transmitted waves induced by earthquakes and detected in the zero displacement plane of *P* waves produce shorter and simpler records than compressional waves.

In mountains, the surface relief may strongly affect the general aspect of the seismogram in its initial part.

In diapiric media, the principal spurious waves are shear waves (single-mode and converted). The polarization method can identify waves reflected from boundaries in subdiapiric sediments, which fact promises a further extension in the depth range of seismic exploration in previously inaccessible localities.

The polarization method owes its efficiency to several features, namely: (a) wave discrimination at a point based on polarization is combined with wave discrimination according to apparent velocity, which appreciably enhances the capabilities of wave-field analysis; (b) several wave types (compressional, shear, and converted) are utilized concurrently, which fact extends the range of geological tasks that can be handled in addition to improving the accuracy of geological reconstructions, and opens up new possibilities for determining the composition of the section (owing to which the polarization method can be used to advantage in regions where structural problems are tackled by the compressional wave method); and (c) wave polarization is utilized as a direct source of knowledge about the medium.

The complexity of the wave field observed at inner points in the medium and variations in the direction of wave arrival along the line of traverse make the polarization method especially effective in VSP work. In turn, advances in the polarization method and an increase in the number of wave types used concurrently add more value to VSP data in identifying the nature of waves and their stratigraphic correlation.

In cases combining wave discrimination at a point with wave discrimination along the line of traverse, it is advantageous to use the common depth point (CDP) variety of the polarization method. With this method, the choice of the observation systems to be used depends on the complexity of the wave field to be observed.

In the presence of a thick low-velocity layer, analysis of compressional waves by ground observations may often prove insufficient. This is where recourse may be had to the LVL variety of the polarization method. In this case, wave discrimination can be effected in terms of the three variables: frequency, apparent velocity and polarization, thereby enhancing substantially the efficiency of seismic exploration in complex media. Such observations may be helpful under simple platform conditions in the presence of strong multiple waves. The use of the same drill-holes first for observations, then for shooting, and the replacement, where necessary, of seismometer grouping by shot grouping, help to set up the required CDP polarization spreads without an appreciable rise in cost.

In many cases, the polarization method permits drill-hole observations beneath principal marker boundaries used in tackling local structural problems to be replaced by surface or sub-LVL observations having a much higher prospecting value.

The polarization method may be valuable in the study of shallow-water and offshore areas. Pressure sensors cannot discriminate waves according to the direction of particle displacement; also, their use in shallow waters involves certain technical difficulties. In contrast, displacement sensors used in bottom observations utilize all of the capabilities of the polarization method.

In addition to the improved efficiency of wave-field analysis, the polarization method may serve as an additional source of information about the medium. The pattern of variations in the direction of particle displacement in the first waves observed at interior points of the medium can throw more light on variations in the true velocity of wave propagation along a vertical profile.

Of course, the polarization method is still at an early stage of its maturity. This calls for further research with a view to expediting its growth. In this respect, the primary tasks are: (a) theoretical and experimental studies of polarization for the basic types of seismic waves in real media; (b) studies into the effect of various discontinuities in the section, above all the ground surface, low-velocity layers and the top part of the section, on wave polarization; (c) development of software for computer processing of data supplied by the polarization method; and (d) development of equipment with as many as several hundred or more channels specifically adapted to the polarization method, widely combining three-component observations at a point with three-dimensional observations in space.

REFERENCES

1. Alekseev, A. S., Gal'perin, E. I., and Gal'perina, R. M.: 'An attempt to correlate the oscillation vector on a digital computer from VSP seismograms' in: *VSP and Increase in the Efficiency of Seismic Studies*, pp. 39–47, VIEMS Press, Moscow, 1971.
2. Alekseev, A. S., Nersesov, I. L., and Tsibul'chik, G. M.: 'A method of studying the Earth's crust using records of converted waves from distant earthquakes'. *Izv. Akad. Nauk SSSR, Ser. Fiz. Zemli*, No. 8, pp. 35–47, 1972.
3. Amirov, A. N., Merkulov, V. I., and Yakovenko, Yu. N.: 'A device for orienting seismometers in drill-holes'. USSR Pat. No. 314 165. Izobreteniya, prom. obraztsy i tovarnye znaki [Inventions, Commercial Models, and Trade Marks], No. 27, 171, 1971.
4. Nersesov, I. L. *et al.*: 'The Alma-Ata high-sensitivity seismographic radio-telemetering test ground'. *Izv. Akad. Nauk SSSR, Ser. Fiz. Zemli*, No. 7, pp. 72–77, 1974.
5. Andreev, S. S.: 'A study of the deep-seated construction of the crust using converted *PS* waves recorded during earthquakes'. *Izv. Akad. Nauk SSSR, Ser. Geofiz.*, No. 1, pp. 21–29, 1957.
6. Vakharevskaya, T. M. and Brodov, L. Yu.: 'The recognition of low-amplitude faults from data on converted reflected *PS* waves' in: *Transverse and Converted Waves in Seismic Prospecting*, pp. 219–226, Nedra Press, Moscow, 1967.
7. Bergodkov, V. A. and Orlov, V. S.: 'Recognition of waves with different polarization during operations employing the method of converted earthquake-waves'. *Izv. Akad. Nauk Turkm. SSR, Ser. Fiz.-Tekhn., Khim. i Geol. Nauk*, No. 1, pp. 97–100, 1971.
8. Berdennikova, N. I.: 'Some cases of anisotropy in a layered medium experienced during work with transverse waves' in: *Problems of the Dynamic Theory of Propagation of Seismic Waves*, Vol. 2, pp. 187–196, Leningr. Gos. Univ. Press, Leningrad, 1959.
9. Berdennikova, N. I. and Kulichikhina, T. N.: 'A study of the kinematic and dynamic characteristics of transverse waves in drill-holes'. *Trudy Inst. Geol. Geofiz., Sib. Otd. Akad. Nauk SSSR* **16**, pp. 31–63, Novosibirsk, 1962.
10. Bereza, G. V.: 'A method of recording seismic oscillations in seismic prospecting'. USSR Patent No. 98804. [Inventions, etc.] No. 8, p. 24, 1954.
11. Berzon, I. S.: *High-Frequency Seismics*, Akad. Nauk SSSR Press, Moscow, 1957.
12. Berzon, I. S.: 'Seismic wave-fields in various models of the true Earth' in: *The State and Problems of Exploration Geology*, pp. 464–468, Nedra Press, Moscow, 1970.
13. Bobrovnik, I. I. and Monastyrëv, V. K.: 'Method of submerged seismometers'. *Geologiya i Geofiz. Novosibirsk*, No. 8, pp. 92–101, 1968.
14. Bondarev, V. I.: 'Assessment of accuracy of determining particle displacements in azimuthal seismic observations as a function of the set's angular errors'. *Izv. Akad. Nauk SSSR, Ser. Geofiz.*, No. 3, pp. 374–377, 1964.
15. Bondarev, V. I.: "Method of determining the parameters of an elliptically-polarized oscillation from line-ups of azimuthal seismograms'. *Izv. Akad. Nauk SSSR, Ser. Fiz. Zemli*, No. 5, pp. 82–93, 1965.
16. Bondarev, V. I.: 'Method of determining the directions of particle motion from azimuthal observation data'. *Prikladn. Geofiz.* **45**, pp. 77–82, 1965.
17. Brodov, L. Yu.: 'Some problems involving the excitation of transverse waves' in: *Transverse and Converted Waves in Seismic Prospecting*, pp. 81–83, Nedra Press, Moscow, 1967.
18. Bulin, N. K.: 'Dynamic characteristics of deep seismic *PS* waves in Middle Asia'. *Izv. Akad. Nauk SSSR, Ser. Fiz. Zemli*, No. 6, pp. 117–125, 1967.

19. Bulin, N. K.: 'Determination of depth to folded basement with the aid of transmitted *PS* waves recorded during earthquakes'. *Izv. Akad. Nauk SSSR, Ser. Geofiz.*, No. 6, pp. 781–786, 1960.
20. Vakulin, G. P.: 'A seismic polarization analyser'. USSR Patent No. 389480. [Inventions, etc.] No. 29, p. 169, 1973.
21. Vasil'ev, Yu. I.: 'Converted refracted waves in seismic prospecting' in: *The State and Prospects of Development of Geophysical Prospecting and Exploration for Useful Minerals*, pp. 236–240, Gostoptekhizdat, Moscow, 1961.
22. Vinogradov, F. V.: 'Some test results with a three-component automatically-oriented seismometer' in: *Transverse and Converted Waves in Seismic Prospecting*, pp. 141–145, Nedra Press, Moscow, 1967.
23. Voronin, Yu. A. and Zhadin, V. V.: 'The frequency distortions of seismic signals recorded by a three-component drill-hole seismometer cluster'. *Geologiya i Geofiz. Novosibirsk*, No. 3, pp. 154–155, 1964.
24. Gal'perin, E. I.: *The Azimuthal Method of Seismic Observations*, Gostoptekhizdat, Moscow, 1955.
25. Gal'perin, E. I.: 'Azimuthal deviations of seismic rays'. *Izv. Akad. Nauk SSSR, Ser. Geofiz.*, No. 11, pp. 1283–1293, 1956.
26. Gal'perin, E. I.: 'Solution of direct three-dimensional problems of geometrical seismics for multi-layered media with interfaces of arbitrary shape'. *Izv. Akad. Nauk SSSR. Ser. Geofiz.*, No. 4, pp. 391–403, 1956.
27. Gal'perin, E. I.: 'Pattern technique of the first kind and a method of obtaining multi-component azimuthal seismograms'. *Izv. Akad. Nauk SSSR, Ser. Geofiz.*, No. 9, pp. 81–98, 1957.
28. Gal'perin, E. I.: 'A method of processing space-probe observations'. *Prikladn. Geofiz.* **17**, pp. 67–75, 1957.
29. Gal'perin, E. I.: 'The variation in particle-displacement directions in seismic waves passing through a low-velocity layer'. *Izv. Akad. Nauk SSSR, Ser. Geofiz.*, No. 5, pp. 585–594, 1962.
30. Gal'perin, E. I.: 'The study of displacement-vector directions in seismic waves in drill-hole observations'. *Izv. Akad. Nauk SSSR, Ser. Geofiz.*, No. 2, pp. 278–292, 1963.
31. Gal'perin, E. I.: 'The determination of the stratigraphic elements of a refracting interface from the displacement direction in first events observed in drill-holes'. *Izv. Akad. Nauk SSSR, Ser. Geofiz.*, No. 4, pp. 513–524, 1963.
32. Gal'perin, E. I.: 'The effect of the surface and the upper part of a section on the nature and structure of a seismogram' in: *Problems of the Dynamic Theory of Seismic Wave Propagation.* Chapter 7, pp. 201–214, Leningrad Gos. Univ. Press, Leningrad, 1964.
33. Gal'perin, E. I.: 'Detailed studies of a velocity model of the upper part of a section under conditions of weak velocity differentiation'. *Izv. Akad. Nauk SSSR, Ser. Geofiz.*, No. 4, pp. 456–474, 1964.
34. Gal'perin, E. I.: 'Processing three-component and azimuthal observations' in: *Handbook of Geophysics*, Vol. 4, *Seismic Prospecting*, pp. 602–606, Nedra Press, Moscow, 1966.
35. Gal'perin, E. I.: 'Seismic observations in drill-holes and their prospecting potentialities' in: *Methodology, Techniques, and Results of Geophysical Exploration*, pp. 151–167, Nedra Press, Moscow, 1967.
36. Gal'perin, E. I.: *Vertical Seismic Profiling*, Nedra Press, Moscow, 1971 [English translation: Soc. Explor. Geophysicists, Tulsa, Oklahoma, USA, 1974].
37. Gal'perin, E. I.: Studies of polarization of seismic waves aimed at improving efficiency of observations in complex media' in: *Problems of Ore Geophysics in Kazakhstan*, Vol. 7, pp. 5–20, Kaz. Fil. VIRG Press, Alma-Ata, 1974.
38. Gal'perin, E. I.: 'Polarization of seismic waves and potentialities for improving the efficiency of seismic prospecting'. *Izv. Akad. Nauk SSSR, Ser. Fiz. Zemli*, No. 2, pp. 107–121, 1975.
39. Gal'perin, E. I.: 'Polarization-positional correlation of seismic waves'. *Dokl. Akad. Nauk SSSR* **223**, No. 2, pp. 336–334, 1975.
40. Gal'perin, E. I.: 'Means and possibilities of improving the efficiency of seismic studies of complex media' in: *Regional, Exploration, and Industrial Geophysics*, ser. 9, No. 9, pp. 1–28, VIEMS Press, Moscow, 1975.

41. Gal'perin, E. I. and Frolova, A. V.: 'Azimuthal and phase correlation of elliptically-polarized waves'. *Izv. Akad. Nauk SSSR, Ser. Geofiz.*, No. 2, pp. 195–208, 1960.
42. Gal'perin, E. I. and Frolova, A. V.: 'Three-component seismic observations in drill-holes'. *Izv. Akad. Nauk SSSR, Ser. Geofiz.*, No. 6, pp. 793–809, 1961.
43. Gal'perin, E. I. and Frolova, A. V.: 'Three-component seismic observations in drill-holes. Studies of displacement directions'. *Izv. Akad. Nauk SSSR, Ser. Geofiz.*, No. 7, pp. 977–993, 1961.
44. Gal'perin, E. I. and Frolova, A. V.: 'The direction of seismic waves refracted at a plane interface'. *Prikladn. Geofiz.* **48**, pp. 37–40, 1966.
45. Gal'perin, E. I. and Frolova, A. V.: 'A method of processing space-probe data using converted reflected waves'. *Razved. Geofiz.* **26**, pp. 29–33, 1968.
46. Gal'perin, E. I., Amirov, A. N., and Troitsky, P. A.: 'A method of selecting waves in vertical seismic profiling'. *Izv. Akad. Nauk SSSR, Ser. Fir. Zemli*, No. 6, pp. 92–95, 1970.
47. Gal'perin, E. I., Aksenovich, G. I., and Frolova, A. A.: 'Method and instrumentation for obtaining multi-component polar seismograms and their orientation in space' in: *Problems of Ore Geophysics in Kazakhstan* 7, pp. 33–45, Kaz. Fil. VIRG Press, Alma-Ata, 1974.
48. Gal'perin, E. I., Nersesov, I. L., and Gal'perina, R. M.: 'The problem of the effect of surface relief on the structure of seismograms'. *Dokl. Akad. Nauk SSSR* **217**, No. 3, pp. 554–557, 1974.
49. Gal'perin, E. I., Frolova, A. V., and Khairutdinov, R. N.: 'Study of polarization of seismic waves during VSP in ore deposits in Central Kazakhstan – experiment and results' in: *Problems of Ore Geophysics in Kazakhstan* 7, pp. 67–86, Kaz. Fil. VIRG Press, Alma-Ata, 1974.
50. Gal'perina, R. M.: 'The problem of the physical bases and prospecting potentialities of the converted-wave method from data obtained in vertical seismic profiling'. *Dokl. Akad. Nauk SSSR* **182**, No. 2, pp. 334–336, 1968.
51. Gamburtsev, G. A.: 'A new kind of phase correlation in seismic observations'. *Dokl. Akad. Nauk SSSR* **87**, No. 1, pp. 37–40, 1952.
52. Gamburtsev, G. A.: 'The determination of the azimuth at an epicentre in the process of recording local earthquakes'. *Dokl. Akad. Nauk SSSR* **87**, No. 2, pp. 105–106, 1952.
53. Gamburtsev, G. A.: 'Correlation methods of studying earthquakes'. *Dokl. Akad. Nauk SSSR* **92**, No. 4, pp. 747–749, 1953.
54. Gamburtsev, G. A. and Gal'perin, E. I.: 'A procedure for employing the correlation method of studying earthquakes'. *Izv. Akad. Nauk SSSR, Ser. Geofiz.*, No. 1, pp. 3–10, 1954.
55. Gamburtsev, G. A. and Gal'perin, E. I.: 'Azimuthal seismic observations with inclined seismometers'. *Izv. Akad. Nauk SSSR, Ser. Geofiz.*, No. 2, pp. 184–189, 1954.
56. Gamburtsev, G. A. and Gal'perin, E. I.: 'Studies of weak local earthquakes in the Khait epicentral zone in the Tadzhik SSR' in: *Published Works of G. A. Gamburtsev*, pp. 400–426, Akad. Nauk SSSR Press, Moscow, 1960.
57. Gamburtsev, G. A. and Gal'perin, E. I.: 'Results of studies of weak local earthquakes in South-western Turkmenia with the aid of the correlation method' in: *Published Works of G. A. Gamburtsev*, pp. 390–393, Akad. Nauk SSSR Press, Moscow, 1960.
58. Ginodman, A. G.: 'The selection of transverse waves based on polarization of oscillations'. *Razved. Geofiz.* **23**, pp. 3–6, 1967.
59. Golitsyn, B. B.: *Proc. Standing Seismic Commission of the Academy of Sciences, USSR*, No. 7 (1), 1915.
60. Gogonenkov, G. R.: 'Some results of three-component inverted microseism logging'. *Razved. Geofiz.* **20**, pp. 40–45, 1967.
61. Gol'din, S. A.: *Linear Transformations of Seismic Signals*, Nedra Press, Moscow, 1974.
62. Gurvich, I. I.: *Seismic Prospecting*, Nedra Press, Moscow, 1970.
63. Gurvich, I. I. and Yanovsky, A. K.: 'Seismic pulses from blasts in a homogeneous absorbing medium'. *Izv. Akad. Nauk SSSR, Ser. Fiz. Zemli*, No. 10, pp. 14–24, 1967.
64. Gurvich, I. I., Zaudel'son, I. I., and Daderko, Yu. R.: 'A procedure for combining observation and shot holes in the reflected-wave method of seismic prospecting'. *Razved. Geofiz.* **58**, pp. 24–29, 1973.
65. Dantsig, L. G. and Obolentzeva, I. R.: 'Results of theoretical studies of space polarization of

transmitted longitudinal and converted waves in the case of oblique interfaces'. *Geologiya i Geofiz. Novosibirsk*, No. 4, pp. 93–102, 1973.

66. Karus, E. V. *et al.*: 'Twelve-channel instrumentation for studying seismic wave fields in deep drill-holes', pp. 93–123, VNIIYaGG Press, Moscow, 1970.
67. Berzon, I. S., Epinatyeva, A. M., Pariiskaya, G. N., and Starodubrovskaya: *Dynamic Characteristics of Seismic Waves in Real Media*. Akad. Nauk SSSR Press, Moscow, 1962.
68. Egorkina, G. V.: 'A study of the anisotropy of the Earth's crust on the basis of records of converted seismic waves'. *Izv. Akad. Nauk SSSR, Ser. Fiz. Zemli* 9, pp. 40–50, 1969.
69. Egorkina, G. V.: 'The polarization of transmitted converted (*PS*) waves'. *Prikladn. Geofiz.* **55**, pp. 61–69, Nedra Press, Moscow, 1969.
70. Egorkina, G. V.: 'Velocity anisotropy of elastic waves in armenia'. *Prikladn. Geofiz.* **78**, pp. 106–118, Nedra Press, Moscow, 1975.
71. Egorkina, G. V.: 'Phase shifts in transmitted converted (*PS*) waves'. *Prikladn. Geofiz.* **70**, pp. 74–85, Nedra Press, Moscow, 1973.
72. Epinatyeva, A. M.: *Physical Fundamentals of Seismic Prospecting*. Moscow University Press, 1970.
73. Ershova, T. N. and Kats, S. A.: 'Two-component self-tuning system for the detection of linearly-polarized waves with a specified direction of polarization' in: *Seismic Waves in Thin-Layered Media*, pp. 182–190, Nedra Press, Moscow, 1973.
74. Zhadin, V. V.: 'Three-component measurements of amplitudes and velocities of compressional and shear waves in deep drill-holes'. *Geologiya i Geofiz. Novosibirsk* **10**, pp. 129–136, 1960.
75. Gal'perin, E. I., Aksenovich, G. I., and Pokidov, V. L.: 'Study of particle motion paths with a view to enhancing the efficiency of VSP' in: *VSP and Improvement in the Efficiency of Seismic Exploration*, pp. 1–12. VIEMS Press, Moscow, 1971.
76. Kanareikin, D. V., Pavlov, N. V., and Potëkhin, R. A.: *Polarization of Radar Signals*. Soviet Radio Press, Moscow, 1966.
77. Kerimov, I. G.: 'Determination of the earth's construction beneath a station from the polarization of seismic waves'. *Dokl. Akad. Nauk Azerb. SSR* **25**, No. 6, pp. 56–62, 1969.
78. Klem-Musatov, K. D., Obolentzeva, I. R., and Aisenberg, A. M.: 'Computation of elastic wave fields for a model of an anisotropic medium' in: *Dynamic Characteristics of Seismic Waves*, pp. 73–98, Nauka Press, Moscow, 1973.
79. Aksenovich, G. I. Galperin, Y. I., Kuzmenko, G. P.: 'Complete equipment for three-component observations and data processing' in: *Problems of Ore Geophysics in Kazakhstan* 7, pp. 46–52, Kaz. Fil. VIRG Press, Alma-Ata, 1974.
80. Kostenich, V. I. and Nikolayev, A. V.: 'Calculation of particle motion paths on a digital computer' in: *Problems of Ore Geophysics in Kazakhstan* 7, pp. 87–138, Kaz. Fil. VIRG Press, Alma-Ata, 1974.
81. Krauklis, L. A., Moiseeva, L. A., and Gel'chinsky, B. Ya.: 'A program for computer correlation of seismic waves and experience in its use' in: *Problems of the Dynamic Theory of Seismic Wave Propagation*, Nauka Press, Moscow–Leningrad, 1966.
82. Lebedeva, G. N., Lebedev, K. A., and Puzyrev, N. N.: 'Seismic-wave discrimination based on polarization for sources with horizontal trends' in: *Seismic Prospecting Techniques*, pp. 127–135, Nauka Press, Moscow, 1965.
83. Lin'kov E. M., Tripol'sky, V. P., and Sabantsev, S. B.: 'Polarization seismic arrays'. *Uchen. Zap. Lenigr. Gos. Univ., Ser. Fiz. Geol.* **303, 13**, pp. 135–137, 1962.
84. Malinovskaya, L. N.: 'The calculation of theoretical seismograms of interference-produced waves' in: *Problems of the Dynamic Theory of Seismic Wave Propagation*, pp. 357–377, Leningrad Gos. Univ. Press, 1959.
85. Manukov, V. S.: 'The use of position and azimuth observations of compressional and converted waves in determining the velocity of shear waves'. *Ekspress. Inform.*, VIEMS Press, Moscow, 1975.
86. Gal'perin, E. I., Krasilshchikova, G. A., Mironova, V. I., and Frolova, A. V.: 'Techniques and procedures for stereographic projection in solving direct spatial problems of geometric seismics'. *Prikladn. Geofiz.* **18**, pp. 3–29, 1957.

87. Gal'perin, E. I., Aksenovich, G. I., Gal'perina, R. M., and Erenburg, M. S.: 'Techniques, equipment, and experience of three-component seismic observations for the detection of transmitted converted waves by polar correlation'. *Izv. Akad. Nauk SSSR, Ser. Fiz. Zemli* 7, pp. 74–85, 1975.
88. Meshbei, V. I.: 'Some features of the wave field in the diapir zone of the Taman Peninsula'. *Prikladn. Geofiz.* **48**, pp. 63–77, 1966.
89. Meshbei, V. I. and Muzyka, I. M.: 'The use of directional sources in the detection of converted grazing-incidence diffracted head ($S_{12}P_{21}$) and grazing-incidence diffracted ($S_{12}P_1$) waves'. *Izv. Akad. Nauk SSSR, Ser. Geofiz.* 9, pp. 1324–1339, 1963.
90. Brodov, L. Vu., Kulichikhina, T. N., Korzheva, L. V., and Yevstigneyev, V. I.: 'Some procedural problems of three-component drill-hole observations of shear waves' in: *Proceedings of the Saratov Seminar on the Use of Shear and Converted Waves*, VIEMS Press, Moscow, 1974.
91. Nedashkovsky, I. Yu.: 'A method for seismic-wave discrimination based on the direction of particle displacement'. USSR Patent No. 251842 [Inventions, etc.], No. 4, pp. 97–106, 1971.
92. Obolentzeva, I. R.: 'Polarization of reflected shear waves excited by a directional source on a horizontal observation plane in the case of inclined interfaces' in: *Techniques of Seismic Exploration*, pp. 41–54, Nauka Press, Moscow, 1969.
93. Obolentzeva, I. R.: 'Ray velocities and polarization of seismic waves as a function of effective elastic parameters of thin-layered periodic media'. *Geologiya Geofiz. Novosibirsk* **12**, pp. 79–94, 1974.
94. Obolentzeva, I. R. and Dantsig, L. G.: 'Polarization of transmitted compressional and converted waves in the case of inclined boundaries'. *Geologiya i geofizika Geofiz. Novosibirsk* **4**, pp. 97–106, 1971.
95. Dantsig, L. G., Degachev, A. A., Zhadin, V. V., and Senyukov, V. A.: 'Detection of converted waves on distant earthquake records' in: *Techniques of Seismic Exploration*, pp. 132–142. Nauka Press, Moscow, 1969.
96. Oznobikhin, Yu. V.: 'The arrival polarity of converted and multiple waves induced by earthquakes and a technique for computing their intensity in some models of media'. *Trudy Zap. Sib. Nauch-no-Issled. Geol. Neft. Inst.* **64**, pp. 97–108, 1972.
97. Gal'perin, E. I, Kovalskaya, I. Y., Krakshina, R. M., and Frolova, A. V.: 'The nature of spurious waves recorded over large adyr Areas in Middle Asia' in: *Techniques and Results of Integrated Deep Geophysical Exploration*, pp. 223–236. Nedra Press, Leningrad, 1969.
98. Pod"yapolsky, G. S.: 'An approximate expression for displacement in the vicinity of the main front in the case of a small angle between the ray and the interface'. *Izv. Akad. Nauk SSSR, Ser. Geofiz.* **12**, pp. 1761–1763, 1959.
99. Pokidov, V. L., Frolova, A. V., and Chastnaya, T. G.: 'Features of wave fields in the ore deposits of Central Kazakhstan based on VSP data' in: *Problems of Ore Geophysics in Kazakhstan* 7, pp. 53–66, 1974.
100. Pomerantseva, I. V.: *The Detection and Polarization of PS Waves Recorded by the 'Zemlya' Station*. VNIIGeofizika Press, Moscow, 1973.
101. Pomerantseva, I. V., Barskova, L. P., and Mozzhenko, A. N.: 'Crustal models derived on the basis of data supplied by the 'Zemlya' station for some platforms, platform troughs, and fore-deeps differing in age' in: *The Deep Construction of the Earth's Crust*, pp. 49–60. Nedra Press, Moscow, 1975.
102. Puzyrev, N. N. (ed.): *Shear and Converted Waves in Seismic Prospecting*. Nedra Press, Moscow, 1967.
103. Potap'ev, S. V.: 'Locating a refracting boundary from three-component observations of *PP* waves' in: *Regional Geophysical Studies in Inaccessible Areas*, Nauka Press, Novosibirsk, 1974.
104. Puzyrev, N. N., Lebedev, K. A., and Lebedeva, G. N.: 'Excitation of shear seismic waves by explosions'. *Geologiya Geofiz. Novosibirsk* **2**, pp. 88–99, 1966.
105. Puzyrev, N. N. and Obolentseva, I. R.: 'Polarization of compressional and converted reflected waves on a horizontal observation plane in the case of inclined interfaces' in: *Shear and Converted Waves in Seismic Prospecting*, pp. 171–202, Nedra Press, Moscow, 1967.

106. Puzyrev, N. N., Kefeli, A. S., and Lebedev, K. A.: 'Excitation of shear waves in drill-holes'. *Geologiya Geofiz. Novosibirsk* **10**, pp. 82–92, 1968.
107. Puzyrev, N. N. and Brodov, L. Y., 'Efficiency of shear-wave excitation'. *Geologiya i Geofiz. Novosibirsk* **5**, pp. 81–88, 1969.
108. Puzyrev, N. N., Trigubov, A. V., and Lebedeva, G. N.: 'Three-component recording of various seismic waves from symmetrical and asymmetrical sources' in: *Experimental and Theoretical Studies of Reflected Waves*, pp. 5–28. Nauka Press, Novosibirsk, 1975.
109. Razumovsky, N. K.: *Stereographic Projection*, Gorn. Inst. Press, Leningrad, 1932.
110. Riznichenko, Yu. V.: 'Seismic quasi-anisotropy'. *Izv. Akad. Nauk SSSR, Ser Georg. Geofiz.* **6**, pp. 518–544, 1949.
111. Rudnitsky, V. P.: 'Three-component measurement of amplitudes and seismic velocities' in: *Geophysics and Astronomy* **9**, pp. 18–23. Naukova Dumka Press, Kiev, 1966.
112. Rudnitsky, V. P.: 'Polarization properties of converted seismic waves'. *Geofiz. Sb. Akad. Nauk UKr. SSR* **44**, pp. 54–65, 1971.
113. Ryabinkin, L. A. and Znamensky, V. V.: 'The detection of low-intensity waves by the CDR method'. *Prikladn. Geofiz.* **34**, pp. 3–24, Gostoptekhizdat, 1962.
114. Savarenskaya, Ye. F. and Kirnos, D. N.: *Elements of Seismology and Seismometry*. Gostoptekhizdat, Moscow, 1955.
115. Aksenovich, G. I., Gal'perin, E. I., and Pokidov, V. L.: 'Method of Vertical Seismic Profiling'. USSR Patent No. 369528. [Inventions, etc.] No. 10, 1973.
116. Gal'perin, E. I.: *Method and Apparatus for Producing Oriented Records in Three-Component Observations by the VSP Method*. VIEMS Press, Moscow, 1974.
117. Aksenovich, G. I.: 'Method for producing three-component seismic records'. USSR Patent No. 366431. [Inventions, etc.] No. 7, p. 88, 1973.
118. Aksenovich, G. I.: 'Method of seismic prospecting in drill-holes'. USSR Patent No. 36643. [Inventions, etc.] No. 7, p. 89, 1973.
119. Teplitsky, V. A.: 'Method of reflected-wave inverse travel-time curves' in: *New Technical and Methodological Developments in Seismic Prospecting*, pp. 87–89, VIEMS Press, Moscow, 1969.
120. Aksenovich, G. I.: 'Three-component clusters and sensing their alignment in drill-holes' in: *Problems of Ore Geophysics in Kazakhstan* 7, pp. 21–32, Kaz. Fil. VIRG Press, Alma-Ata, 1974.
121. Trigubov, A. V.: 'The technique of combined observation of converted, shear, and compressional reflected waves' in: *Experimental and Theoretical Studies of Reflected Waves*, pp. 57–64. Nauka Press, Novosibirsk, 1975.
122. Trigubov, A. V., Fanenkov, V. N., and Pavtovets, L. E.: 'A study of low-amplitude tectonic disturbances by various types of reflected waves' in: *Experimental and Theoretical Studies of Reflected Waves*, pp. 64–77. Nauka Press, Novosibirsk, 1975.
123. Shcherbakova, B. E.: 'Conditions of excitation and recording of deep converted waves' in: *Proceedings of Jubilee Scientific and Technical Conference on Geophysics*, pp. 73–93. Ashkhabad, 1972.
124. Gal'perin, E. I.: 'A device for sensing the path of particle motion during propagation of seismic waves'. USSR Patent No. 372529. [Inventions, etc.] No. 13, p. 113, 1973.
125. Aksenovich, G. I.: 'A device for producing oriented seismic records in drill-holes'. USSR Patent No. 376739. [Inventions, etc.] No. 17, p. 145, 1973.
126. Aksenovich, G. I.: 'A device for polarization-position correlation of seismic waves'. USSR Patent No. 456241. [Inventions, etc.] No. 1, p. 110, 1975.
127. Tsybul'chik, G. M.: 'Analysis of distant earthquake seismograms'. *Geologiya i Geofiz. Novosibirsk* **4**, pp. 73–86, 1969.
128. Chubov, P. G.: 'A device for mathematical processing of three-component seismograms'. USSR Patent No. 173047. [Inventions, etc.] No. 14, p. 92, 1965.
129. Shchepin, V. D. and Ruch, G. S.: 'A device for directional reception of elastic waves'. USSR Patent No. 157516. [Inventions, etc.] No. 18, p. 64, 1963.
130. Bakharevskaya, T. M.: 'Experimental study of polarization of converted *PS* waves reflected from an inclined interface' in: *Shear and Converted Waves in Seismic Prospecting*, pp. 203–209. Nedra Press, Moscow, 1967.

131. Puzyrev, N. N. (ed.): *Experimental Studies of Shear and Converted Waves*. Akad. Nauk SSSR, Sib. Otdel. Press, Novosibirsk, 1962.
132. Yudina, R. I.: 'The pattern of variations in the velocity of shear waves in geological media' in: *Shear and Converted Waves in Seismic Prospecting*, pp. 157–168. Nedra Press, Moscow, 1967.
133. Barr Jr., Frederick, J. 'Method and apparatus for increasing seismic signal-to-noise ratio'. *Petty Geophysical Engineering Co*. US Patent No. 3736556, Class 340–15, 5, EC.
134. Burg, J. P. 'Three-dimensional filtering with an array of seismometers'. *Geophysics* **29**, pp. 693–713, 1964.
135. Cassinis, R.: 'Application of three-component seismic systems to the evaluation of soil properties'. *Bol. Geofis. Teorica ed Applicata* **9**, No. 30, pp. 285–303, 1967.
136. Cook, K., Algermissen, S., and Costain, J.: 'The status of *PS* converted waves in crystal studies. *Geophys. Rev.* **67**, pp. 4769–4778, 1962.
137. Flinn, E. A.: 'Data processing techniques for the detection and interpretation of teleseismic signals'. *Proc. IEEE* **53**, pp. 1860–1884, 1965.
138. Flinn, E. A., Archambeau, C. B., and Lambert, D. G.: 'Detection, analysis and interpretation of teleseismic signals. Part I. Compressional phases from the Salmon Event'. *J. Geophys. Res.* **71**, pp. 3433–3501, 1966.
139. Flinn, E. A., Cohen, T. I., and McCowan, D. E.: 'Detection and analysis of multiple seismic events'. *Bull, Seismo. Soc. Am.* **63**, pp. 1921–1936, 1973.
140. Galitsin, B. B. 'Zur Frage der Bestimmung des Azimuts des Epizentrums eines Bebens'. *Izv. Imper. Akad. Nauk*, Ser. 6, 1909.
141. Gamburtsev, G. A.: 'Some new methods of of seismological research'. *Publ. Bur. Centr. Seism. Int. Ser. a, Trav. Scient.* **19**, pp. 373–381, Toulouse, 1956.
142. Ganguli, D. K. and Gupta, I. N.: 'Variation of ground motion characteristics with nature of media in the vicinity of small explosions'. *J. Sci. and Eng. Res.* **10**, pp. 337–344, 1966.
143. Geyer, R. L. and Martner, S. T.: '*SH* waves from explosive sources'. *Geophysics* **34**, pp. 893–905, 1969.
144. Griffin, J. N.: 'Remode: signal-noise tests in polarized noise'. *Seismic Data Laboratory Report*, No. 162, pp. 86–92, Teledyne Inc., Alexandria, 1966.
145. Gupta, I. N.: 'Premonitory variations in *S* wave anisotropy before earthquakes in Nevada'. *Science, N.Y.* **182**, pp. 1129–1132, 1973.
146. Jolly, R. N.: 'Investigation of shear waves'. *Geophysics* **21**, pp. 905–938, 1956.
147. Key, F. A.: 'Signal-generated noise recorded at the Eskdalemuir seismometer array station'. *Bull, Seism. Soc. Am.* **57**, pp. 27–39, 1967.
148. Key, F. A.: Some observations and analysis of signal-generated noise'. *Geophys. J. R. Astron. Soc.* **15**, pp. 377–392, 1968.
149. Mercado, E. J.: 'Linear phase filtering of multicomponent seismic data'. *Geophysics* **33**, pp. 926–935, 1968.
150. Mines, C. H. and Sax, R. Z.: 'Rectilinear motion detection (Remode)'. *Seismic Data Laboratory Report* **118**, pp. 51–58, Teledyne Inc., Alexandria, 1965.
151. Montalbetty, L. F. and Kanashevich, E. R.: 'Enhancement of teleseismic body phase with a polarization filter'. *Geophys. J. R. Astron. Soc.* **21**, pp. 119–129, 1970.
152. Shimshoni, M. and Smith, S. W.: 'Seismic signal enhancement with three-component detectors'. *Geophysics* **2**, pp. 664–671, 1964.
153. Strobach, K.: 'Stereoskopische Vektorregistrierung'. *Z. Geophys.* **23**, pp. 306–315, 1957.
154. Strobach, K.: 'Morfologische Untersuchung microseismischer Bodenbewegungen nach stereoskopischen Vektorregistrierung'. *Z. Geophys.* **24**, pp. 369–379, 1958.
155. Sutton, G. H. and Pomery, P. W.: 'Analog analysis of seismograms recorded on magnetic tape'. *J. Geophys. Res.* **68**, pp. 2791–2815, 1963.
156. Tatel, H. E. and Tuve M. A.: 'Note on the nature of a seismogram'. *J. Geophys. Res.* **59**, pp. 287–288, 1954.
157. Walzer J.: 'Polarization of oscillation planes of microseisms and signal-to-noise ratio augmentation'. *Pure and Appl. Geophy.* **82**, No. 5, pp. 19–25. 1970.
158. White, J. E.: Heaps, S. H. and Lawrence, P. Z.: 'Seismic waves from a horizontal force'. *Geophysics* **21**, pp. 715–723, 1956.

159. White J. E.: 'Motion product seismograms. *Geophysics* **29**, pp. 288–298, 1964.
160. White, J. E.: *Seismic Waves, Radiation, Transmission and Attenuation*. McGraw-Hill, New York, 1965.
161. Whitecomb, Z. H.: 'Shear wave detection in near-surface seismic refraction studies'. *Geophysics* **31**, pp. 981–983, 1966.
162. Wuenschel, P. C.: 'Dispersive body waves. An experimental study'. *Geophysics* **30**, pp. 539–551, 1965.
163. Zoltan, A.: 'Some data processing results for a vertical array of seismometers'. *Geophysics* **35**, pp. 337–343, 1970.

APPENDIX

GLOSSARY

Term	Definition
1. Azimuthal seismogram.	*See*: polar seismogram
2. Azimuthal method of seismic observations.	A method of seismic studies based on the use of correlational principles involving observations at a single point when vibrations of the earth are registered by a group of over three seismometers constituting an azimuthal set.
3. Azimuthal seismometer set.	A device comprising more than three seismometers whose maximum sensitivity axes are arranged at an equal angle to the horizon, but at different azimuths, the axes constituting the generatrices of a cone (conical sets). Was in use mainly before CDR–I has been developed.
4. Azimuthal-phase correlation.	The tracking of waves or of their different phases as functions of the azimuth of spatial oscillation components (a particular case of *polar correlation*).
5. Azimuthal observations in boreholes.	A term sometimes erroneously applied to three-component observations in boreholes (PM VSP).
6. Analyzer (polarization)	A device used to obtain a set of *fixed components* definitely oriented in space from the original signals of a three component set.
7. Grouping of the 1st kind.	A method of obtaining a directivity diagram arbitrarily oriented in space by summing the signals of seismometers constituting three-componet sets.
8. Three-component set orientation gauge (DOU).	A device used to determine the actual position of a three-component set in a borehole (the direction of the set's axis and the azimuth of one of the seismometers): The following types are to be distinguished. DOU–M utilizes the magnetic field of the Earth and is used only in uncased boreholes. DOU–G utilizes the gyroscopic effect and is used both in cased and uncased holes.

Term	Definition
9. Directivity diagram of a seismometer with respect to displacements (diagram of the 1st kind).	The sensitivity of a seismometer with its maximum sensitivity axis fixed in space as a function of the direction in space of constant-amplitude displacements. The directivity diagram of a seismometer is represented by two tangent spheres.
10. Line of 'vision'.	A line on the stereographic projection described by the direction of the displacement (velocity) vector in the course of observations along a vertical profile from one shot-point or along a level profile from a set of shot-points.
11. Continuous seismic control of instrumentation in the course of PM observations.	A method of control based on the comparison of recordings of an oscillation component that does not coincide with any of the set's axes obtained from direct observations or as a result of summing signals of a three-component set. In symmetrical sets the vertical oscillation component is conveniently used as the control component. To this end the set is equipped with a fourth seismometer.
12. Off-line vertical seismic profiling.	One of possible VSP observation systems in which the source of oscillations is located at a distance from the borehole's entrance. Sometimes erroneously classified as a separate observation method.
13. Three-dimensional observation system.	An observation system, which combines PM area observations on the surface and observations along a vertical profile with the aim of studying all parameters of the wave field in the space surrounding the borehole.
14. Space surrounding the borehole.	In *production seismic exploration* the term implies the space defined by a truncated cone with its axis coinciding with the borehole axis that can be studied with the aid of a three-dimensional observation system. The radius of the cone's base depends on seismologic conditions and on the goals of the studies. In mining geology it usually does not exceed a few hundred meters. In oil geology it may be as large as a few kilometers. The cone's altitude may exceed the depth of the hole. The space surrounding the borehole should be distinguished from the space contagious with the borehole, which is studied by different production geophysical methods (various modifications of the electric, radioactive and sonic logging). The radius of the space contagious with the borehole does not exceed 5–10 m.
15. Orientation analyzer.	A device used to obtain a three-component recording with a specified orientation of oscillation

Term	Definition
	components in space from the signals of the original three-component set. It is based on the grouping of the 1st kind and is a modification of the polarization analyzer.
16. Orientation of the components.	The production of seismograms with components having a specified orientation in space. Is executed both in analog and in digital form.
17. Polarization parameters.	A quantitative characteristic of the trajectory of particle motion. Parameters of a *linearly-polarized oscillation*. 1. The displacement direction determined by two angles one in the horizontal and one in the vertical plane (φ and ω, respectively). 2. The amplitude A. Of an *elliptically-polarized oscillation*. 1. The orientation of the polarization plane in space as determined by the direction of its normal (φ_n, ω_n). 2. The shape of the ellipse in the polarization plane as determined by the ratio of its semi-axes $\epsilon = a/A$. 3. The orientation of the ellipse in the polarization plane as determined by the direction of one of its axes φ_A, ω_A.
18. PM data digital processing subsystem.	A library of computer programs for digital processing of various modifications of the polarization method. The following subsystems are available: 5 CS–3 PM, 'Azimuth', 'Polar'. They are matched with corresponding universal seismic exploration data digital processing systems.
19. Fields of displacement (velocity) vectors.	A set of displacement (velocity) vectors obtained in observations along a vertical profile from several shot points. A particular case of the vector field is a field of only the displacement directions.
20. Polarization filtering.	A method of three-dimensional filtering based on the polarizational properties of seismic waves. Employed both to discriminate waves with different types of polarization and to determine their polarization parameters. The polarization filtering is executed at the digital processing stage.
21. Polar-positional correlation (PPC).	A method of three-dimensional filtering in which the waves are discriminated and tracked not according to fixed (e.g. vertical) oscillation components, but according to tracking (optimum) components for which the signal-to-noise ratio is at its maximum.

Term	Definition
22. Polarization method of seismic studies (PM).	A method of seismic studies based on the utilization of all the parameters of the wave field (including polarization) for its analysis and for the joint processing of waves of different types (the longitudinal, converted and transverse) to obtain information about the earth. The method is applicable to various sorts of observations. Up- and down-hole modifications of the polarization method are being developed: the reflection (PM RW); the common depth point (PM CDP); the refraction and the curved-ray refraction; the method of regional studies employing vertical seismic profiling.
23. Wave polarization.	A space-and-time characteristic of a seismic wave. *Polarization* type defines the path of particle motion in the process of a seismic wave passing through a fixed point in space. The following types are distinguished: *linear* – the particles move in a rectilinear path. *plane* – the path of particle motion lies in a plane. The *elliptical* and the *circular* are particular cases of plane polarization. *spatial* – the particles move in a path that does not lie in a plane.
24. Polar correlation.	The tracing of peculiar properties of a recording as determined by oscillation components oriented in space. It is executed with the aid of a multicomponent polar seismogram obtained for a point with differently oriented seismometer axes (a particular case of polar correlation is the azimuthal). The polar correlation can be conventionally regarded as the result of polarization filtering having a cosine directivity diagram.
25. Polar seismogram.	A set of traces corresponding to oscilation components definitely oriented in space. The following are to be distinguished: *overall* – with a uniform distribution of components in space. *planar* – in which the components are located in the vicinity of a plane orthogonal to the oscillations in an unwanted wave; *azimuthal* – whose components differ only in their azimuths.
26. Production seismic exploration.	A trend in seismic exploration useful not only at the prospecting stages, but also at the stage of exploitation of a field as well as for the solution of drilling technology problems. The production

Term	Definition
	seismic exploration is based on the studies of the space surrounding the borehole with the aid of combined PM VSP and surface PM CDP methods involving an extensive use of production geophysical data. Production seismic exploration is executed in close association with drilling.
27. Controlled directional reception of the 1st kind (CDR–I).	A method of discriminating waves of a point according to the directions of particle motion at the point based on the selectivity of the seismometer's directivity diagram. It is similar to the traditional CDR–II method, which discriminates waves according to their propagation direction along the line of the profile.
28. Coordinate systems used in PM.	PM makes use of orthogonal and spherical coordinate systems. The following orthogonal coordinate systems are to be distinguished: The *original* whose axes coincide with the maximum sensitivity axes of the seismometers of a three-component set that has been used in field observations. The *fixed-in-space* the orientation of whose axes does not change along the line of observations (e.g. an *XYZ* or a symmetrical I, II, III, three-component set whose *X* or I axis points North or along the line of observations); The local (*PRT*) coordinate system the orientation of whose axes is associated with the plane of incidence of the primary wave at the given point. The positive direction of the *P* axis is chosen to coincide with the direction of particle motion in the first longitudinal wave, the other two axes (*R* and *T*) lying in the plane of the *P* wave's front. The *T* axis is horizontal and orthogonal to the plane of rays. The positive direction of the axes is determined with the aid of the left-hand rule. The local coordinate system is convenient, because in homogenous and axially-symmetrical media all waves are polarized in the coordinate planes.
29. Oscillation component.	The projection of the full oscillation vector on a given direction in space specified by the parameters: the angle with the vertical (φ) and the aximuth (ω). The following components are distinguished in PM: The *original*, which coincides with the direction of one of the seismometer axes of a three-component set and was recorded directly in field conditions; The *fixed* calculated from the original whose orientation does not change inside the entire registration interval and along the whole profile

Term	Definition
	or in the whole plane of observations (original components are particular cases of the fixed component); The *tracking* the typical feature of which is the maximum signal-to-noise ratio recorded on a specified basis inside a specified time interval. The orientation of the tracking component in a wave may change along the line of observations (along a profile) and depends on the direction of the wave's full vector and on the state of the unwanted waves.
30. Spectral polarizational analysis.	The study of polarization parameters as functions of the frequency.
31. Stereographic projection.	The projection of a sphere onto a plane.
32. Stereo net.	The projection onto a plane of a grade net. There is a direction in space to correspond to every point on the net. It is very convenient for the display of three-dimensional trajectories and for the graphical solution of problems in the determination of directions and in the measurement of angles in space. May be used to analyse and study the paths of particle motion.
33. Three-component seismometer set.	A set comprising three mutually-orthogonal seismometers. The following sets are to be distinguished. The *XYZ* consisting of two horizontal and a vertical seismometer; the *symmetrical* (uniform) I, II, III comprising three similar seismometers whose axes are inclined at an identical angle of 35°20′ to the horizon, the axes of adjacent seismometers making angles of 120° with one another. Possesses considerable advantages in design and methodology over the traditional *XYZ* set. PM normally utilizes the symmetrical sets.
34. Level observations.	A feasible system of observations in VSP in which the depth of the seismometer remains fixed, the variable parameter being the location of the source of oscillations (such observations became known in the literature as the inverted travel-time method). A more appropriate term for such a travel-time curve is a level travel-time curve.

INDEX OF SUBJECTS

NAME INDEX